Single Charge Tunneling

Coulomb Blockade Phenomena In Nanostructures

Edited by

Hermann Grabert

Universität Essen
Essen, Germany

and

Michel H. Devoret

Centre d'Etudes de Saclay
Gif-sur-Yvette, France

Plenum Press
New York and London
Published in cooperation with NATO Scientific Affairs Division

Proceedings of a NATO Advanced Study Institute
on Single Charge Tunneling,
held March 5–15, 1991,
in Les Houches, France

NATO-PCO-DATA BASE

The electronic index to the NATO ASI Series provides full bibliographical references (with key-words and/or abstracts) to more than 30,000 contributions from international scientists published in all sections of the NATO ASI Series. Access to the NATO-PCO-DATA BASE is possible in two ways:

—via online FILE 128 (NATO-PCO-DATA BASE) hosted by ESRIN, Via Galileo Galilei, I-00044 Frascati, Italy.

—via CD-ROM "NATO-PCO-DATA BASE" with user-friendly retrieval software in English, French, and German (© WTV GmbH and DATAWARE Technologies, Inc. 1989)

The CD-ROM can be ordered through any member of the Board of Publishers or through NATO-PCO, Overijse, Belgium.

Library of Congress Cataloging in Publication Data

Single charge tunneling: Coulomb blockade phenomena in nanostructures /
edited by Hermann Grabert and Michel H. Devoret.
    p.   cm.—(NATO ASI series. Series B, Physics; v. 294)
    "Proceedings of a NATO Advanced Study Institute on Single Charge Tunneling, held March 5–15, 1991, in Les Houches, France"—Verso t.p.
    "Published in cooperation with NATO Scientific Affairs Division."
    Includes bibliographical references and index.
    ISBN 0-306-44229-9
    1. Tunneling (Physics)—Congresses. 2. Coulomb potential—Congresses. 3. Nanostructures—Congresses. I. Grabert, Hermann, date. II. Devoret, Michel H. III. NATO Advanced Study Institute on Single Charge Tunneling (1991: Les Houches, Haute-Savoie, France) IV. North Atlantic Treaty Organization. Scientific Affairs Division. V. Series.
QC176.8.T8S56   1992   92-13436
530.4′16—dc20   CIP

ISBN 0-306-44229-9

© 1992 Plenum Press, New York
A Division of Plenum Publishing Corporation
233 Spring Street, New York, N.Y. 10013

Printed in the United States of America

Preface

The field of single charge tunneling comprises of phenomena where the tunneling of a microscopic charge, usually carried by an electron or a Cooper pair, leads to macroscopically observable effects. The first conference entirely devoted to this new field was the NATO Advanced Study Institute on Single Charge Tunneling held in Les Houches, France, March 5-15, 1991. This book contains a series of tutorial articles based on lectures presented at the meeting. It was intended to provide both an introduction for nonexperts and a valuable reference summarizing the state of the art of single charge tunneling. A complementary publication with contributions by participants of the NATO Advanced Study Institute is the Special Issue on Single Charge Tunneling of Zeitschrift für Physik B, Vol. 85, pp. 317-468 (1991). That issue with original papers provides a snapshot of the leading edge of current research in the field.

The success of the meeting and the publication of this volume was made possible through the generous support of the NATO Scientific Affairs Division, Brussels, Belgium. The Centre de Physique des Houches has provided a superbly situated conference site and took care of many local arrangements. Both for the preparation of the conference and the handling of some manuscripts the support of the Centre d'Etudes de Saclay was essential. The editing of the proceedings volume would not have been possible without the dedicated efforts of Dr. G.-L. Ingold, who tailored a LaTeX style-file to the needs of this book, and the assistance of J. Ankerhold and P. Wyrowski from the staff of the Fachbereich Physik of the Universität Essen, who prepared the manuscripts in camera-ready form. The editors are most appreciative of the efforts made by the colleagues who accepted the burden of writing a chapter covering a specific aspect of single charge tunneling in this volume, and they thank all participants of the conference as well as the supporting staff for their help in making the Advanced Study Institute so successful.

Michel H. Devoret
Groupe Quantronique
Centre d'Etudes de Saclay
France

Hermann Grabert
Fachbereich Physik
Universität Essen
Germany

Introduction

This book is divided into 9 chapters, which, except for chapter 1, deal with a particular aspect of single charge tunneling (SCT) phenomena. Chapter 1 by DEVORET and GRABERT provides an introduction to the other chapters, each of which is to a large extent self-contained. In recent years it has become increasingly clear that the more fundamental objects in the theory of SCT phenomena should be "islands", that is metallic regions with small total capacitance into which electrons can tunnel, rather than individual small capacitance tunnel junctions. Chapter 1 is devoted mainly to the discussion of this idea.

Chapter 2 by INGOLD and NAZAROV deals with the dynamics of ultrasmall junction circuits and teaches one how to compute the tunneling rate of electrons and Cooper pairs across a particular junction in the circuit. Starting with single junction systems, progressively more complex cases are treated. In the Appendix, a microscopic justification for the starting hypotheses of the calculations is given. Chapter 3 by ESTEVE explains the basic ideas on which the manipulation of electrons one-by-one in metallic junction circuits are founded. The primary focus is thus on normal junction circuits, although the manipulation of Cooper pairs is briefly discussed. Particular emphasis is put on the accuracy of devices in which electrons are clocked by an external frequency, such as the "turnstile" and the "electron pump". These devices may lead one day to experiments with metrological accuracy.

Chapter 4 by TINKHAM treats the superconducting single junction system. It gives a survey of phenomena observed in low-capacitance tunnel junctions and also provides a link with the usual Josephson effects. Chapter 5 by VAN HOUTEN, BEENAKKER, and STARING deals with semiconductor systems, mostly GaAs/AlGaAs heterostructures, in which the metallic islands consist of quasi-isolated regions of the 2D electron gas. In these systems, the interplay between the Coulomb effects and both the Quantum Hall effect and the resonant tunneling effect leads to exciting phenomena. While in the first five chapters single charge tunneling phenomena are analysed mainly in terms of charge tunneling across one junction at a time, Chapter 6 by AVERIN and NAZAROV is a detailed discussion of the process of co-tunneling in which several tunneling events across different junctions occur coherently. Co-tunneling causes fundamental limitations of the accuracy of the charge transferring devices described in Chapter 3.

Chapters 7 and 8 deal with systems containing a large number of junctions but which nevertheless display great simplicity because of their translational symmetry properties. Chapter 7 by DELSING treats one-dimensional arrays, with a particular emphasis on RF excitation revealing the self-correlations between tunnel events that exist in sufficiently long arrays. A link with the turnstile of Chapter 3 is made. Chapter 8 by MOOIJ and

SCHÖN treats two-dimensional arrays, both in the vortex and the charge regimes. The two-dimensional character of these arrays allows for the possibility of a phase transition. The approximate duality between the vortex and charge regimes is elucidated. Finally, Chapter 9 by AVERIN and LIKHAREV handles the delicate task of predicting the future of the field. In particular, it discusses the bold and tantalizing claim that digital electronics, in its ultimate stage, will be based on SCT phenomena.

Contents

4 Josephson Effect in Low-Capacitance Tunnel Junctions
by M. Tinkham

139

5 Coulomb-Blockade Oscillations in Semiconductor Nanostructures
by H. van Houten, C. W. J. Beenakker, and A. A. M. Staring

167

Chapter 1

Introduction to Single Charge Tunneling

M. H. DEVORET

Groupe Quantronique, Service de Physique de l'Etat Condensé,
Centre d'Etudes de Saclay, 91191 Gif-sur-Yvette, France

and

H. GRABERT

Fachbereich Physik, Universität–GH Essen, 4300 Essen, Germany

1. Basic ingredients of single charge tunneling phenomena

Consider a charge transport experiment in which a voltage difference is applied to two electrodes (a "source" and a "drain", see Fig. 1) separated by an insulating gap. In the middle of the gap lies a third metallic electrode, which we call an "island" since it is surrounded by an insulator. To travel from the source to the drain the electrons must go through the island. We assume that the conduction of electrons through the insulating gaps between the source and the island and between the island and the drain occurs by quantum tunneling. This process is so fast that we can consider that the electrons are traversing the insulating gaps one at a time. Nevertheless, the successive tunnel events across a particular junction are uncorrelated and constitute a Poisson process. The key point is that, during its journey from the source to the drain, the electron necessarily makes the charge of the island vary by e. This is a tiny amount of charge if we consider

Figure 1. The quantum tunneling of electrons between a "source" and a "drain" electrode through an intermediate "island" electrode can be blocked if the electrostatic energy of a single excess electron on the island is large compared with the energy of thermal fluctuations.

Single Charge Tunneling, Edited by H. Grabert and
M.H. Devoret, Plenum Press, New York, 1992

ordinary electronic devices: each charge packet in a charge coupled device (CCD), for example, is composed of about 10^6 electrons [1]. However, if the island is small enough, the variation of the island potential due to the presence of an excess electron can be large enough to react back on the tunneling probabilities. The existence of such a feedback effect was proposed several decades ago [2]–[6].

At that time, the effect could only be observed in granular metallic materials. It was realized that the hopping of electrons from grain to grain could be inhibited at small voltages if the electrostatic energy $e^2/2C$ of a *single* excess electron on a grain of capacitance C was much greater than the electron thermal energy $k_B T$. The interpretation of these pioneering experiments, in which single electron effects and random media properties interplay, was complicated by the limited control over the structure of the sample. Nowadays, with modern nanofabrication techniques, it is possible to design metallic islands of known geometry separated by well controlled tunnel barriers [7]. Bias leads can impose a voltage across the whole set of barriers and the charge distribution of the islands can be acted upon by leads connected to small gate capacitors. In these nanoscale tunnel junction systems, a fully developed "Coulomb gap" arises [8, 9] which can be exploited to control a current by means of a single charge on a gate [10] and to transfer single charges from one island to another in a controlled way [11, 12]. The mechanism underlying these systems exploits the feedback effect of the Coulomb interaction energy of a charge with the other charges in the system. More generally, this feedback effect characterizes what we call *single charge tunneling* (SCT) phenomena. They can take place not only in normal metal and semiconductor junction systems in which the individual charge carriers are electrons and holes but also in superconducting systems in which the charge carriers are Cooper pairs.

The introductory remarks above indicate the basic requirements for single charge tunneling phenomena to occur in nanoscale junction systems. Leaving aside for the moment the special case of a single tunnel junction which will be discussed in the following section, these conditions are as follows. Firstly, the system must have metallic islands that are connected to other metallic regions only via tunnel barriers with a tunneling resistance R_T that exceeds the resistance quantum $R_K = h/e^2 \simeq 25.8$ kΩ, i.e.,

$$R_T \gg R_K. \tag{1}$$

The tunneling resistance is a phenomenological quantity which is defined in the situation where a fixed voltage difference V is imposed to the two electrodes on either side of the tunnel barrier. The tunneling rate Γ of an electron through the barrier is then proportional to V : $\Gamma = V/eR_T$. The tunneling resistance can be expressed in terms of the microscopic quantity $\mathcal{T}$, which is the barrier transmission coefficient at the Fermi energy: $R_T^{-1} = 4\pi N \mathcal{T} R_K^{-1}$, where N is the number of independent electron channels through the barrier. Condition (1) is obtained by requiring that for an excess charge on the island the energy uncertainty associated with the lifetime due to tunneling $\tau_r = R_T C$ is much smaller than the Coulomb energy $E_c = e^2/2C$. Essentially, condition (1) ensures that the wave function of an excess electron or Cooper pair on an island is localized there. It is generally believed that in systems with tunneling resistances that are small on the scale provided by R_K charging effects will be suppressed since delocalized states in which electrons flow through an island without charging it are available for charge transport

[13, 14], although the exact circumstances are not precisely known at the time of this writing.

Secondly, the islands must be small enough and the temperature low enough that the energy E_c required to add a charge carrier to an island far exceeds the available energy of thermal fluctuations, i.e.,

$$E_c \gg k_B T. \tag{2}$$

In practice, only islands having capacitances not much below a fF can be reliably designed, thus imposing experiments done at a few tens of mK, now routinely attainable with a dilution refrigerator. Conditions (1) and (2) ensure that the transport of charge from island to island is governed by the Coulomb charging energy. With the use of externally applied gate voltages, the charging energy of the various islands can be sequentially lowered or increased in order to manipulate single charge carriers (see Chap. 3).

At present, two main types of systems where single charge tunneling phenomena arise are being explored. The majority of experiments carried out so far have used metallic (mostly Al) thin film systems in which the lithographically patterned islands are separated by oxide layer tunnel barriers (see Chaps. 3, 4, 7 and 8). In this case, three-dimensional electron gases confined to small regions are coupled by the tunnel effect through the oxide layer which is only about 10Å thick. The capacitances of the tunnel junctions thus make up for the most part of the capacitance of an island. These systems also allow one to explore charging effects involving Cooper pairs since the metals used to fabricate the circuits are superconductors at the temperatures required to satisfy (2). As a matter of fact, one must apply a magnetic field to keep the metals in the normal state.

Single charge tunneling effects also occur when the two-dimensional electron gas of a GaAs/AlGaAs heterostructure is confined to small islands by means of Schottky gates. In that case, covered in Chap. 5, electrons can tunnel through the depleted region between islands and the tunnel resistances of the "junctions" can be tuned by changing the confining gate voltages. Further, the islands may be reduced to quantum dots with a discrete energy spectrum for single electron wave functions. This situation yields interesting phenomena combining charging effects and resonant tunneling. It is remarkable that even for such dots containing only several tens of electrons, charging effects can usually be described through an effective island capacitance in close relationship with its geometry. The fact that the electrostatic capacitance remains a useful concept for such small islands calls for an explanation: The applicability of the notion of an island capacitance depends on the ratio between the screening length and the size of the island. This ratio, which is of order 10^{-4} in metallic systems, is still comfortably smaller than 1 in the case of quantum dots.

Apart from granular films and lithographically patterned systems covered in this book, single charge tunneling phenomena are observed in a number of other cases like small metal particles embedded in an oxide layer or disordered quantum wires. Also, one of the tunneling barriers may be formed by a scanning tunneling microscope. Detailed lists of references to these studies can be found in the review articles by Averin and Likharev [8] and Schön and Zaikin [9], and also in the Special Issue of Zeitschrift für Physik B – Condensed Matter on SCT [15].

2. Single current biased junction

In the preceding section we have stressed that the basic system in which single charge tunneling phenomena occur is a metallic island connected to electron reservoirs through at least *two* tunnel barriers. What would happen with only *one* small capacitance tunnel junction? Physicists are attracted to simple systems. Historically, it is this question which started the field a few years ago. Several new effects due to the quantization of charge were predicted to arise in an ultrasmall tunnel junction, both in the superconducting and the normal state [16]–[19]. Likharev and coworkers [17, 19] gave a major thrust to this new area of low temperature physics by making detailed predictions of Coulomb blockade phenomena in a single junction and by proposing various applications of the new effects.

The theory of Likharev and coworkers considers a tunnel junction which is biased by a current I and whose voltage V is measured by a very high impedance voltmeter. The junction is characterized by two parameters: its capacitance C and tunnel resistance R_T. The state of the junction is described by two degrees of freedom whose different nature is crucial. The first degree of freedom is the charge Q on the junction capacitance. It is a continuous variable since it describes the bodily displacement of the electron density in the electrodes with respect to the positive ionic background. In fact, Q can be an arbitrarily small fraction of the charge quantum. The second degree of freedom is the discrete number n of electrons (or Cooper pairs if the electrodes are in the superconducting state) which have passed through the tunnel barrier. The key hypothesis in the theory is that Q and n are classical variables with a well defined value at every instant of time t. Charge conservation is imposed by the relation $\dot{Q}(t) + e^* \dot{n}(t) = i(t)$, where $i(t)$ is the current flowing in the leads of the junction and e^* is the charge e (normal state) or $2e$ (superconducting state) of the carriers. Since the current bias is assumed to be ideal we have $i(t) = I$. During a tunneling event, the charge Q must thus discontinuously jump by the elementary charge e^*. The resulting change in electrostatic energy of the junction is

$$\Delta E = \frac{Q^2}{2C} - \frac{(Q - e^*)^2}{2C} = \frac{e^*(Q - e^*/2)}{C}. \tag{3}$$

At zero temperature, tunneling can only occur if ΔE is positive. This has two consequences. Firstly, the $I - V$ characteristic should have an $I = 0$ branch

$$-\frac{e^*}{2C} < V < \frac{e^*}{2C} \quad \text{for} \quad I = 0 \tag{4}$$

where the particular value of V is determined by the history of the current in the junction leads: $CV = \int_{-\infty}^{t} i(t') \mathrm{d}t'$ modulo e^*. This is the Coulomb blockade for single junctions. Secondly, when a non-zero current is imposed through a junction in the normal state, the junction capacitor charge Q will increase linearly until the threshold charge $e/2$ is reached. Then, a tunneling event occurs, making Q jump to $-e/2$ and a new charging cycle starts again. This leads to single electron tunneling (SET) sawtooth oscillations of the junction voltage with the fundamental frequency

$$f_{\text{SET}} = I/e. \tag{5}$$

By a similar kind of reasoning, one predicts for a superconducting junction (Josephson junction) the so-called Bloch oscillations with the frequency

$$f_{\text{Bloch}} = I/2e. \tag{6}$$

The difference between the SET and Bloch oscillations is that in the normal state the charge tunnels irreversibly as Q goes beyond $e/2$ because it is accompanied by quasiparticle excitations whereas in the superconducting state the charge tunnels reversibly at $Q = e$ because Cooper pairs have no kinetic degrees of freedom.

This analysis rests on Q and n being classical variables. The classical nature of the variable n is solely determined by the properties of the junction. We can safely assume that condition (1), which is a statement about the tunnel barrier and which translates directly in terms of junction fabrication, is a sufficient condition. However, the classical nature of the variable Q depends on the junction electromagnetic environment and the original predictions concerning Coulomb blockade phenomena did not make very explicit statements about what the characteristics of this environment should be. In this theoretical void, two questions concerning the observability of Coulomb blockade and SET or Bloch oscillations arose:

Question A: The pads on the junction chip which are needed to make connections to the $I - V$ measuring apparatus have parasitic capacitances in the pF range. How should the junction environment be designed for these parasitic capacitances not to shunt the junction capacitance, which needs to be kept in the fF range to observe the charging effects?

Question B: Each mode in the environment is coupled to the charge Q and its zero point energy induces $i(t)$ and $Q(t)$ to fluctuate. How should the environment be designed for these quantum mechanical fluctuations not to affect the Coulomb blockade? In other words, how perfect does the current biasing need to be?

We will see below that these two questions are in fact closely related and that their answer can be obtained by a fully quantum mechanical analysis of the influence of the junction environment on the tunneling probability. Before presenting the results of this analysis, which is essential to the understanding of single charge tunneling phenomena, we have to discuss the various time scales of the problem, both the time scales pertaining to the junction itself and those pertaining to its environment.

The junction is characterized by three time scales. The two longer ones can be deduced from quantities we have already mentioned. The longest time scale is set by the tunneling resistance and the capacitance: $\tau_r = R_T C$. It is the reciprocal of the rate of tunnel events for a junction biased at the Coulomb voltage e/C. The intermediate time scale is the uncertainty time associated with the Coulomb energy $\tau_c = h/(e^2/C) = R_K C$. The shortest time scale is the tunneling time τ_t of the junction which is given by

$$\tau_t = \hbar \left(\frac{\partial \ln \mathcal{T}(E)}{\partial E} \right)_{E=E_F} \tag{7}$$

where, as previously, $\mathcal{T}(E)$ is the transmission probability through the tunnel barrier of an electron with energy E. This tunneling time, whose importance has been stressed by Büttiker and Landauer [20] (see also [21] and references therein), can be loosely described as the time spent by the tunneling electron under the barrier. In metallic

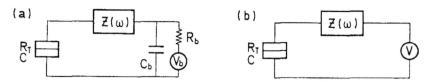

Figure 2. Lumped element model of the electromagnetic environment of a current-biased tunnel junction, which is represented by a double box symbol. The capacitance and the tunnel resistance of the junction are C and R_T, respectively. The impedance $Z(\omega)$ models the high frequency response of the environment which is dominated by the effect of the leads attached to the junction. The environmental low frequency response, which is dominated by the bias circuitry, is modelled in (a) by a resistance R_b, a capacitance C_b, and a voltage source V_b. In all practical cases $C_b \gg C$, and one can use the simplified model (b) in which the junction is biased by an effective voltage source V which is a function of the time-averaged current through the junction.

tunnel junctions it is of the order of 10^{-15} s. Here, it may be worthwhile pointing out that electron tunneling in a metallic junction is in fact a complex process, at least much more complex than what elementary textbooks might lead one to suppose: the electrons in the metallic electrodes travel as quasiparticles, i.e., bare electrons dressed by a positive cloud of charge. When a tunneling quasiparticle impinges on the insulating barrier it has to undress, leaving the positive charge cloud behind as it travels through the barrier. When this bare electron arrives in the other electrode it attracts a new cloud of positive charge and dresses again to form a quasiparticle. The characteristic time for the undressing and dressing processes is the inverse of the plasma frequency. These processes have to be taken into account in the computation of the effective tunneling time, which is the one of interest here. The tunneling rate of a quasiparticle will be quite different from the tunneling rate of a bare electron if the effective tunneling time is notably longer than the inverse of the plasma frequency [22].

Let us now discuss the time scales of the environment. We first have to indicate how one should model the junction electromagnetic environment which includes not only the $I - V$ characteristic measuring apparatus at high temperature but also the leads close to the junction. A priori, we need to consider the response of the environment up to the frequency τ_t^{-1}. Although this natural cut-off provided by the tunneling time is a frequency in the optical domain, the junction is small enough to be treated as a lumped element since its dimensions have to be of the order of 100 nm or less to ensure a capacitance in the fF range. The electromagnetic environment as seen from the location of the junction can thus be completely described in electrical engineering terms by the relationship between the complex amplitudes $v(\omega) = Q(\omega)/C$ and $i(\omega)$ at frequency ω of the voltage across the junction and the current in the first few hundred nanometers of its leads. Assuming that the environment is linear, we arrive for the electromagnetic environment as seen from the location of the junction at the general lumped element model of Fig. 2a. The bias circuitry, which includes the room temperature electronics, the filters, and the leads down to the pads on the junction chip, is modelled by a bias resistor $R_b \gg R_T$ in series with a voltage source V_b. There is also, in parallel with the resistor and the source, a capacitance C_b which models the parasitic capacitances in the bias circuitry. This three element model of the bias circuitry accounts for the low frequency response of the environment and is placed in series with a complex impedance $Z(\omega)$ which represents the impedance of the last few mm of leads on the junction chip. The impedance $Z(\omega)$ accounts for the high frequency response of the environment.

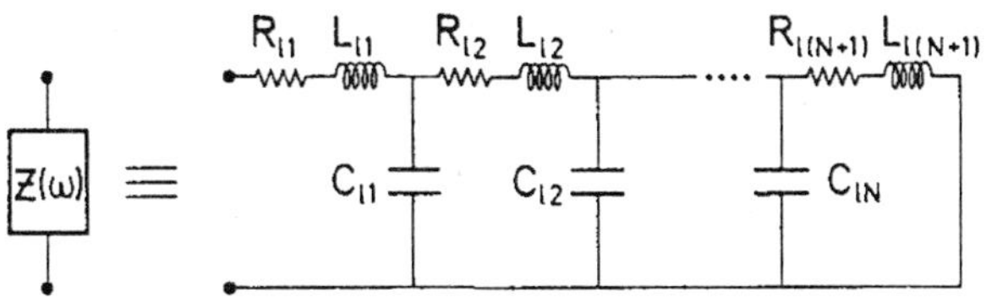

Figure 3. Resistive transmission line model for the impedance $Z(\omega)$.

This general model of the environment of the junction can be somewhat simplified, however. One has to note that the parasitic capacitance C_b is larger than the junction capacitance C by orders of magnitude (typically $C_b \simeq 10^4 C$). This means that the voltage on the capacitance C_b is essentially time independent although the current through the junction is composed of pulses corresponding to the tunnel events. One can thus replace the model of Fig. 2a by the model of Fig. 2b in which the impedance $Z(\omega)$ is simply in series with a voltage source V. Of course, V has to be determined self-consistently from the time-averaged current $I(V)$ through the junction by the relation $V = V_b - R_b I(V)$, but this is not a problem. If we know how to calculate the $I - V$ characteristic of the junction for the model of Fig. 2b, the junction voltage as a function of the bias current V_b/R_b for the model of Fig. 2a can be reconstructed.

An important remark is now in order. The value of the impedance $Z(\omega)$ at moderately high frequencies can be made large by making the leads on the junction chip very narrow and by using a resistive material like NiCr. However, at frequencies corresponding to micron wavelengths, no matter how careful one is in the fabrication of the leads, the impedance $Z(\omega)$ will be dominated by radiation phenomena and will be of the order of the impedance of free space $Z_V = (\mu_0/\epsilon_0)^{1/2} \simeq 377\,\Omega$. The modulus $|Z(\omega)|$ of the impedance is thus a decreasing function of frequency. This behavior can be crudely understood by considering the parasitic capacitance between the leads, whose shorting effect on the junction becomes more and more pronounced as the frequency gets higher.

A more precise understanding of the frequency dependence of the environment is provided by a resistive transmission line model of the function $Z(\omega)$. This model is analogous to the model described by Martinis and Kautz for experiments on the phase diffusion of a small Josephson junction [23]. The resistive transmission line can be thought of as a ladder of discrete components R_{ln}, C_{ln} and L_{ln}, as shown on Fig. 3. The total capacitance and resistance of the transmission line are $C_l = \sum_{n=1}^{N} C_{ln}$ and $R_l = \sum_{n=1}^{N} R_{ln}$, respectively, while the characteristic impedance of the line is $Z_l = (L_{ln}/C_{ln})^{1/2}$. In practice, Z_l is always a fraction of the vacuum impedance Z_V while C_l, which one tries to get as small as possible, is not much below $0.1\,\mathrm{pF}$. As we mentioned above, by using very narrow leads made from NiCr, values of the order of $100\,\mathrm{k\Omega}$ can be obtained for R_l (we refer to this as the "extreme" case). If no special effort is put in making high resistance lead resistors, typical values for R_l are in the $100\,\Omega$–$1\,\mathrm{k\Omega}$ range (hereafter referred to as the "standard" case). A log-log plot of the function $|Z(\omega)|$ is shown schematically in Fig. 4. If one is in the standard case where the lead total resistance R_l is comparable to the line impedance Z_l, the environment behaves as a resistor Z_l. If one is in the extreme case where the lead total resistance is much higher than the line impedance, the environment behaves at low frequencies as a resistor R_l until a roll-off frequency given by $1/R_l C_l$ is reached. One then enters an RC line regime where the leads

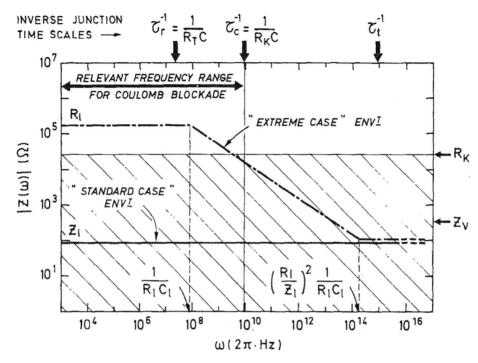

Figure 4. Schematic behavior of the modulus $|Z(\omega)|$ of the environmental impedance as a function of the frequency ω. We have shown for comparison (i) the inverse junction time scales on the frequency axis and (ii) the resistance quantum $R_K = h/e$ and the impedance of the vacuum $Z_V = \sqrt{\mu_0/\epsilon_0}$ on the resistance axis.

behave as an impedance with equal reactive and dissipative parts falling off as $\omega^{-1/2}$. Finally, at the saturation frequency $\omega_s = (R_l/Z_l)^2/(R_lC_l)$ one recovers the frequency independent behavior of the standard case. Note that the saturation frequency ω_s is independent of the length of the leads provided that $C_{ln} = C_l/N$ and $R_{ln} = R_l/N$, which is a realistic assumption. As an example, for leads with distributed resistance, capacitance, and inductance of $100\,\Omega/\mu$m, $0.5 \times 10^{-16}\,$F$/\mu$m, and $0.5 \times 10^{-12}\,$H$/\mu$m, respectively, one finds that $\omega_s/2\pi = 1.6 \times 10^{13}\,$Hz. These values correspond to NiCr resistors that are 60 nm thick and 1 μm wide.

The imperfection of the current biasing scheme, on the one hand, and the parasitic capacitances with which the environment shunts the junction, on the other hand, are thus just two aspects of the properties of the function $Z(\omega)$. Questions A and B can now be unified into a single one: What values should $|Z(\omega)|$ have at the junction characteristic frequencies in order to observe a Coulomb blockade? It is clear that in the spirit of (1) a *sufficient* condition to ensure that Q is a classical variable reads

$$|Z(\omega)| \gg R_K \quad \text{for} \quad \omega < \tau_t^{-1}, \tag{8}$$

but this is impossible to satisfy. There is a fundamental limitation since the ratio between the impedance of free space Z_V, which controls the asymptotic behavior of $Z(\omega)$ at high frequencies, and the resistance quantum R_K is equal to twice the fine structure constant $1/137$. Thus one cannot avoid the problem of finding the tunneling rate as a function of V for a junction coupled to an arbitrary $Z(\omega)$. The environment needs to be treated quantum mechanically, since in most of the relevant frequency range thermal fluctuations are smaller than quantum fluctuations, i.e. $\hbar\omega \gg k_BT$. Details on the way the problem is dealt with can be found in Chap. 2. In the following we will just emphasize the important features of the theory on the effect of the electromagnetic environment [24]–[26].

The theory first assumes a clear separation of time scales

$$\tau_t \ll \tau_c \ll \tau_r. \tag{9}$$

The first inequality states that the tunneling time is negligible while the second one states the classical nature of n. The theory then considers the modes of the linear circuit formed by the environmental impedance $Z(\omega)$ in series with the junction capacitance C. Of course, for a dissipative environment, the mode frequencies form a continuous spectrum. It is assumed that before a tunnel event the environmental modes are in their equilibrium state. A tunnel event can excite them. This process is described by a function $P(E)$, which gives the probability that the tunneling electron transfers the energy E to the distribution of modes of the circuit [25]. One finds, from a quantum calculation, that $P(E)$ is a distribution function which is determined in terms of the density of environmental modes given by the real part of the total circuit impedance $Z_t(\omega) = 1/[iC\omega + Z(\omega)^{-1}]$. The probability $P(E)$ is given by

$$P(E) = \frac{1}{2\pi\hbar} \int_{-\infty}^{+\infty} dt \, \exp[J(t) + iEt/\hbar] \tag{10}$$

with

$$J(t) = 2 \int_0^\infty \frac{d\omega}{\omega} \frac{\text{Re}[Z_t(\omega)]}{R_K} \left(\coth(\beta\hbar\omega/2)[\cos(\omega t) - 1] - i \sin(\omega t) \right) \tag{11}$$

where $\beta = 1/k_B T$ is the inverse temperature. Finally, the tunneling rate in the direction imposed by V is computed from $P(E)$ using

$$\Gamma = \frac{1}{2\tau_r E_c} \int_{-\infty}^{+\infty} dE \int_{-\infty}^{+\infty} dE' \, f(E)[1 - f(E')]P(E + eV - E') \tag{12}$$

which reflects the fact that only a part of the energy eV of the voltage source is used to excite the environment, the rest being used to excite one hole and one electron on either side of the barrier. At thermal energies much lower than the Coulomb energy, i.e. $\beta E_c \gg 1$, one draws the following conclusions:

i) For impedances $Z(\omega)$ such that $|Z(\omega)| \ll R_K$ for all frequencies, the function $P(E)$ is sharply peaked at $E = 0$, i.e. $P(E) \simeq \delta(E)$, and we find from (12) a straight $I - V$ characteristic with no Coulomb blockade. This result means that most tunneling transitions leave the environmental modes undisturbed except those near $\omega = 0$. The charge transferred through the junction is thus removed instantaneously by the voltage source constituted by the pads even though it is physically located a few mm away. In a way, for most tunneling events, *the environment acts as a perfect voltage source*. This is a purely quantum mechanical effect. It seems to defy locality since one would expect the charge to propagate at the speed of light from the reservoir of charge to the junction. This expectation, which seemed based on good relativistic common sense, was actually the basis for an argument in favor of the existence of Coulomb blockade phenomena for tunnel junctions in a low impedance environment [27]. There is in fact no contradiction between locality and the perfect "quantum rigidity" of the charge along the leads suggested by quasi-elastic tunneling. In this case of a low impedance environment, one calculates that the zero point motion of the environmental modes induces quantum fluctuations of the charge Q which are much larger than e. Remember that the junction and the environment behave as a whole quantum mechanically coherent unit. One can thus assume that the transferred charge is entirely provided by the zero point charge fluctuations of the environment. Crudely speaking, even though the charge is removed instantaneously by the voltage source, the source cannot tell when a tunnel event occurs because the charge pulse associated with a tunnel event is buried in the quantum fluctuations. This "charge-less" transfer of charge through the junction is analogous to the Mössbauer effect. Gamma rays can be emitted from a nucleus in a solid without exciting the phonon modes. The conservation of momentum is not violated because the recoil momentum of the nucleus is transferred to the whole crystal ("recoil-less" emission). One can think of our function $P(E)$ as being equivalent to the gamma ray energy spectrum.

ii) For impedances such that $|Z(\omega)| \gg R_K$ for all frequencies $\omega \leq \tau_c^{-1}$, the tunneling electrons are well coupled to the environmental modes. One finds at zero temperature that the function $P(E)$ is sharply peaked at E_c, i.e. $P(E) \simeq \delta(E - E_c)$. Hence, like in the classical case, an electron can only tunnel when it gains at least E_c from the applied voltage, which leads to a Coulomb blockade of tunneling. The problem is that this limit is very difficult to achieve experimentally. We have represented in Fig. 4 the domain where $|Z(\omega)| < R_K$ by a shaded area. The $\omega^{-1/2}$ roll-off of the impedance must cross the impedance quantum R_K at a frequency high enough on the scale of the Coulomb frequency τ_c^{-1}. In practice this means that the on-chip lead resistors must have a saturation frequency ω_s as high as possible. This requirement is unfortunately in conflict with the requirement of no heating in the resistor and a compromise has to be found. This has been achieved by Cleland et al. [28] for normal junctions and by Kuzmin et al. [29] for superconducting junctions.

The theory we have outlined can easily be adapted to the tunneling of Cooper pairs [30]. The function $P(E)$, modified slightly to take into account the charge $2e$ of Cooper pairs, yields directly the $I - V$ characteristic if no quasiparticles are present. Schön and coworkers [31, 32] have worked out detailed predictions taking into account finite quasiparticle tunneling. As in the normal state, theory predicts that no "Cooper pair gap" exists for a single junction if the environmental impedance is less than R_K. This result explains why no Cooper pair gap was found in the Harvard group experiments described in Chap. 4. The extension of the theory to Josephson junctions is also discussed in Chap. 2.

3. Single island circuits

The preceding section showed that a current biased single tunnel junction is not, after all, a particularly simple system as far as single charge tunneling is concerned. It is certainly of interest for the foundations of the field, but it is not suited for practical applications, since the requirements for a clear-cut Coulomb blockade are so difficult to realize experimentally. We have seen that the smallness of the fine structure constant imposes the magnitude of the charge fluctuations on the junction to be much greater than e in standard cases.

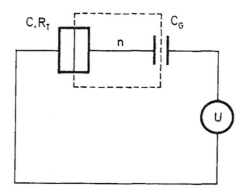

Figure 5. The single electron box, consisting of a tunnel junction in series with a capacitor. The number n of excess electrons on the island is controlled by a gate voltage U.

Another point of view can be adopted to understand why it is difficult to observe Coulomb blockade in a current biased single junction. This other point of view will make clear why islands are necessary for the occurrence of fully developed single charge tunneling phenomena. Let us compute the total *equilibrium* electrostatic energy of the circuit of Fig. 2a as a function of the number n of electrons that went through the junction. In making this calculation one assumes that the junction behaves as a perfect capacitor after the last electron has gone through it and one takes into account the work performed by the voltage source, the whole circuit having relaxed to equilibrium. One then finds

$$E_{\text{eq}} = -neV + \text{ terms independent of } n. \tag{13}$$

It is thus always energetically favorable for an electron to tunnel. Coulomb blockade in a current biased single junction is, at best, just a dynamical effect in which one is trying to slow down the tunneling rates as much as possible by making the environmental impedance as high as possible.

Consider now the circuit of Fig. 5 which is the simplest tunnel junction circuit containing at least one island. The island lies between a nanoscale tunnel junction with capacitance C and a gate capacitance C_G whose order of magnitude is close to that of C. This junction-capacitor combination has been nicknamed the *single electron box* [33]. The box is controlled by a voltage source U which closes the circuit. We shall restrict ourselves in the sequel to the standard case of low impedance leads. Then, as described above, the charges Q and Q_G on the junction and gate capacitances undergo large quantum fluctuations. However, these two capacitances in series couple to the leads like one capacitance $C_s = CC_G/(C + C_G)$ carrying the charge $Q_s = (C_G Q + C Q_G)/(C + C_G)$. Only this linear combination of Q and Q_G is affected by the electromagnetic environment [34]. The other linear combination, which is the island charge $Q_i = Q - Q_G$ decouples from the leads. The charge $Q_i = -ne$ is quantized in units of the elementary charge, and n, the number of "electrons in the box", is the number of excess electrons on the island. To change n an electron has to tunnel through the junction. As in (13), one can calculate the equilibrium electrostatic energy of the circuit as a function of n. It is given by

$$E_{\text{eq}} = \frac{(C_G U - ne)^2}{2(C + C_G)} + \text{terms independent of } n. \tag{14}$$

There is now a big difference with the current biased junction. The equilibrium energy change

$$\Delta E_{\text{eq}} = \frac{e(C_G U - ne - 1/2)}{C + C_G} \tag{15}$$

that accompanies a transition from n to $n + 1$ can now be *positive*, hence ensuring an *equilibrium* Coulomb blockade. The expression (15) is in fact similar to (3), except that the junction charge is replaced by the island charge $Q_i = -ne$ which is shifted by the gate voltage. Despite the fact that ΔE_{eq} depends on the entire circuit, it can be written in terms of the average charge $\langle Q \rangle$ on the tunnel junction. Of course, $\langle Q \rangle$ depends on the applied voltage U. Using simple electrostatics one finds

$$\Delta E_{\text{eq}} = \frac{e}{C}(\langle Q \rangle - Q_c), \tag{16}$$

where

$$Q_c = \frac{C}{C + C_G} \frac{e}{2} \tag{17}$$

is the so-called critical charge of the junction [11, 34] which is less than $e/2$. For the case of a low impedance environment assumed here, the tunneling rate is given by

$$\Gamma = \frac{1}{e^2 R_T} \frac{\Delta E_{\text{eq}}}{1 - \exp[-\Delta E_{\text{eq}}/k_B T]}, \tag{18}$$

which at zero temperature reduces to

$$\Gamma = \begin{cases} \Delta E_{\text{eq}}/e^2 R_T & \text{for} \quad \Delta E_{\text{eq}} > 0 \\ \\ 0 & \text{for} \quad \Delta E_{\text{eq}} < 0. \end{cases} \tag{19}$$

As soon as $\langle Q \rangle$ exceeds Q_c an electron can tunnel onto the island. According to formula (18), the tunneling rate is determined by the change (15) of the equilibrium electrostatic energy of the entire system caused by the transition. This so-called global rule rate [35, 36] is found to be very accurate when the tunneling resistance R_T satisfies (1) and the electromagnetic environment is of low impedance [34]. From (19) we see that at zero temperature a transition from n to $n + 1$ only occurs for $C_G U > e(n + \frac{1}{2})$. Likewise, one finds for transitions from n to $n - 1$ the condition $C_G U < e(n - \frac{1}{2})$. Hence, for sufficiently low temperatures and gate voltages U in the interval

$$e(n - \frac{1}{2}) < C_G U < e(n + \frac{1}{2}), \tag{20}$$

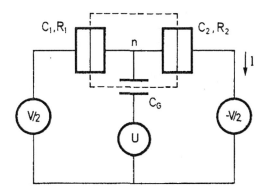

Figure 6. The SET transistor consists of two tunnel junctions in series forming an island the electrostatic potential of which is acted upon by the gate voltage U through the capacitance C_G. The transport voltage V induces a net flow of charge through the device, the value of current I being controlled by the gate voltage U.

the state with n electrons in the box is stable. By changing U electrons can thus be added one-by-one to the box. Hence, the single electron box is a simple device allowing for the manipulation of a single charge. Further details on this system are given by Lafarge et al. [33] and in Chap. 3. At the time of this writing, a box for Cooper pairs could not be operated successfully. Since for quasiparticles the threshold voltage for tunneling is always lower than that for Cooper pairs, an island between a Josephson junction and a capacitance will behave like an electron box even when the density of quasiparticles is small.

Another basic device with just one island is the double junction driven by a transport voltage V. Very often a gate capacitor with a gate voltage U is coupled to the island between the junctions. This is the single electron tunneling (SET) transistor [8] with the circuit diagram depicted in Fig. 6. The first observation of single charge tunneling in microfabricated samples by Fulton and Dolan [10] was made with this device. SET transistors are also part of more elaborate devices fabricated with oxide layer tunnel junctions (see Chap. 3), and most of the studies on charging effects in semiconductors have used this type of circuit (see Chap. 5).

In the SET transistor the island can be charged by tunneling across one junction and discharged by tunneling across the other junction, which leads to a net current through the device. If the tunneling resistances of both junctions satisfy (1), the electron transfer rates are again determined by the change of the equilibrium electrostatic energy of the circuit. In semiconductor devices with a small number of electrons in the segment between the junctions, the island may form a quantum dot with an energy level separation that exceeds $k_B T$. Then the energy difference between the Fermi level of the dot and the next available state has to be considered when calculating the energy change. This case is discussed in Chap. 5. When the discreteness of the spectrum of electronic states on the island can be disregarded, the energy change due to a transition from n to $n+1$ excess electrons as a consequence of tunneling across the first junction [cf. Fig. 3] is found to be (we drop the subscript eq)

$$\Delta E_1 = \frac{e[(C_2 + \frac{1}{2}C_G)V + C_G U + ne - \frac{e}{2}]}{C_\Sigma},$$ (21)

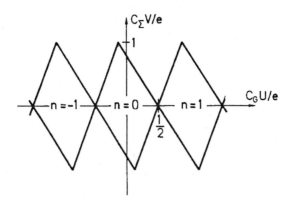

Figure 7. The stability diagram of a SET transistor with $2C_2 = 10C_G = C_1$. The transistor conducts only outside the rhombic-shaped regions. Inside these regions, there is a constant number n of electrons on the island.

where

$$C_\Sigma = C_1 + C_2 + C_G \tag{22}$$

is the capacitance of the island. In (21) the gate voltage U only appears in the combination $C_G U - ne$, which leads for all measurable quantities to a periodicity in U with period e, since the integer part of $C_G U/e$ can always be absorbed in n. In semiconductor devices this strict periodicity is usually not met because the gate voltage U influences the tunneling resistances of the junctions and because the gate capacitance is weakly dependent on U.

A straightforward analysis of the rate formula (19) shows that at zero temperature the state with n electrons on the island of the SET transistor is stable with respect to tunneling across the first and second junctions for voltages satisfying

$$e(n - \frac{1}{2}) < C_G U + (C_2 + \frac{1}{2}C_G)V < e(n + \frac{1}{2})$$

$$e(n - \frac{1}{2}) < C_G U - (C_1 + \frac{1}{2}C_G)V < e(n + \frac{1}{2}), \tag{23}$$

respectively. Hence, in the UV–plane there are rhombic–shaped regions along the U–axis within which the transistor island is charged with a fixed number of excess electrons [cf. Fig. 7]. Inside these rhombi all transitions are suppressed by a Coulomb blockade and no current flows through the device.

For example, near $U = V = 0$ the state $n = 0$ is stable. The inequalities (23) show that whenever the system leaves the stability region of $n = 0$ at a point in the UV–plane with $V \neq 0$ and a tunneling transition, say, to $n = 1$ occurs, the new state is not stable with respect to tunneling across the other junction. Hence, shortly after the first tunneling event an electron leaves the island through the other junction and the system returns to $n = 0$, where the cycle can start again. As a net effect, a current flows through the device. The second tunneling transition of this cycle occurs for $Q > Q_c$ and part of the change of electrostatic energy is left as kinetic energy of the tunneling electron. Therefore, the transistor is a dissipative element, in contrast to the single electron box discussed above which is reversible when the gate voltage is changed slowly.

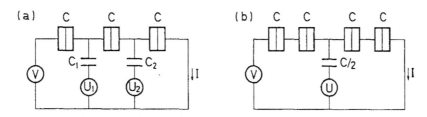

Figure 8. The circuit diagrams of the single electron pump (a) and turnstile (b). An appropriate rf modulation applied to the gate voltage(s) of these devices transfers precisely one electron through them per cycle.

For voltages V of order e/C the current I is very sensitive to U. A small change of the polarization charge $C_G U$ by a fraction of the elementary charge e can change the current from zero to values of the order of E_c/eR_T. This is why the SET transistor can be used as a highly sensitive electrometer [10, 33]. It also could serve as a low-noise amplifier of analog signals. Finally, the SET transistor is the basic active element of digital and other applications proposed for single charge tunneling (see Chap. 9). A detailed discussion of the SET transistor is given in Chap. 2. Since the transistor is a dissipative element, Cooper pairs can usually be transferred only in combined processes involving quasiparticles or environmental modes or both. The complex behavior of the superconducting device is discussed by Maassen van den Brink et al. [32].

At zero temperature and for voltages within the intervals (23), the state with n electrons on the transistor island is stable with respect to tunneling across either junction. However, in the presence of an applied voltage V the state can only be metastable. In fact, Averin and Odintsov [37] have pointed out that second order transitions always lead to a finite current in the presence of an applied voltage. In these co-tunneling events an electron tunnels onto the island while a second electron simultaneously leaves the island across the other junction. Since the charge on the island is only changed virtually, there is no Coulomb barrier for this process. The co-tunneling rate is proportional to $(R_K/R_T)^2$ and hence is a factor R_K/R_T smaller than the rate for first order processes. Accordingly, in more complicated multijunction circuits there are co-tunneling events involving N junctions with rates proportional to $(R_K/R_T)^N$. Of course, co-tunneling mainly arises when the inequality (1) is only poorly satisfied. However, since the time scale of all single charge tunneling processes is proportional to R_K/R_T, very large tunneling resistances severely reduce the speed of devices. That is why a detailed understanding of co-tunneling is of essential importance to the field. A survey of the theory is given in Chap. 6.

4. Circuits with several islands

More sophisticated multijunction circuits can be built using the single electron box or the single electron tunneling transistor as basic units. Fig. 8 shows the circuit diagram of the "pump" and "turnstile" devices fabricated recently by the Saclay and Delft groups. The pump designed by Pothier et al. [12] can be seen as two boxes connected by a tunnel junction. The boxes allow for a control of the input and output of electrons by means of the gate voltages. When the appropriate ac voltages with frequency f are applied to the gates, precisely one electron is transferred per cycle through the device, giving a current

$$I = ef. \tag{24}$$

This relation is the basis of high precision SCT current sources. The difference between (5) and (24) is that in (24) f is an externally imposed frequency. Since the pump principle employs only reversible processes, it can also be used to transfer Cooper pairs [38].

The turnstile designed by the Delft and Saclay groups [11] can be seen as two double junctions connected by a common island. The charging and discharging of this island is controlled by a gate voltage. Again, a current obeying (24) can be generated by means of an ac voltage. In a semiconductor version of the turnstile fabricated by Kouwenhoven et al. [39], the tunneling resistances are modulated via the voltages applied to the Schottky gates. Deviations from (24) mainly arise from finite temperature effects, electron heating, co-tunneling, and moving background charges. These effects must be reduced to achieve a current standard with metrological accuracy. A detailed introduction into the art of manipulating electrons one-by-one is given in Chap. 3.

Another class of multi-island circuits are one-dimensional and two-dimensional arrays (see Fig. 9). In these wonderful man-made crystals, the tolerable dispersion of "microscopic" parameters is more than compensated for by the fact that one can tune them to explore effects that would not exist in the natural world. In the one-dimensional arrays (Fig. 9a) of the Göteborg group (see Chap. 7) extended charge solitons are created by an external voltage which then makes them drift through the array. Only one soliton can exist at a time in the array and the reciprocal of the average time the soliton takes to travel along the array is given by (5). In the two-dimensional arrays (Fig. 9b) of the Delft group (see Chap. 8) many charge solitons of limited size exist at the same time and an interesting cooperative behavior similar to the Kosterlitz-Thouless transition arises. The two-dimensional arrays in the superconducting state are particularly fascinating, since, depending on the junction and island parameters, one can have either charge solitons (one extra Cooper pair on an island) or flux solitons (vortices sitting on the plaquette defined by four junctions). Under proper experimental conditions it seems possible that the motion of charge and flux vortices could be quantum mechanical. The duality relationship between the charge and flux solitons can be exploited to predict new quantum effects [40] discussed in Chap. 8.

5. Conclusions

Single charge tunneling has only recently developed into a field investigated in many laboratories world-wide, partly due to the progress in nanoscale fabrication techniques. Yet this area of research is about to leave its infancy. The main problems still to be overcome have been identified, and it might be appropriate to conclude by speculating about possible applications.

As mentioned above, the single electron tunneling transistor serves as a highly sensitive electrometer. The sensitivity of existing prototypes already exceeds those of other

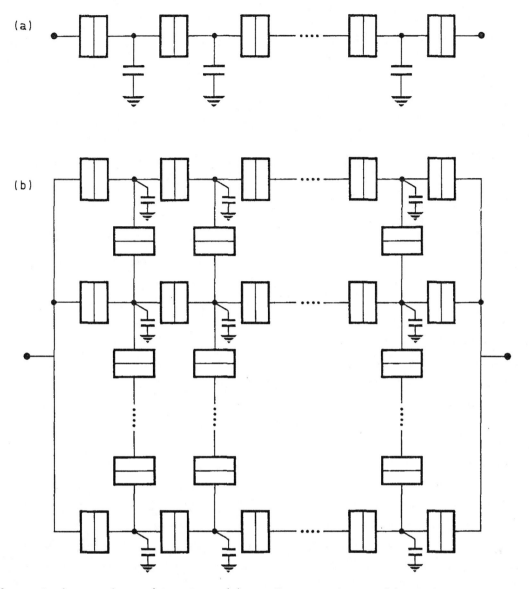

Figure 9. Arrays of tunnel junctions: (a) one-dimensional array, (b) two-dimensional array.

electrometers by 6 orders of magnitude, and the performance can certainly be improved further. The maximal sensitivity attainable is presently not known, but it should be at least $10^{-5}e/\sqrt{\mathrm{Hz}}$. Despite this very high precision, the SET electrometer is for the measurement of electrical charges not quite as revolutionary as the SQUID was for the measurement of magnetic flux. Since there is no analogue of the superconducting flux transformer, the very small input capacitance of the single electron tunneling electrometer might limit its usefulness.

The pump and turnstile devices demonstrate that single charge tunneling can be employed to construct frequency-controlled current sources. The relative uncertainty of the current produced by existing devices is below 10^{-2}, but the error sources are being analysed and improved designs are under investigation. Whether metrological accuracy of 10^{-8} is really achievable is not known, although the prospects are rather promising. Since the ampère is presently derived from the kilogram, a closure of the quantum metrological triangle could ultimately revolutionize the metrological system, and perhaps do away with the last relic of the Bureau International des Poids et Mesures in Sèvres, i.e., the standard kilogram.

Much of the fascination of single charge tunneling derives from the idea that, in the future, a single bit in an information flow might possibly be represented by a single electron (see Chap. 9). Although ingenious designs are being proposed, the less-than-unity voltage gain of the single electron tunneling transistor remains at present a fundamental engineering problem. A complementary logic analogous to the CMOS recently proposed by Tucker [41], is an attempt to overcome this problem. It is important to note that in the single electron tunneling transistor the modulation of the flow of electrons by the gate ceases as soon as the bias voltage exceeds the Coulomb gap, whereas in the FETs used in digital circuits the modulation of the source-drain current by the gate only saturates at large bias voltages [1]. This latter feature ensures enough voltage gain to compensate for the dispersion in device parameters and makes robust integrated circuit designs possible. It may happen that the real impact of single charge tunneling phenomena on nanoelectronics will be to show how, in the next generation of FET devices, one can make Coulomb charging effects reinforce the dominant Fermi effects upon which FETs are based, instead of spoiling them. Not only is this a very valuable goal, but along the way, other striking advances in fundamental science are likely to be made, like perhaps the controlled transfer of fractional charges in semiconducting systems in the quantum Hall effect regime.

ACKNOWLEDGEMENTS. We should like to thank the participants of the NATO Advanced Study Institute on SCT in Les Houches for many inspiring discussions, in particular D. Esteve, G.-L. Ingold, P. Lafarge, H. Pothier, and C. Urbina. This work was started during an extended stay by one of us (H.G.) at the Centre d'Etudes de Saclay whose kind hospitality is gratefully acknowledged. Financial support was provided through grants from NATO, the European Community, and the Deutsche Forschungsgemeinschaft.

References

[1] D. A. Fraser, *The Physics of Semiconductor Devices*, (Clarendon, Oxford, 1986).

[2] C. J. Gorter, Physica **17**, 777 (1951).

[3] C. A. Neugebauer and M. B. Webb, J. Appl. Phys. **33**, 74 (1962).

[4] I. Giaever and H. R. Zeller, Phys. Rev. Lett. **20**, 1504 (1968).

[5] J. Lambe and R. C. Jaklevic, Phys. Rev. Lett. **22**, 1371 (1969).

[6] I. O. Kulik and R. I. Shekhter, Zh. Eksp. Teor. Fiz. **68**, 623 (1975) [Sov. Phys. JETP **41**, 308 (1975)].

[7] G. J. Dolan and J. H. Dunsmuir, Physica B **152**, 7 (1988).

[8] K. K. Likharev, IBM J. Res. Dev. **32**, 144 (1988); D. V. Averin and K. K. Likharev, in: *Quantum Effects in Small Disordered Systems*, ed. by B. L. Altshuler, P. A. Lee, and R. A. Webb (Elsevier, Amsterdam, 1991).

[9] G. Schön and A. D. Zaikin, Phys. Rep. **198**, 237 (1990).

[10] T. A. Fulton and G. J. Dolan, Phys. Rev. Lett. **59**, 109 (1987).

[11] L. J. Geerligs, V. F. Anderegg, P. A. M. Holweg, J. E. Mooij, H. Pothier, D. Esteve, C. Urbina, and M. H. Devoret, Phys. Rev. Lett. **64**, 2691 (1990).

[12] H. Pothier, P. Lafarge, P. F. Orfila, C. Urbina, D. Esteve, and M. H. Devoret, Physica B **169**, 573 (1991); Europhys. Lett. **17**, 249 (1992).

[13] K. A. Matveev, Zh. Eksp. Teor. Fiz. **99**, 1598 (1991) [Sov. Phys. JETP **72**, 892 (1991)].

[14] W. Zwerger and M. Scharpf, Z. Phys. B **85**, 421 (1991).

[15] *Special Issue on Single Charge Tunneling*, Z. Phys. B **85**, 317–468 (1991).

[16] A. Widom, G. Megaloudis, T. D. Clark, H. Prance, and R. J. Prance, J. Phys. A **15**, 3877 (1982).

[17] K. K. Likharev and A. B. Zorin, J. Low Temp. Phys. **59**, 347 (1985).

[18] E. Ben-Jacob and Y. Gefen, Phys. Lett. A **108**, 289 (1985).

[19] D. V. Averin and K. K. Likharev, J. Low Temp. Phys. **62**, 345 (1986).

[20] M. Büttiker and R. Landauer, Phys. Rev. Lett. **49**, 1739 (1982).

[21] E. H. Hauge and J. A. Støvneng, Rev. Mod. Phys. **61**, 917 (1989).

[22] B. N. J. Persson and A. Baratoff, Phys. Rev. B **38**, 9616 (1988).

[23] J. M. Martinis and R. L. Kautz, Phys. Rev. Lett. **63**, 1507 (1989).

[24] Yu. V. Nazarov, Pis'ma Zh. Eksp. Teor. Fiz. **49**, 105 (1989) [JETP Lett. **49**, 126 (1989)].

[25] M. H. Devoret, D. Esteve, H. Grabert, G.-L. Ingold, H. Pothier, and C. Urbina, Phys. Rev. Lett. **64**, 1824 (1990).

[26] S. M. Girvin, L. I. Glazman, M. Jonson, D. R. Penn, and M. D. Stiles, Phys. Rev. Lett. **64**, 3183 (1990).

[27] M. Büttiker and R. Landauer, IBM J. Res. Dev. **30**, 451 (1986).

[28] A. N. Cleland, J. M. Schmidt, and J. Clarke, Phys. Rev. Lett. **64**, 1565 (1990).

[29] L. S. Kuzmin, Yu. V. Nazarov, D. B. Haviland, P. Delsing, and T. Claeson, Phys. Rev. Lett. **67**, 1161 (1991).

[30] D. V. Averin, Yu. V. Nazarov, and A. A. Odintsov, Physica B **165&166**, 945 (1990).

[31] G. Falci, V. Bubanja, and G. Schön, Europhys. Lett. **16**, 109 (1991); Z. Phys. B **85**, 451 (1991).

[32] A. Maassen van den Brink, A. A. Odintsov, P. A. Bobbert, and G. Schön, Z. Phys. B **85**, 459 (1991).

[33] P. Lafarge, H. Pothier, E. R. Williams, D. Esteve, C. Urbina, and M. H. Devoret, Z. Phys. B **85**, 327 (1991).

[34] H. Grabert, G.-L. Ingold, M. H. Devoret, D. Esteve, H. Pothier, and C. Urbina, Z. Phys. B **84**, 143 (1991).

[35] K. K. Likharev, N. S. Bakhvalov, G. S. Kazacha, and S. I. Serdyukova, IEEE Trans. Magn. **25**, 1436 (1989).

[36] U. Geigenmüller and G. Schön, Europhys. Lett. **10**, 765 (1989).

[37] D. V. Averin and A. A. Odintsov, Phys. Lett. A **140**, 251 (1989).

[38] L. J. Geerligs, S. M. Verbrugh, P. Hadley, J. E. Mooij, H. Pothier, P. Lafarge, C. Urbina, D. Esteve, and M. H. Devoret, Z. Phys. B **85**, 349 (1991).

[39] L. P. Kouwenhoven, A. T. Johnson, N. C. van der Vaart, A. van der Enden, C. J. P. M. Harmans, and C. T. Foxon, Z. Phys. B **85**, 381 (1991).

[40] B. J. van Wees, Phys. Rev. B **44**, 2264 (1991).

[41] J. R. Tucker, to be published.

Chapter 2

Charge Tunneling Rates in Ultrasmall Junctions

GERT-LUDWIG INGOLD

Fachbereich Physik, Universität-GH Essen, 4300 Essen, Germany

and

YU. V. NAZAROV

Nuclear Physics Institute, Moscow State University
Moscow 119899 GSP, USSR

1. Introduction

1.1. Ultrasmall tunnel junctions

With the advances of microfabrication techniques in recent years it has become possible to fabricate tunnel junctions of increasingly smaller dimensions and thereby decreasing capacitance C. Nowadays one can study tunnel junctions in a regime where the charging energy $E_c = e^2/2C$ is larger than the thermal energy $k_B T$. Then charging effects play an important role and this has been the subject of a by now large body of both theoretical and experimental work.

To study ultrasmall tunnel junctions, that is tunnel junctions with capacitances of 10^{-15} F or less, one may either use metal-insulator-metal tunnel junctions or constrictions in a two-dimensional electron gas formed by a semiconductor heterostructure. Due to the very different density of charge carriers in metals and semiconductors the physics of metallic tunnel junctions and two-dimensional electron gases with constrictions differ. In this chapter we restrict ourselves to metallic tunnel junctions while Chap. 5 is devoted to single charging effects in semiconductor nanostructures.

Metal-insulator-metal tunnel junctions produced by nanolithography are widely studied today. Other systems where metallic tunnel junctions are present include granular films, small metal particles embedded in oxide layers, crossed wires, and the scanning tunneling microscope. The general setup consists of two pieces of metal separated by a

Single Charge Tunneling, Edited by H. Grabert and
M.H. Devoret, Plenum Press, New York, 1992

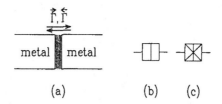

Figure 1. (a) Schematic drawing of a metal tunnel junction. The arrows indicate forward and backward tunneling through the barrier. (b) Symbol for an ultrasmall tunnel junction. The capacitor-like shape emphasizes the role of the charging energy. (c) Symbol for an ultrasmall superconducting tunnel junction.

thin insulating barrier as shown in Fig. 1. The metal may either be normal or superconducting at low temperatures. In the latter case, one may study Josephson junctions with ultrasmall capacitances as well as normal tunnel junctions if a sufficiently high magnetic field is applied.

Classically, there is no electrical transport through the barrier and the junction will act like a capacitor of capacitance C. By connecting a single junction to an external circuit it may be charged with a charge $Q = CV$ where V is the voltage applied to the junction. The charge Q is an influence charge created by shifting the electrons in the two metal electrodes with respect to the positive background charge. A very small shift of the electrons will lead to a small change in Q. Therefore, the charge Q is continuous even on the scale of the elementary charge. One finds that for such small charges the interaction between the two charge distributions on either side of the barrier may still be described by a charging energy $Q^2/2C$.

Taking into account quantum effects there is a possibility of charge transport through the barrier by tunneling of electrons indicated by the arrows in Fig. 1. In contrast to the charge motion in the electrodes this transport process involves discrete charges since only electrons as an entity may tunnel. The typical change in energy for such a process is therefore the charging energy $E_c = e^2/2C$.

This energy scale has to be compared with the other energy scale present in the system, namely k_BT. Since our interest is in the discussion of charging effects we require $e^2/2C \gg k_BT$. Otherwise thermal fluctuations will mask these effects. As an example let us note that a junction with an area of about $0.1 \times 0.1 \mu m^2$ and a typical oxide layer thickness of 10Å has a capacitance of about 10^{-15} F corresponding to a temperature of about 1K. For decreasing capacitance which requires decreasing dimensions of the junction this restriction for temperature becomes more relaxed.

1.2. Voltage-biased tunnel junction

In the following two sections we present two different pictures for the behavior of tunnel junctions. We will not give very detailed derivations at this point since similar calculations will be presented in subsequent sections.

Let us first consider a tunnel junction coupled to an ideal voltage source V. In order to determine the current-voltage characteristic one needs to calculate electron tunneling rates in both directions through the junction taking into account the external bias. The tunneling process is described by a tunneling Hamiltonian [1]

$$H_T = \sum_{kq\sigma} T_{kq} c_{q\sigma}^{\dagger} c_{k\sigma} + \text{H.c.} \tag{1}$$

where the term written out explicitly describes the annihilation of an electron with wave vector k and spin σ on the left electrode and the creation of an electron with the same spin but wave vector q on the right electrode thereby transferring an electron from the left to the right side. The corresponding matrix element is given by T_{kq}. The Hermitian conjugate part describes the reverse process. We note that what we call electrons here for simplicity, are really quasiparticles of the many-electron system in the electrodes.

As a result of a golden rule calculation treating H_T as a perturbation and assuming an elastic tunneling process one finds after some calculation for the average current

$$I(V) = \frac{1}{eR_T} \int dE \big\{ f(E)\left[1 - f(E + eV)\right] - \left[1 - f(E)\right] f(E + eV) \big\}. \tag{2}$$

Here, $f(E) = 1/[1 + \exp(\beta E)]$ is the Fermi function at inverse temperature $\beta = 1/k_B T$. The integrand of (2) can easily be understood. The first term gives the probability to find an electron with energy E on the left side and a corresponding empty state on the right side. The difference in the Fermi energies on both sides of the barrier is accounted for in the argument of the second Fermi function. This difference is assumed to be fixed to eV by the ideal voltage source. An analogous interpretation may be given for the second term in the integrand which describes the reverse process. The tunneling matrix element and the densities of state were absorbed in R_T which is proportional to $1/|T_{kq}|^2$. The integral in (2) will be calculated explicitly in Sec. 3.2. where we will find independent of temperature

$$I(V) = V/R_T. \tag{3}$$

Since this gives a current proportional to the applied voltage, the current-voltage characteristic is of the same form as for an Ohmic resistor. It is therefore suggestive to call R_T a tunneling resistance. We stress, however, that the tunneling resistance should not be confused with an Ohmic resistance because of the quite different nature of charge transport through a tunnel junction and an Ohmic resistor. This becomes apparent for example in the different noise spectra.[2]

1.3. Charging energy considerations

In the previous section we have discussed a tunnel junction coupled to an ideal voltage source. As a consequence, the charge on the junction capacitor is kept fixed at all times. Now, we consider a different case where an ideal external current I controls the charge Q on the junction. At zero temperature a tunneling process leading from Q to $Q - e$ is only possible if the difference of charging energies before and after the tunneling process is positive

$$\Delta E = \frac{Q^2}{2C} - \frac{(Q - e)^2}{2C} > 0. \tag{4}$$

This condition is satisfied if $Q > e/2$ or the voltage across the junction $U > U_c = e/2C$. Note, that we distinguish between the voltage U across the junction and an externally applied voltage V.

Assuming an ideal current-biased junction, the following picture results. Starting at a charge $|Q| < e/2$, the junction is charged by the external current. At a charge $Q > e/2$ an electron may tunnel thereby decreasing the charge on the junction below the threshold $e/2$. Then the cycle starts again. This process which occurs with a frequency $f = I/e$ only determined by the external bias current I is called SET oscillation.[3]–[5] One can show that the average voltage across the junction is proportional to $I^{1/2}$.

We may also use an ideal voltage source and feed a current to the junction through a large resistor. Its resistance is assumed to be smaller than the tunneling resistance of the junction but large enough to inhibit a fast recharging of the capacitor after a tunneling event. According to the argument given above there will be no current if the external voltage is smaller than $e/2C$. Beyond this voltage one finds at zero temperature that the average current is determined by an Ohmic current-voltage characteristic with resistance R_T shifted in voltage by $e/2C$. This shift in the current-voltage characteristic is called the Coulomb gap and the phenomenon of suppression of the current below U_c is referred to as Coulomb blockade. For the energy consideration presented in this section it was important that the charge on the capacitor is well defined and continuous even on the scale of an elementary charge. Only a junction charge less than $e/2$ together with the fact that tunneling always changes this charge by e gave rise to the possibility of a Coulomb gap.

1.4. Local and global view of a single tunnel junction

Comparing the discussions in the two previous sections we find that there are different energy differences associated with the tunneling process, namely eV and E_c. As we will argue now these two cases can be viewed as a local and a global description [4]–[6] of a single tunnel junction coupled to an external circuit, at least at zero temperature. In Sec. 1.3. we used the energy difference (4) which gives the difference in charging energy of the junction before and immediately after the tunneling process. This is called the local view since it only considers the junction through which the electron is tunneling and ignores its interaction with the rest of the world.

In contrast, in Sec. 1.2. the energy changes in the circuit were viewed globally. After the tunneling process a nonequilibrium situation occurs since the charge $Q - e$ on the junction and the charge $Q = CV$ imposed by the voltage source are different. To reestablish equilibrium the voltage source transfers an electron and recharges the junction capacitor to the charge Q. In the end there is no change in charging energy. However, the work done by the voltage source which amounts to eV has to be taken into account. This is indeed the case in (2) where eV appears as the difference between the Fermi energies of the two electrodes.

Now, the question arises which one of these two descriptions, if any, is correct. This problem cannot be solved by treating the single junction as decoupled from the rest of the world or by replacing its surroundings by ideal current or voltage sources. The discussion of the local and global view rather suggests that one has to consider the junction embedded in the electrical circuit.[7]–[9] A junction coupled to an ideal

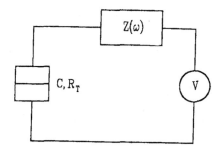

Figure 2. An ultrasmall tunnel junction with capacitance C and tunneling resistance R_T coupled to a voltage source V via the external impedance $Z(\omega)$.

voltage source, for example, will always behave according to the global rule since after the tunneling process charge equilibrium is reestablished immediately. This is not the case if the voltage source is attached to the junction with a large resistor in series. In the sequel we will discuss the influence of the electrodynamic environment on electron tunneling rates in ultrasmall tunnel junctions. We will learn that the local and global rules are just limiting cases of the so-called orthodox theory and find out to what kind of circuit we have to couple the junction in order to observe Coulomb blockade phenomena.

2. Description of the environment

2.1. Classical charge relaxation

In this section we discuss the coupling of a tunnel junction to the external circuit classically. That means that we forget about tunneling for the moment and consider the junction just as a capacitor of capacitance C carrying the charge $Q = CU$ where U is the voltage across the junction. The junction is attached to the external circuit which we describe by its impedance

$$Z(\omega) = \frac{V(\omega)}{I(\omega)}. \tag{5}$$

The impedance gives the ratio between an alternating voltage of frequency ω applied to the circuit and the current which then is flowing through it if the junction capacitor is replaced by a short. The external circuit shall contain an ideal dc voltage source V in series with the impedance as shown in Fig. 2. As discussed in Chap. 1 and [8, 10], the assumption of a voltage source is reasonable even if in a real experiment a current source is used. Generally, the leads attached to the junction generate a capacitance which is so large compared to the junction capacitance that a current source will charge this large capacitor which then acts as an effective voltage source.

In equilibrium the average charge on the junction $Q_e = CV$ is determined by the external voltage source. Let us assume now that at some initial time the equilibrium is disturbed and the charge on the junction is Q_0. In the following we will derive the relaxation from this initial condition back to equilibrium. The information on the charge relaxation which depends only on the external impedance and the junction capacitance will later be needed to describe the influence of the environment.

To solve this initial value problem it is convenient to work in the Laplace space. Then the Laplace transform of the voltage across the impedance follows from (5) as

$$\hat{V}(p) = \hat{Z}(p)\hat{I}(p) \tag{6}$$

where the hat denotes the Laplace transform and

$$\hat{Z}(p) = Z(-ip). \tag{7}$$

Applying the rule for the Laplace transformation of a derivative, one finds for the current in terms of the charge

$$\hat{I}(p) = p\hat{Q}(p) - Q_0. \tag{8}$$

Since the Laplace transform of the constant external voltage V is V/p, we get for the voltage balance in the circuit

$$\frac{Q_e}{pC} = \frac{\hat{Q}(p)}{C} + \hat{Z}(p)\left(p\hat{Q}(p) - Q_0\right). \tag{9}$$

Solving this equation for $\hat{Q}(p)$, doing the inverse Laplace transformation, and rewriting the final result in terms of the original impedance $Z(\omega)$ we get for the relaxation of the charge

$$Q(t) = Q_e + (Q_0 - Q_e)R(t). \tag{10}$$

Here, the Fourier transform of the charge relaxation function

$$\int_0^\infty dt e^{-i\omega t} R(t) = CZ_t(\omega) \tag{11}$$

is related to the total impedance

$$Z_t(\omega) = \frac{1}{i\omega C + Z^{-1}(\omega)} \tag{12}$$

of the circuit consisting of the capacitance C in parallel with the external impedance $Z(\omega)$. It will become clear in Sec. 6.2. that $Z_t(\omega)$ is the effective impedance of the circuit as seen from the tunnel junction.

2.2. Quantum mechanics of an *LC*-circuit

In the previous section we have found that the classical relaxation of charge in a circuit can be described in terms of its impedance. We will now make a first step towards the quantum mechanical treatment of a tunnel junction coupled to an external circuit by discussing the most simplest case where the environmental impedance of the circuit shown in Fig. 2 is given by an inductance $Z(\omega) = i\omega L$. In the next section this will turn out to be the fundamental building block for a general description of the environment.

For the following it is convenient to introduce the phase [11]

$$\varphi(t) = \frac{e}{\hbar} \int_{-\infty}^{t} dt' U(t') \tag{13}$$

where $U = Q/C$ is the voltage across the junction. The definition (13) becomes the Josephson relation for a superconducting tunnel junction if we replace the electron charge e by the charge of Cooper pairs $2e$. In the superconducting case this phase is of course of great importance as the phase of the order parameter.

To derive the Hamiltonian of a voltage-biased LC-circuit let us first write down the Lagrangian

$$\mathcal{L} = \frac{C}{2} \left(\frac{\hbar}{e} \dot{\varphi} \right)^2 - \frac{1}{2L} \left(\frac{\hbar}{e} \right)^2 \left(\varphi - \frac{e}{\hbar} Vt \right)^2. \tag{14}$$

The first term represents the charging energy of the capacitor which can easily be verified by means of the definition of the phase (13). The magnetic field energy of the inductor is given by the second term since up to a factor $\hbar/e$ the flux through the inductor is given by the phase difference across the inductor. The latter relation is obtained from the requirement that the phase differences at the capacitor and inductor should add up to the phase difference $(e/\hbar)Vt$ produced by the voltage source according to (13).

Switching to the Hamilton formalism we find that the charge Q on the junction is the conjugate variable to $(\hbar/e)\varphi$. In a quantum mechanical description this results in the commutation relation

$$[\varphi, Q] = ie. \tag{15}$$

Now, the phase φ, the voltage U across the junction, and the charge Q are operators. Note that there is no problem with phase periodicity when constructing the phase operator φ. Due to the continuous charge Q the spectrum of the phase operator is continuous on the interval from $-\infty$ to $+\infty$.

From the Lagrangian (14) we immediately get the Hamiltonian

$$H = \frac{Q^2}{2C} + \frac{1}{2L} \left(\frac{\hbar}{e} \right)^2 \left(\varphi - \frac{e}{\hbar} Vt \right)^2. \tag{16}$$

According to the equations of motion derived either from (14) or (16) one finds that the average phase evolves in time like $(e/\hbar)Vt$. The average charge on the capacitor is given by CV. It is therefore convenient to introduce the variables

$$\tilde{\varphi}(t) = \varphi(t) - \frac{e}{\hbar} Vt \tag{17}$$

and

$$\tilde{Q} = Q - CV \tag{18}$$

describing the fluctuations around the mean value determined by the external voltage.

The commutator between the new variables is again

$$[\tilde{\varphi}, \tilde{Q}] = ie. \tag{19}$$

Substituting $\varphi - (e/\hbar)Vt$ by $\tilde{\varphi}$ in the Hamiltonian (16) amounts to going into a "rotating reference frame". This transformation results in an extra contribution $-(i/\hbar)QV$ to the time derivative in the time-dependent Schrödinger equation. We thus obtain, up to a term depending only on the external voltage,

$$H = \frac{\tilde{Q}^2}{2C} + \frac{1}{2L}\left(\frac{\hbar}{e}\tilde{\varphi}\right)^2 \tag{20}$$

which demonstrates the equivalence between an LC-circuit and a harmonic oscillator. Note that the influence of an external voltage is entirely accounted for by the definitions (17) and (18).

2.3. Hamiltonian of the environment

The special environment of the previous section did not give rise to dissipation so that there were no problems in writing down a Hamiltonian. On the other hand, in general, an impedance $Z(\omega)$ will introduce dissipation. At first sight, this is in contradiction to a Hamiltonian description of the environment. However, after realizing that dissipation arises from the coupling of the degrees of freedom Q and φ, in which we are interested, to other degrees of freedom our task is not as hopeless anymore. We will introduce a Hamiltonian for the system coupled to the environment which, after elimination of the environmental degrees of freedom, describes a dissipative system. One approach would be to start from a microscopic model. This will be discussed in the appendix. Here, we represent the environment by a set of harmonic oscillators which are bilinearly coupled to φ and which may be viewed as LC-circuits. These harmonic oscillators may in some cases be justified microscopically. In most cases, however, this representation of the environment is introduced phenomenologically. It then has to fulfill the requirement that in the classical limit the reduced dynamics is described correctly. We now can write down the Hamiltonian for the environmental coupling

$$H_{\text{env}} = \frac{\tilde{Q}^2}{2C} + \sum_{n=1}^{N}\left[\frac{q_n^2}{2C_n} + \left(\frac{\hbar}{e}\right)^2\frac{1}{2L_n}(\tilde{\varphi} - \varphi_n)^2\right] \tag{21}$$

which is expressed in terms of the variables $\tilde{\varphi}$ and $\tilde{Q}$ defined in (17) and (18) thereby accounting for an external voltage source. The first term describes the charging energy of the junction capacitor. In the second term we sum over the environmental degrees of freedom represented by harmonic oscillators of frequency $\omega_n = 1/\sqrt{L_n C_n}$ which are bilinearly coupled to the phase of the tunnel junction. In order to describe an effectively dissipative environment the number N of environmental degrees of freedom has to be rather large. Usually, in practice the limit $N \to \infty$ has to be performed. The model Hamiltonian (21) is not new. Hamiltonians of this form have been used in quantum optics for several decades.[12] More recently, Caldeira and Leggett [13] introduced this description of the environment in the context of macroscopic quantum tunneling.

Table I. Correspondence between electrical and mechanical quantities

Electrical quantity	Mechanical quantity
charge Q	momentum p
voltage $U = Q/C$	velocity $v = p/M$
capacitance C	mass M
phase φ	coordinate x
$[\varphi, Q] = ie$	$[x, p] = i\hbar$
inductance L	spring constant k
LC-circuit	harmonic oscillator

To derive the reduced dynamics of the system described by (21) we write down the Heisenberg equations of motion for the operators $\tilde{Q}, \tilde{\varphi}, q_n$, and φ_n. It is easy to solve for q_n and φ_n by considering $\tilde{\varphi}$ as a given function of time. After substituting the result into the equations of motion for $\tilde{Q}$ and $\tilde{\varphi}$ and solving these we obtain after a partial integration

$$\dot{\tilde{Q}}(t) + \frac{1}{C} \int_0^t ds\, Y(t - s)\tilde{Q}(s) = I_N(t). \tag{22}$$

Here,

$$Y(t) = \sum_{n=1}^{N} \frac{1}{L_n} \cos(\omega_n t). \tag{23}$$

Note that an arbitrary function $Y(t)$ can be described in this way by an adequate choice of the model parameters L_n and C_n. In general, for a given $Y(t)$ the sum has to be replaced by an integral over a continuous distribution of harmonic oscillators. The Fourier transform of $Y(t)$ is the admittance $Y(\omega) = 1/Z(\omega)$. The inhomogeneity $I_N(t)$ in (22) is the quantum mechanical noise current and depends on the initial conditions at $t = 0$. By Laplace transforming the left-hand side of (22) we recover (9) which describes the classical relaxation of the junction charge according to the total impedance $Z_t(\omega)$ introduced in (12). Therefore, the Hamiltonian (21) gives us an equivalent description of the environment which enables us to treat a tunnel junction coupled to the external circuit quantum mechanically.

Sometimes it is useful to use a mechanical analogue of the model presented above. The correspondence between the electrical and mechanical quantities is given in table I. At zero bias the Hamiltonian (21) may then be interpreted as describing a free particle coupled to N harmonic oscillators forming the heat bath and (22) is indeed the equation of motion describing such a system.

3. Electron tunneling rates for single tunnel junctions

3.1. Tunneling Hamiltonian

In the previous section we have treated the tunnel junction as a capacitor thereby neglecting the fact that electrons may tunnel through the junction. We will now include tunneling to allow for a current through the junction. The quasiparticles in the two metal electrodes are described by the Hamiltonian

$$H_{qp} = \sum_{k\sigma} \epsilon_k c_{k\sigma}^\dagger c_{k\sigma} + \sum_{q\sigma} \epsilon_q c_{q\sigma}^\dagger c_{q\sigma} \tag{24}$$

where the first and second sum correspond to the left and right electrode, respectively. ϵ_k and ϵ_q are the energies of quasiparticles with wave vector k and q while σ denotes their spin.

Tunneling is introduced by the Hamiltonian [8, 14, 15]

$$H_T = \sum_{kq\sigma} T_{kq} c_{q\sigma}^\dagger c_{k\sigma} e^{-i\varphi} + \text{H.c.} \tag{25}$$

This is the tunneling Hamiltonian (1) presented in Sec. 1.2. apart from the operator $\exp(-i\varphi)$. Using the mechanical analogue of Sec. 2.3. the latter operator would correspond to a momentum shift operator. Indeed according to

$$e^{i\varphi} Q e^{-i\varphi} = Q - e \tag{26}$$

which follows from the commutator (15) this new operator acts as a "translation" operator changing the charge on the junction by an elementary charge e. In the Hamiltonian (25) we use operators $c^\dagger$ and c representing quasiparticles and in addition the phase φ which is conjugate to the charge Q. These operators may be expressed in terms of true electron creation and annihilation operators. In the following, we will assume that the quasiparticle operators commute with the charge and phase operators since a large number of quasiparticle states contribute to these operators and the contribution of a single state is negligible. This is closely related to the assumption of linearity of the electrodynamics in the system under consideration. Although quasiparticle and charge operators commute, the tunneling Hamiltonian (25) now establishes a coupling between the tunneling electron and the environment which "sees" the junction charge. It will be shown below that this coupling makes the current-voltage characteristic of the junction nonlinear.

As in Sec. 2.2. we may introduce the phase $\tilde{\varphi}$ defined in (17) into H_T. This will help to clarify the relation between the two tunneling Hamiltonians (1) and (25). Exploiting the relation

$$e^{-i\alpha c^\dagger c} c e^{i\alpha c^\dagger c} = c e^{i\alpha} \tag{27}$$

we perform a time-dependent unitary transformation with

$$U = \prod_{k\sigma} \exp\left[i\frac{e}{\hbar}Vtc_{k\sigma}^{\dagger}c_{k\sigma}\right].$$ (28)

The new tunneling Hamiltonian then reads

$$\tilde{H}_T = U^{\dagger}H_T U = \sum_{kq\sigma} T_{kq}c_{q\sigma}^{\dagger}c_{k\sigma}e^{-i\tilde{\varphi}} + \text{H.c.}$$ (29)

Since the transformation (28) is time-dependent is also shifts quasiparticle energies on the left electrode and we obtain from (24)

$$\tilde{H}_{\text{qp}} = U^{\dagger}H_{\text{qp}}U - i\hbar U^{\dagger}\frac{\partial}{\partial t}U$$

$$= \sum_{k\sigma}(\epsilon_k + eV)c_{k\sigma}^{\dagger}c_{k\sigma} + \sum_{q\sigma}\epsilon_q c_{q\sigma}^{\dagger}c_{q\sigma}.$$ (30)

In the absence of an environment, the operator $\exp(-i\tilde{\varphi})$ in (29) has no effect on the tunneling process. The Hamiltonian (29) then becomes identical to the Hamiltonian (1). The phase factor $\exp[-(i/\hbar)eVt]$ which was present in (25) has vanished. Instead, the energy levels on the left and right electrodes are shifted by eV relative to each other. This shift was taken into account in the result (2).

In the following we will use the tunneling Hamiltonian in the form (29). Before starting with the calculation let us collect the Hamiltonians describing the whole system. The total Hamiltonian

$$H = \tilde{H}_{\text{qp}} + H_{\text{env}} + \tilde{H}_T$$ (31)

contains the contributions of the quasiparticle Hamiltonian (30) for the two electrodes, the Hamiltonian (21) describing the environment including the charge degree of freedom, and finally the tunneling Hamiltonian (29) which couples the first two parts.

3.2. Calculation of tunneling rates

3.2.1. Perturbation theory

Starting from the Hamiltonian (31) we now calculate rates for tunneling through the junction. First we make two important assumptions. The tunneling resistance R_T shall be large compared to the resistance quantum $R_K = h/e^2$ which is a natural resistance scale (for Josephson junctions one often uses $R_Q = h/4e^2$ to account for the charge $2e$ of Cooper pairs). Since R_T is inversely proportional to the square of the tunneling matrix element, this implies that the states on the two electrodes only mix very weakly so that the Hamiltonian (30) is a good description of the quasiparticles in the electrodes. We then may consider the tunneling Hamiltonian $\tilde{H}_T$ as a perturbation. Here, we will restrict ourselves to the leading order, i.e. we calculate the tunneling rate within the golden rule approximation. We further assume that charge equilibrium is established before a tunneling event occurs. This defines the states to be used in the perturbation theoretical calculation as equilibrium states. On the other hand, it means that the time between two tunneling processes should be larger than the charge relaxation time.

As mentioned above we will calculate the tunneling rates by means of the golden rule

$$\Gamma_{i\to f} = \frac{2\pi}{\hbar}|\langle f|\tilde{H}_T|i\rangle|^2\delta(E_i - E_f) \tag{32}$$

which gives the rate for transitions between the initial state $|i\rangle$ and the final state $|f\rangle$. In the absence of the tunneling Hamiltonian we may write the total state as a product of a quasiparticle state and a charge state which in the following we call reservoir state because it is connected with the coupling to the environment. Specifically, we set $|i\rangle = |E\rangle|R\rangle$ and $|f\rangle = |E'\rangle|R'\rangle$ where $|E\rangle$ and $|E'\rangle$ are quasiparticle states of respective energy and $|R\rangle$ and $|R'\rangle$ are reservoir states with energies E_R and E'_R. The matrix element in (32) then becomes

$$\langle f|\tilde{H}_T|i\rangle = \langle E'|H_T^e|E\rangle\langle R'|e^{-i\tilde{\varphi}}|R\rangle + \langle E'|H_T^{e\dagger}|E\rangle\langle R'|e^{i\tilde{\varphi}}|R\rangle \tag{33}$$

with

$$H_T^e = \sum_{kq\sigma} T_{kq}c_{q\sigma}^{\dagger}c_{k\sigma} \tag{34}$$

being the part of the tunneling Hamiltonian acting in the quasiparticle space.

To calculate the total rate for electron tunneling from left to right we have to sum over all initial states weighted with the probability to find these states and over all final states. We thus have to evaluate

$$\vec{\Gamma}(V) = \frac{2\pi}{\hbar}\int_{-\infty}^{+\infty}dEdE'\sum_{R,R'}|\langle E'|H_T^e|E\rangle|^2\,|\langle R'|e^{-i\tilde{\varphi}}|R\rangle|^2$$

$$\times P_\beta(E)P_\beta(R)\delta(E + eV + E_R - E' - E'_R). \tag{35}$$

Let us consider one term $T_{kq}c_{q\sigma}^{\dagger}c_{k\sigma}$ contained in H_T^e. The only possible states with nonvanishing matrix element $\langle E'|c_{q\sigma}^{\dagger}c_{k\sigma}|E\rangle$ are $|E\rangle = |\ldots,1_{k\sigma},\ldots,0_{q\sigma},\ldots\rangle$ and $|E'\rangle = |\ldots,0_{k\sigma},\ldots,1_{q\sigma},\ldots\rangle$ where this notation means that in $|E\rangle$ a quasiparticle with wave vector k and spin σ on the left side of the barrier is present while the state with wave vector q and spin σ on the right side is unoccupied. The occupation of states with other quantum numbers is arbitrary. Since $P_\beta(E)$ factorizes we then obtain

$$\vec{\Gamma}(V) = \frac{2\pi}{\hbar}\int_{-\infty}^{+\infty}d\epsilon_k d\epsilon_q\sum_{kq\sigma}|T_{kq}|^2 f(\epsilon_k)[1 - f(\epsilon_q)]$$

$$\times \sum_{R,R'}|\langle R'|e^{-i\tilde{\varphi}}|R\rangle|^2 P_\beta(R)\,\delta(\epsilon_k + eV + E_R - \epsilon_q - E'_R). \tag{36}$$

Here, the probability to find the initial state is given by the product of Fermi functions and $\epsilon_k + eV - \epsilon_q$ is the difference of quasiparticle energies associated with the tunneling process since the occupation of the other states remains unchanged. Note that eV does not appear in $f(\epsilon_k)$ since the Fermi level on the left side is shifted by this amount. If the applied voltage is such that eV is much smaller than the Fermi energy we may assume that all quasiparticle states involved have energies close to the Fermi energy. Taking the tunneling matrix element to be approximately independent of ϵ_k and ϵ_q we may replace $\sum_{kq\sigma} |T_{kq}|^2$ by an averaged matrix element $|T|^2$ which also accounts for the density of states at the Fermi energy. We collect all constant terms in the tunneling resistance R_T. The rate expression (36) then becomes

$$\vec{\Gamma}(V) = \frac{1}{e^2 R_T} \int_{-\infty}^{+\infty} \mathrm{d}E \mathrm{d}E' f(E)[1 - f(E')]$$

$$\times \sum_{R,R'} |\langle R|e^{-i\tilde{\varphi}}|R'\rangle|^2 P_\beta(R)\, \delta(E + eV + E_R - E' - E'_R) \quad (37)$$

where we have renamed the energies ϵ_k and ϵ_q into E and E'. The justification for calling R_T a tunneling resistance will be given below when we calculate current-voltage characteristics.

3.2.2. Tracing out environmental states

We now have to do the sum over R and R' in (37). The probability of finding the initial reservoir state $|R\rangle$ is given by the corresponding matrix element

$$P_\beta(R) = \langle R|\rho_\beta|R\rangle \quad (38)$$

of the equilibrium density matrix

$$\rho_\beta = Z_\beta^{-1} \exp(-\beta H_{\mathrm{env}}) \quad (39)$$

of the reservoir at inverse temperature β. Here,

$$Z_\beta = \mathrm{Tr}\Big\{ \exp(-\beta H_{\mathrm{env}})\Big\} \quad (40)$$

is the partition function of the environment. To proceed, it is useful to rewrite the delta function in (37) in terms of its Fourier transform

$$\delta(E + eV + E_R - E' - E'_R)$$

$$= \frac{1}{2\pi\hbar} \int_{-\infty}^{+\infty} \mathrm{d}t \exp\left(\frac{i}{\hbar}(E + eV + E_R - E' - E'_R)t\right) \quad (41)$$

and to use the part containing the reservoir energies to introduce the time dependent phase operator in the Heisenberg picture. We thus obtain

$$\vec{\Gamma}(V) = \frac{1}{e^2 R_T} \int_{-\infty}^{+\infty} \mathrm{d}E \mathrm{d}E' \int_{-\infty}^{+\infty} \frac{\mathrm{d}t}{2\pi\hbar} \exp\left(\frac{i}{\hbar}(E - E' + eV)t\right) f(E)[1 - f(E')]$$

$$\times \sum_{R,R'} P_\beta(R)\langle R|e^{i\tilde{\varphi}(t)}|R'\rangle\langle R'|e^{-i\tilde{\varphi}(0)}|R\rangle. \quad (42)$$

Since the reservoir states form a complete set we can do the sum over R'. Together with the definition (38) of $P_\beta(R)$ we find that the reservoir part in the rate formula is given by the equilibrium correlation function

$$
\langle e^{i\tilde{\varphi}(t)} e^{-i\tilde{\varphi}(0)} \rangle = \sum_R \langle R| e^{i\tilde{\varphi}(t)} e^{-i\tilde{\varphi}(0)} |R\rangle P_\beta(R)
$$

$$
= \frac{1}{Z_\beta} \sum_R \langle R| e^{i\tilde{\varphi}(t)} e^{-i\tilde{\varphi}(0)} e^{-\beta H_{\text{env}}} |R\rangle \tag{43}
$$

so that we get from (42)

$$
\vec{\Gamma}(V) = \frac{1}{e^2 R_T} \int_{-\infty}^{+\infty} \mathrm{d}E \mathrm{d}E' f(E)[1 - f(E')]
$$

$$
\times \int_{-\infty}^{+\infty} \frac{\mathrm{d}t}{2\pi\hbar} \exp\left(\frac{i}{\hbar}(E - E' + eV)t\right) \langle e^{i\tilde{\varphi}(t)} e^{-i\tilde{\varphi}(0)} \rangle. \tag{44}
$$

3.2.3. Phase-phase correlation function

The correlation function defined in (43) may be simplified. According to (38) the probability $P_\beta(R)$ of the unperturbed reservoir is given by the equilibrium density matrix of the environmental Hamiltonian (21). Since this Hamiltonian is harmonic, the equilibrium density matrix in the $\tilde{\varphi}$-representation is a Gaussian and therefore determined only by its first and second moments. Hence, it should be possible to express the correlation function (43) in terms of phase correlation functions of at most second order. This goal may be achieved by exploiting the generalized Wick theorem for equilibrium correlation functions [16]

$$
\langle \psi_1 \psi_2 \ldots \psi_n \rangle = \langle \psi_1 \psi_2 \rangle \langle \psi_3 \psi_4 \ldots \psi_n \rangle + \langle \psi_1 \psi_3 \rangle \langle \psi_2 \psi_4 \ldots \psi_n \rangle + \ldots
$$

$$
+ \langle \psi_1 \psi_n \rangle \langle \psi_2 \psi_3 \ldots \psi_{n-1} \rangle. \tag{45}
$$

This theorem applies if the Hamiltonian of the system for which the thermal average is performed may be represented in terms of independent harmonic oscillators and if the operators ψ_i are linear combinations of creation and annihilation operators. The first condition is fulfilled since the Hamiltonian (21) may in principle be diagonalized. Due to the linearity of the equation of motion, $\tilde{\varphi}(t)$ is a linear combination of creation and annihilation operators and thus also the second condition holds. After expanding the exponentials on the right hand side of

$$
\frac{\mathrm{d}}{\mathrm{d}\alpha} \langle e^{i\alpha\tilde{\varphi}(t)} e^{-i\alpha\tilde{\varphi}(0)} \rangle = i\left[\langle \tilde{\varphi}(t) e^{i\alpha\tilde{\varphi}(t)} e^{-i\alpha\tilde{\varphi}(0)} \rangle - \langle e^{i\alpha\tilde{\varphi}(t)} \tilde{\varphi}(0) e^{-i\alpha\tilde{\varphi}(0)} \rangle\right] \tag{46}
$$

we may apply the generalized Wick theorem. The resulting sums may again be expressed in terms of exponentials and we find

$$
\frac{\mathrm{d}}{\mathrm{d}\alpha} \langle e^{i\alpha\tilde{\varphi}(t)} e^{-i\alpha\tilde{\varphi}(0)} \rangle = 2\alpha \langle [\tilde{\varphi}(t) - \tilde{\varphi}(0)]\tilde{\varphi}(0) \rangle \langle e^{i\alpha\tilde{\varphi}(t)} e^{-i\alpha\tilde{\varphi}(0)} \rangle \tag{47}
$$

where we made use of $\langle \tilde{\varphi}(t)^2 \rangle = \langle \tilde{\varphi}(0)^2 \rangle$ which is a consequence of the stationarity of equilibrium correlation functions. The differential equation (47) may easily be solved and we obtain with the correct initial condition at $\alpha = 0$ the result for $\alpha = 1$

$$\langle e^{i\tilde{\varphi}(t)} e^{-i\tilde{\varphi}(0)} \rangle = e^{\langle [\tilde{\varphi}(t) - \tilde{\varphi}(0)]\tilde{\varphi}(0) \rangle}. \tag{48}$$

For later convenience we introduce the abbreviation

$$J(t) = \langle [\tilde{\varphi}(t) - \tilde{\varphi}(0)]\tilde{\varphi}(0) \rangle \tag{49}$$

for the phase-phase correlation function. In view of (44) it is useful to introduce the Fourier transform of the correlation function (48)

$$P(E) = \frac{1}{2\pi\hbar} \int_{-\infty}^{+\infty} dt \exp \left[J(t) + \frac{i}{\hbar} Et \right]. \tag{50}$$

3.2.4. Tunneling rate formula

Using the definition of $P(E)$ we may now rewrite the expression (44) for the forward tunneling rate in the form

$$\overrightarrow{\Gamma}(V) = \frac{1}{e^2 R_T} \int_{-\infty}^{+\infty} dE dE' \, f(E)[1 - f(E' + eV)]P(E - E') \tag{51}$$

which allows for a simple physical interpretation. As already pointed out in Sec. 1.2. the Fermi functions describe the probability of finding an occupied state on one side and an empty state on the other side of the barrier. The difference in the Fermi energies due to the applied voltage is taken into account in the argument of the second Fermi function. In the discussion of Sec. 1.2. we had assumed an ideal voltage bias and no environmental modes were present. Therefore, the energy conservation condition in the golden rule (32) applied directly to the tunneling electron. The expression (51) is more general and takes into account the possibility of energy exchange between the tunneling electron and the environment. We may interpret $P(E)$ as the probability to emit the energy E to the external circuit. Correspondingly, $P(E)$ for negative energies describes the absorption of energy by the tunneling electron.

To further simplify (51) we first calculate the integral over Fermi functions

$$g(x) = \int_{-\infty}^{+\infty} dE \, [f(E) - f(E + x)] \tag{52}$$

which will also be of use later on. The derivative of g with respect to x can easily be evaluated yielding $dg(x)/dx = f(-\infty) - f(\infty) = 1$. Integration with the initial condition $g(0) = 0$ then gives the formula

$$\int_{-\infty}^{+\infty} dE \, [f(E) - f(E + x)] = x. \tag{53}$$

By means of the relation

$$f(E)[1 - f(E + x)] = \frac{f(E) - f(E + x)}{1 - e^{-\beta x}} \tag{54}$$

we find for the integral which we need in order to simplify (51)

$$\int_{-\infty}^{+\infty} dE \, f(E)[1 - f(E + x)] = \frac{x}{1 - e^{-\beta x}}.$$ (55)

This together with (51) finally gives for the forward tunneling rate through a single junction

$$\vec{\Gamma}(V) = \frac{1}{e^2 R_T} \int_{-\infty}^{+\infty} dE \, \frac{E}{1 - \exp(-\beta E)} P(eV - E).$$ (56)

A corresponding calculation can be done for the backward tunneling rate. However, it is rather obvious from the symmetry of a voltage biased single junction that

$$\overleftarrow{\Gamma}(V) = \vec{\Gamma}(-V)$$ (57)

which is indeed the result one obtains from redoing the calculation.

For a further discussion of the tunneling rates and the current-voltage characteristic of a single tunnel junction it is useful to know more about the function $P(E)$ and the correlation function $J(t)$ by which it is determined. We will be able to derive some general properties from which we will deduce a few facts about rates and current-voltage characteristics. Further, we will discuss the limits of very low and very high impedance. For a realistic environment one usually has to evaluate $P(E)$ numerically. We will present several examples for impedances from which we learn more about how $P(E)$ is related to properties of the environmental circuit.

3.3. Phase-phase correlation function and environmental impedance

In Sec. 2.3. we presented the Hamiltonian (21) to describe the electrodynamic environment and derived the operator equation of motion (22) for the junction charge $\tilde{Q}$. According to (13) the phase is proportional to the time derivative of the charge so that we immediately get the equation of motion for the phase

$$C\ddot{\tilde{\varphi}} + \int_0^t ds \, Y(t - s)\dot{\tilde{\varphi}}(s) = \frac{e}{\hbar} I_N(t)$$ (58)

where again $I_N(t)$ is the quantum mechanical noise current. In terms of our mechanical analogue introduced in Table I, (58) can be interpreted as the equation of motion of a free Brownian particle.

The effect of the environmental degrees of freedom on the charge and phase degrees of freedom is twofold. They produce a damping term which depends on the admittance $Y(\omega)$ and is responsible for the relaxation of the charge into equilibrium. The relaxation of the mean charge is described by a dynamical susceptibility which for this linear system is the same in the classical and the quantum case due to the Ehrenfest theorem. From our results in Sec. 2.1. we obtain for the dynamical susceptibility describing the response of the phase to the conjugate force $(e/\hbar)I(t)$

$$\chi(\omega) = \chi'(\omega) - i\chi''(\omega) = \left(\frac{e}{\hbar}\right)^2 \frac{Z_t(\omega)}{i\omega}.$$ (59)

The second effect of the environment manifests itself in the noise current $I_N(t)$ and appears in correlation functions as the one introduced in (49). Since damping and fluctuations have the same microscopic origin they are not independent of each other and in fact the so-called fluctuation-dissipation theorem [17]

$$\tilde{C}(\omega) = \frac{2\hbar}{1 - e^{-\beta\hbar\omega}} \chi''(\omega) \tag{60}$$

relates the absorptive part $\chi''(\omega)$ of the dynamical susceptibility (59) to the Fourier transform

$$\tilde{C}(\omega) = \int_{-\infty}^{+\infty} dt\, e^{-i\omega t} \langle \tilde{\varphi}(0)\tilde{\varphi}(t) \rangle \tag{61}$$

of the equilibrium phase-phase correlation function. The fluctuation-dissipation theorem may be proven in the framework of linear response theory which becomes exact if a linear system is treated as is the case here. From (60) and (61) together with the stationarity of equilibrium correlation functions we then immediately get

$$\langle \tilde{\varphi}(t)\tilde{\varphi}(0) \rangle = 2 \int_{-\infty}^{+\infty} \frac{d\omega}{\omega} \frac{\mathrm{Re}Z_t(\omega)}{R_K} \frac{e^{-i\omega t}}{1 - e^{-\beta\hbar\omega}}. \tag{62}$$

Since in general the real part of the impedance Z_t at $\omega = 0$ does not vanish, the correlation function (62) does not exist due to an infrared divergence. This can easily be understood within our mechanical picture of a free Brownian particle.[18] In the absence of a confining potential the variance of the position of the particle in equilibrium should diverge. There are, however, no problems with the correlation function $J(t)$ in which according to its definition (49) the diverging static correlation $\langle \tilde{\varphi}^2 \rangle$ is subtracted off. Since the Fourier transform of the impedance has to be real, the real part $\mathrm{Re}Z_t(\omega)$ of the total impedance is even and together with the identity

$$\frac{1}{1 - e^{-\beta\hbar\omega}} = \frac{1}{2} + \frac{1}{2}\coth\left(\frac{1}{2}\beta\hbar\omega\right) \tag{63}$$

we finally get for the correlation function appearing in the definition (50) of $P(E)$

$$J(t) = 2\int_0^\infty \frac{d\omega}{\omega}\frac{\mathrm{Re}Z_t(\omega)}{R_K}\left\{\coth\left(\frac{1}{2}\beta\hbar\omega\right)[\cos(\omega t) - 1] - i\sin(\omega t)\right\}. \tag{64}$$

3.4. General properties of $P(E)$

With the expression (64) for the correlation function $J(t)$ derived in the last section one may calculate the probability $P(E)$ for energy exchange between the tunneling electron and the environment once the external impedance is known. In general it is not possible to calculate $P(E)$ analytically for a given impedance except for some special cases which we will discuss later. On the other hand, there are general properties of $P(E)$ which are independent of the actual impedance.

Recalling the definition (50) of $P(E)$ we find a first sum rule

$$\int_{-\infty}^{+\infty} dE\, P(E) = e^{J(0)} = 1 \tag{65}$$

since $J(0) = 0$ which follows directly from the definition (49). Eq. (65) confirms our interpretation of $P(E)$ as a probability. A second sum rule is obtained by taking the time derivative of $\exp[J(t)]$ resulting in

$$\int_{-\infty}^{+\infty} \mathrm{d}E\, EP(E) = i\hbar J'(0) = E_c. \tag{66}$$

To prove this relation we have to calculate $J'(0)$ which can be done by a short time expansion of (64) which yields

$$J'(0) = -i \int_{-\infty}^{+\infty} \mathrm{d}\omega\, \frac{Z_t(\omega)}{R_K}. \tag{67}$$

Here, we made use of the fact that the imaginary part of the impedance is antisymmetric in ω. The integral in (67) can be evaluated by integrating (11) over ω. Since the response function $R(t)$ jumps from zero to one at $t = 0$ we get

$$\int_{-\infty}^{+\infty} \mathrm{d}\omega\, Z_t(\omega) = \frac{\pi}{C} \tag{68}$$

which together with (67) proves the last equality in (66). It should be noted that we assumed that there is no renormalization of the tunnel capacitance by the environment which means that the first derivative of the charge with respect to time in the equation of motion (22) stems from the charging energy of the tunnel junction. Otherwise, one would have to replace C by an effective tunnel capacitance defined by (68).

Another important property of $P(E)$ concerns a relation between the probabilities to emit and to absorb the energy E. We make use of the two identities

$$\langle e^{i\tilde{\varphi}(t)} e^{-i\tilde{\varphi}(0)} \rangle = \langle e^{-i\tilde{\varphi}(t)} e^{i\tilde{\varphi}(0)} \rangle \tag{69}$$

and

$$\langle e^{i\tilde{\varphi}(t)} e^{-i\tilde{\varphi}(0)} \rangle = \langle e^{-i\tilde{\varphi}(0)} e^{i\tilde{\varphi}(t+i\hbar\beta)} \rangle. \tag{70}$$

One may convince oneself that (69) is correct by substituting $\tilde{\varphi}$ by $-\tilde{\varphi}$ in (48). To prove the second identity one writes the correlation function as a trace

$$\langle e^{i\tilde{\varphi}(t)} e^{-i\tilde{\varphi}(0)} \rangle = \mathrm{Tr}\left(e^{-\beta H} e^{\frac{i}{\hbar}Ht} e^{i\tilde{\varphi}} e^{-\frac{i}{\hbar}Ht} e^{-i\tilde{\varphi}} \right) / \mathrm{Tr}\left(e^{-\beta H} \right) \tag{71}$$

and exploits the invariance of the trace under cyclic permutations. With (69) and (70) one finds from the definition (50) of $P(E)$ the so-called detailed balance symmetry

$$P(-E) = e^{-\beta E} P(E) \tag{72}$$

which means that the probability to excite environmental modes compared to the probability to absorb energy from the environment is larger by a Boltzmann factor. Another consequence is that at zero temperature no energy can be absorbed from the environment. $P(E)$ then vanishes for negative energies.

At zero temperature the asymptotic behavior of $P(E)$ for larges energies may be obtained from an integral equation which is also useful for numerical calculations. We will now derive the integral equation following an idea of Minnhagen [19]. From (62) we find for the phase-phase correlation function at zero temperature

$$J(t) = 2 \int_0^\infty \frac{d\omega}{\omega} \frac{\mathrm{Re} Z_t(\omega)}{R_K} (e^{-i\omega t} - 1). \tag{73}$$

Taking the derivative of $\exp[J(t)]$ with respect to time we get

$$\frac{d}{dt} \exp[J(t)] = -2i \exp[J(t)] \int_0^\infty d\omega \frac{\mathrm{Re} Z_t(\omega)}{R_K} e^{-i\omega t}. \tag{74}$$

Since we are interested in $P(E)$, we take the Fourier transform which on the left hand side results in a term proportional to $EP(E)$ and a convolution integral on the right hand side. Using the fact that $P(E)$ at zero temperature vanishes for negative energies we finally get

$$EP(E) = 2 \int_0^E dE' \frac{\mathrm{Re}\left[Z_t \left(\dfrac{E - E'}{\hbar} \right) \right]}{R_K} P(E'). \tag{75}$$

This enables us to calculate $P(E)$ numerically by starting from an arbitrary $P(0)$ and subsequently normalizing the result. For finite temperatures energy can also be absorbed from the environment. Then an inhomogeneous integral equation may be derived which is more complicated.[20]

We now consider the integral equation (75) for large energies so that the integral on the right hand side covers most of the energies for which $P(E)$ gives a contribution. For these large energies we may neglect E' with respect to E in the argument of the impedance and end up with the normalization integral for $P(E)$. For large energies and zero temperature $P(E)$ therefore decays according to [21]

$$P(E) = \frac{2}{E} \frac{\mathrm{Re} Z_t(E/\hbar)}{R_K} \qquad \text{for } E \to \infty. \tag{76}$$

For the limits of low and high energies one may often approximate the external impedance $Z(\omega)$ by a constant. In this case we can apply the results (111) and (114) which will be derived in Sec. 4.2. for an Ohmic environment.

3.5. General properties of current-voltage characteristics

The detailed balance relation (72) is useful to derive a simple formula for the current-voltage characteristic of a single tunnel junction. The total current through the junction is given by the transported charge e times the difference of the forward and backward tunneling rates

$$I(V) = e(\overrightarrow{\Gamma}(V) - \overleftarrow{\Gamma}(V)). \tag{77}$$

The backward tunneling rate may be obtained from the forward tunneling rate (56) by means of the symmetry (57). Together with the detailed balance relation (72) one obtains

$$I(V) = \frac{1}{eR_T}(1 - e^{-\beta eV}) \int_{-\infty}^{+\infty} dE \, \frac{E}{1 - e^{-\beta E}} P(eV - E). \tag{78}$$

This formula has the property $I(-V) = -I(V)$ as one would expect.

We now consider the limit of zero temperature and assume that $V > 0$. Taking into account that $P(E)$ then vanishes for negative energies, we obtain from (78)

$$I(V) = \frac{1}{eR_T} \int_0^{eV} dE \,(eV - E)P(E). \tag{79}$$

It is no surprise that, in contrast to the finite temperature case, at zero temperature the current at a voltage V depends only on the probability to excite environmental modes with a total energy less than eV since this is the maximum energy at the disposal of the tunneling electron. According to (79) the current at the gap voltage $e/2C$ depends on $P(E)$ at all energies up to the charging energy E_c. In view of the integral equation (75) this means that the environmental impedance up to the frequency $E_c/\hbar$ (which is of the order of 20 GHz for $C = 10^{-15}$ F) is relevant. The general behavior of an impedance up to high frequencies is discussed in Chap. 1. Another consequence of the zero temperature result (79) is that the probability $P(E)$ directly determines the second derivative of the current-voltage characteristic of normal tunnel junctions

$$\frac{d^2 I}{dV^2} = \frac{e}{R_T} P(eV). \tag{80}$$

The sum rules derived in the last section can be used to determine the current-voltage characteristic at very large voltages. We assume that eV is much larger than energies for which $P(E)$ gives a noticeable contribution and that $eV \gg k_B T$. Then the expression (78) becomes

$$I(V) = \frac{1}{eR_T} \int_{-\infty}^{+\infty} dE \,(eV - E)P(E) \tag{81}$$

which together with the sum rules (65) and (66) yields

$$I(V) = \frac{V - e/2C}{R_T} \tag{82}$$

for very large positive voltages. The slope of $I(V)$ confirms the interpretation of R_T as a tunneling resistance. The shift in voltage by $e/2C$ represents the Coulomb gap. In Sec. 4.2. we will discuss in more detail for the Ohmic model how the asymptotic current-voltage characteristic is approached for large voltages.

3.6. Low impedance environment

A special case of an environment is when the impedance is so low that one may effectively set $Z(\omega) = 0$. This will be a good approximation if the impedance is much less than the resistance quantum R_K. Since then the phase fluctuations described by

$J(t)$ vanish, we find $P(E) = \delta(E)$. This corresponds to the fact that in the absence of environmental modes only elastic tunneling processes are possible. From (56) we immediately get for the forward tunneling rate

$$\vec{\Gamma}(V) = \frac{1}{e^2 R_T} \frac{eV}{1 - \exp(-\beta eV)}. \tag{83}$$

According to Sec. 1.4. this is the global rule result which was already discussed in Sec. 1.2. where we introduced the voltage-biased tunnel junction. The appearance of the global rule in this limit is easy to understand. The external voltage source keeps the voltage across the junction fixed at any time. Therefore, after the tunneling process the electron has to be transferred through the circuit immediately to restore the charge on the junction capacitor. The work eV done by the voltage source is thus the only energy which can appear in the rate expressions.

We remark that the second sum rule (66) is violated if the impedance vanishes. As a consequence, in the absence of an external impedance we do not find a Coulomb gap even at highest voltages. On the other hand, for a small but finite impedance the sum rule (66) is valid although the current-voltage characteristic will show a clear Coulomb gap only at very large voltages (cf. Eq. (115)).

3.7. High impedance environment

We now consider the limit of a very high impedance environment, i.e. the impedance is much larger than R_K. Then the tunneling electron may easily excite modes. This situation is described by a spectral density of the environmental modes which is sharply peaked at $\omega = 0$. To check this we consider the case of Ohmic damping, i.e. $Z(\omega) = R$. Then the real part of the total impedance is given by $R/(1 + (\omega RC)^2)$. For very large resistance this becomes $(\pi/C)\delta(\omega)$. The prefactor is consistent with our result (68) for the integral over the total impedance. For the correlation function $J(t)$ this concentration of environmental modes at low frequencies means that the short time expansion

$$J(t) = -\frac{\pi}{CR_K}\left(it + \frac{1}{\hbar\beta}t^2\right) \tag{84}$$

applies for all times. Inserting this result into the definition (50) of $P(E)$ one gets a Gaussian integral which may easily be evaluated yielding

$$P(E) = \frac{1}{\sqrt{4\pi E_c k_B T}} \exp\left[-\frac{(E - E_c)^2}{4E_c k_B T}\right]. \tag{85}$$

This result obviously satisfies the sum rules (65) and (66) derived earlier. For very low temperatures $k_B T \ll E_c$ the probability to excite environmental modes reduces to

$$P(E) = \delta(E - E_c) \tag{86}$$

so that each electron transfers to the environment an amount of energy corresponding to the charging energy E_c. The expression (85) may be used to calculate tunneling rates and current-voltage characteristics in the high impedance limit. The broadening of the Gaussian distribution with respect to the delta function (86) describes the washout of the Coulomb blockade at finite temperatures. For zero temperature the expression (79) for the current together with (86) yields

$$I(V) = \frac{eV - E_c}{eR_T} \Theta(eV - E_c) \tag{87}$$

where $\Theta(E)$ is the unit step function. Since according to (86) a tunneling electron always transfers the energy E_c to the environment, tunneling becomes possible only if the energy eV at disposal exceeds E_c. We thus find the Coulomb gap as we did in Sec. 1.3. by considering only the charging energy of the junction. To make this connection clearer we note that the energy difference (4) of the local rule appears in (87) since

$$\frac{Q^2}{2C} - \frac{(Q - e)^2}{2C} = eV - E_c \tag{88}$$

if V is the voltage across the junction before the tunneling event.

We conclude the discussion of the last two sections by noting that the answer to whether one should use the global or local rule to determine the behavior of a tunnel junction is as follows. In general, neither rule is valid and the rate depends on the external circuit to which the junction is coupled. For impedances very low compared to the resistance quantum we find that the global rule leads to a correct description whereas for a high impedance environment and very low temperatures the local rule is correct. In all other cases $P(E)$ has to be calculated for the specific environment present in order to get the correct current-voltage characteristic. In the following section we will present various examples for external impedances and discuss how they affect tunneling rates and current-voltage characteristics.

4. Examples of electromagnetic environments

4.1. Coupling to a single mode

As a first example let us study the coupling of a tunnel junction to one single environmental mode which comes from a resonance in the lead impedance or might be associated with a molecule in the barrier. This model is so simple that analytical solutions are available for arbitrary temperatures.[8, 22] In addition, the simplicity of the model will allow us to learn important facts about how properties of the environment show up in the probability for energy transfer between the tunneling electron and the external circuit.

The coupling of the tunnel junction to one environmental mode may be accomplished by putting just one inductance L into the external circuit. In this special case our model for the environment introduced in Sec. 2.2. may be taken rather literally. With the impedance $i\omega L$ of an inductor we find for the total impedance

$$Z_t(\omega) = \frac{1}{C} \frac{i\omega}{\omega_s^2 - (\omega - i\epsilon)^2} \tag{89}$$

where we introduced the frequency

$$\omega_s = \frac{1}{\sqrt{LC}} \qquad (90)$$

of the environmental mode. The small imaginary part ϵ is necessary to obtain the correct result for the real part. In the limit $\epsilon \to 0$ we obtain

$$\text{Re} Z_t(\omega) = \frac{\pi}{2C}[\delta(\omega - \omega_s) + \delta(\omega + \omega_s)] \qquad (91)$$

which is what we expected since only the mode with frequency ω_s should be present and the prefactor satisfies (68).

Due to the delta functions in (91) we get the correlation function $J(t)$ simply by substituting ω by ω_s in (64). Inserting the result into (50) we then find

$$P(E) = \frac{1}{2\pi\hbar} \int_{-\infty}^{+\infty} dt \, \exp\left[\rho\left\{\coth(\frac{\beta\hbar\omega_s}{2})(\cos(\omega_s t) - 1) - i\sin(\omega_s t)\right\} + \frac{i}{\hbar} Et\right]. \qquad (92)$$

Here, we have introduced the parameter

$$\rho = \frac{\pi}{CR_K\omega_s} = \frac{E_c}{\hbar\omega_s} \qquad (93)$$

which should be of relevance since it compares the single electron charging energy with the mode excitation energy. This parameter determines the size of charge fluctuations

$$\langle \tilde{Q}(t)\tilde{Q}(0)\rangle = -\left(\frac{\hbar C}{e}\right)^2 \ddot{J}(t). \qquad (94)$$

Using (94) which is obtained from the relation (13) between the phase and the charge together with (17), (18), and the stationarity of equilibrium correlation functions

$$\langle A(t)B(0)\rangle = \langle A(0)B(-t)\rangle, \qquad (95)$$

we find

$$\langle \tilde{Q}^2\rangle = \frac{e^2}{4\rho} \coth(\frac{\beta\hbar\omega_s}{2}) \qquad (96)$$

so that at zero temperature charge fluctuations will only be small compared to the elementary charge if $\rho \gg 1$.

We now proceed with the calculation of the current-voltage characteristic. Using the equality

$$\cos(\omega_s t)\coth\left(\frac{\beta\hbar\omega_s}{2}\right) - i\sin(\omega_s t) = \frac{\cosh\left(\dfrac{\beta\hbar\omega_s}{2} - i\omega_s t\right)}{\sinh\left(\dfrac{\beta\hbar\omega_s}{2}\right)} \qquad (97)$$

we can take advantage of the generating function [23]

$$\exp[\frac{y}{2}(z + \frac{1}{z})] = \sum_{k=-\infty}^{+\infty} z^k I_k(y) \tag{98}$$

of the modified Bessel function I_k for $z = \exp(x)$. Then the integral over time in (92) can easily be done leading to

$$P(E) = \exp\left(-\rho \coth(\frac{\beta \hbar \omega_s}{2})\right)$$

$$\times \sum_{k=-\infty}^{+\infty} I_k\left(\frac{\rho}{\sinh(\frac{\beta \hbar \omega_s}{2})}\right) \exp\left(k \frac{\beta \hbar \omega_s}{2}\right) \delta(E - k \hbar \omega_s). \tag{99}$$

Although this expression for $P(E)$ is rather complicated it has a simple physical origin. This becomes particularly apparent at zero temperature where we find

$$P(E) = e^{-\rho} \sum_{k=0}^{\infty} \frac{\rho^k}{k!} \delta(E - k \hbar \omega_s) = \sum_{k=0}^{\infty} p_k \delta(E - k \hbar \omega_s). \tag{100}$$

Here p_k is the probability to emit k oscillator quanta. Comparing the second and the third expression in (100), one sees that p_k obeys a Poissonian distribution. Therefore, the quanta are emitted independently. The way of reasoning may now be reversed. Making the assumption of independent emission, (100) may of course immediately be obtained. But we also get the expression (99) for finite temperatures. Introducing the Bose factor $N = 1/[\exp(\hbar \beta \omega_s) - 1]$, the probability to emit a quantum is given by $\rho_e = \rho(1 + N)$ and the probability for absorption is $\rho_a = \rho N$. The probability to absorb m quanta and to emit n quanta will then be $\exp[-(\rho_a + \rho_e)]\rho_a^m \rho_e^n/(m!n!)$ so that

$$P(E) = \exp[-(\rho_a + \rho_e)] \sum_{m,n} \frac{\rho_a^m \rho_e^n}{m!n!} \delta(E - (n - m)\hbar \omega_s). \tag{101}$$

Doing the sum over the variable $l = m + n$ and using the ascending series of the modified Bessel function [23]

$$I_k(z) = \left(\frac{z}{2}\right)^k \sum_{l=0}^{\infty} \frac{(z^2/4)^l}{l!(k+l)!} \tag{102}$$

one is left with a sum over the difference $k = n - m$ which is our finite temperature result (99). We note that the argument given here can be generalized to the case of two or three modes and finally to infinitely many modes. The representation (101) of $P(E)$ points clearly to its physical significance. It is apparent that $P(E)$ gives the quantity describing the probability to exchange the energy E with the environment.

With the form (99) for $P(E)$ the convolution integral appearing in the expression (78) for the current-voltage characteristic can easily be evaluated yielding [8, 22]

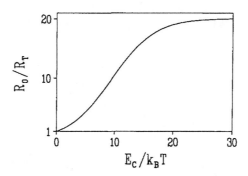

Figure 3. Zero-bias differential resistance as a function of temperature for $\rho = 3$.

$$I(V) = \frac{1}{eR_T} \sinh\left(\frac{\beta eV}{2}\right) \exp\left(-\rho \coth(\frac{\beta \hbar \omega_s}{2})\right)$$

$$\times \sum_{k=-\infty}^{+\infty} I_k\left(\frac{\rho}{\sinh(\frac{\beta \hbar \omega_s}{2})}\right) \frac{\epsilon_k}{\sinh(\frac{\beta \epsilon_k}{2})}. \qquad (103)$$

Here, we introduced the energy

$$\epsilon_k = eV - k\hbar \omega_s \qquad (104)$$

left to the electron after having excited k quanta $\hbar \omega_s$. In the limit of zero temperature and for positive voltages (103) becomes

$$I(V) = \frac{1}{eR_T} e^{-\rho} \sum_{k=0}^{n} \frac{\rho^k}{k!}(eV - k\hbar \omega_s) \qquad (105)$$

where n is the largest integer smaller or equal to $eV/\hbar \omega_s$. This result has a simple interpretation. The sum runs over all possible numbers of excited quanta where the maximum number of modes which can be excited is given by n. The factor $\exp(-\rho)$ determines the slope at zero voltage since at very low voltages only the term with $k = 0$ contributes to the sum. As we expected from our discussion of ρ, this quantity is important for the occurrence of the Coulomb blockade. For small ρ there is no Coulomb blockade and the conductance at zero voltage is about $1/R_T$. Only for large enough ρ a Coulomb blockade becomes apparent in the small factor $\exp(-\rho)$. According to the definition (93), large ρ means that the mode energy $\hbar \omega_s$ is small compared to the charging energy E_c which indicates a high impedance environment as discussed earlier. So again this example shows that Coulomb blockade can only be found if the environmental impedance is large enough. Fig. 3 presents the differential resistance $R_0 = dV/dI$ at zero bias $V = 0$ for $\rho = 3$. For large temperatures the Coulomb blockade is lifted by thermal fluctuations and R_0 is of the order of R_T. As the temperature is decreased a Coulomb gap forms and R_0/R_T approaches $\exp(\rho)$ for vanishing temperature.

So far we have discussed the small voltage behavior. But it is also the current-voltage characteristic at finite voltages which contains information about the environment. Every time the voltage becomes an integer multiple of the mode energy $\hbar \omega_s$ the slope of

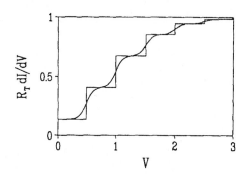

Figure 4. Differential current-voltage characteristic for $\rho = 2$. The voltage is given in units of $e/2C$. The step-like curve corresponds to zero temperature while the smooth curve is for $k_B T = 0.04\, E_c$.

the current-voltage characteristic changes. This becomes even more apparent in the differential current-voltage characteristic where we find steps at voltages $k\hbar\omega_s/e$ when new inelastic channels are opened. This is in agreement with (80) according to which the second derivative at zero temperature is $P(E)$ for which we know that it is a series of delta functions at voltages $k\hbar\omega_s/e$. Thus derivatives of the current-voltage characteristic contain information about the structure of the environment. As an example we show in Fig. 4 the differential current-voltage characteristic for $\rho = 2$. At zero temperature one gets the steps as expected. At finite temperature, however, the sharp resonances in $P(E)$ are washed out and therefore the steps are smoothed.

To end this section let us apply the mechanical analogue of Table I to point out the relation between the Mößbauer effect in solid state physics and the environmental effects on single charge tunneling. For the Mößbauer effect one considers a radioactive nucleus embedded in a crystal. When a γ quant is emitted there are two ways to satisfy momentum conservation. The first possibility is to excite phonons in the crystal, i.e. momentum is transferred to the emitting nucleus and the energy of the γ quant is reduced. In the second possiblity, the so-called Mößbauer transition, the recoil momentum is transferred to the whole crystal. This will be more likely if it is difficult to excite phonons. Due to the large mass of the crystal the energy of the γ quant and, more important for us, the momentum of the nucleus then remain unchanged.

In ultrasmall tunnel junctions the emission of a γ quant corresponds to the tunneling of an electron. According to Table I the momentum of the nucleus is related to the charge of the junction. The question is whether a tunneling process changes the junction charge or not. If this charge is kept fixed we do not find Coulomb blockade. This corresponds to the Mößbauer transition. In both cases no environmental modes are excited. For the occurrence of Coulomb blockade we need a change of the junction charge. This is analogous to a non-Mößbauer transition and requires the excitation of environmental modes. We conclude from this analogy that Coulomb blockade is only possible if there are low frequency environmental modes which are coupled strongly to the tunneling electron, i.e. a high impedance environment is needed. This is in agreement with our previous findings. The analogy with the Mößbauer effect allows us also to interpret the factor $\exp(-\rho)$ in (100) as a Debye-Waller factor giving the possibility for electron tunneling without the excitation of environmental modes.

4.2. Ohmic impedance

For the impedance caused by an external circuit a more realistic choice than the single mode model would be an ideal Ohmic resistor described by the frequency-independent impedance $Z(\omega) = R$. We introduce the dimensionless parameter

$$g = \frac{R_K}{R} \tag{106}$$

which is proportional to the lead conductance.

The mechanical analogue of this problem is very well studied since this special impedance results in a Fourier transform of the admittance which is proportional to a delta function. Reinterpreting (58) we then find the equation of motion for a free Brownian particle which contains a damping term proportional to the velocity of the particle. Without a confining potential the Brownian particle undergoes a diffusive motion. From our knowledge of classical diffusion we conclude that for long times the correlation function $J(t)$ should increase proportional to time t. This classical result holds also for low temperatures. At zero temperature, however, the environment cannot provide the diffusing particle with energy and the correlation function will increase somewhat slower, namely proportional to $\ln(t)$.[18] In general, it is not possible to obtain analytical results for the Ohmic model. We therefore restrict ourselves to the case of zero temperature and consider the limits of low and high energies in $P(E)$. This allows us to find explicit expressions for the current-voltage characteristics at small and large voltages. Numerical calculations bridge the gap between the two limits.

We first discuss the low energy behavior of $P(E)$, which is determined by the long time behavior of the correlation function $J(t)$. This case is of general importance since according to (75) at low voltages and very low temperatures $P(E)$ is governed by the impedance at low frequencies. As long as the impedance at zero frequency is nonvanishing the Ohmic model with $R = Z(0)$ will apply in this regime.

In order to avoid lengthy calculations we determine the low energy behavior of $P(E)$ from the integral equation (75) which is valid at zero temperature. Since this integral equation is homogeneous it will allow us to determine $P(E)$ only up to a multiplicative constant which depends on the behavior of $P(E)$ at all energies. To solve the integral equation we need the real part of the total impedance

$$\frac{\mathrm{Re}Z_t(\omega)}{R_K} = \frac{1}{R_K}\mathrm{Re}\left[\frac{1}{i\omega C + 1/R}\right] = \frac{1}{g}\frac{1}{1 + (\omega/\omega_R)^2} \tag{107}$$

where the frequency

$$\omega_R = \frac{1}{RC} = \frac{g}{\pi}\frac{E_c}{\hbar} \tag{108}$$

describes an effective cutoff for the total impedance due to the junction capacitance. At energies small compared to $\hbar\omega_R$ we may approximate the real part of the total impedance by a constant. Taking the derivative of (75) with respect to energy we get the differential equation

$$\frac{dP(E)}{dE} = \left(\frac{2}{g} - 1\right) \frac{P(E)}{E} \tag{109}$$

which may easily be solved yielding

$$P(E) \sim E^{(2/g-1)} \tag{110}$$

for small positive energies. For negative energies $P(E)$ vanishes since we consider the case of zero temperature. With a more complete analysis of $J(t)$ and $P(E)$ one may determine the normalization constant. One finds [8]

$$P(E) = \frac{\exp(-2\gamma/g)}{\Gamma(2/g)} \frac{1}{E} \left[\frac{\pi}{g} \frac{E}{E_c}\right]^{2/g}, \tag{111}$$

where $\gamma = 0.577\ldots$ is the Euler constant. The factors appearing in (111) may be motivated by the behavior of the correlation function $J(t)$ for large times [18]

$$J(t) = -\frac{2}{g}[\ln(\omega_R t) + i\frac{\pi}{2} + \gamma] \quad \text{for } t \to \infty \tag{112}$$

so that the offset of the logarithmic divergence appears in the result (111). From (79) it is straightforward to calculate the current-voltage characteristic for small voltages

$$I(V) = \frac{\exp(-2\gamma/g)}{\Gamma(2 + 2/g)} \frac{V}{R_T} \left[\frac{\pi}{g} \frac{e|V|}{E_c}\right]^{2/g} \tag{113}$$

which leads to a zero-bias anomaly of the conductance $dI/dV \sim V^{2/g}$.[7, 8, 9, 15, 24] This result remains valid for a more general environment with a finite zero-frequency impedance $Z(0)$. The power law exponent is then given by $2/g = 2Z(0)/R_K$ but the prefactor in (113) depends on the high-frequency behavior of the impedance.

Besides the behavior at low voltages it is also of interest how fast the current-voltage characteristics for finite lead conductance g approach the high impedance asymptote (82). To answer this we need to know the high energy behavior of $P(E)$ which for an Ohmic impedance follows from (76) and (107) as

$$P(E) = \frac{2g}{\pi^2} \frac{E_c^2}{E^3} \quad \text{for } E \to \infty. \tag{114}$$

Inserting this into the expression (79) for the current at zero temperature one finds

$$I(V) = \frac{1}{R_T} \left[V - \frac{e}{2C} + \frac{g}{\pi^2} \frac{e^2}{4C^2} \frac{1}{V}\right] \quad \text{for } V \to \infty. \tag{115}$$

As expected the corrections to (82) for finite lead conductance are positive and for a given voltage they become smaller with decreasing conductance of the external resistor. The voltage at which the corrections become negligible will increase with $\sqrt{g}$ as the lead conductance is increased. In the limit $g \to \infty$ no crossover will occur and the Ohmic current-voltage characteristic (3) will be correct for all voltages.

In Fig. 5 we present a $P(E)$ for zero temperature and an Ohmic lead conductance

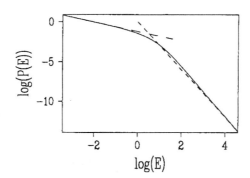

Figure 5. Log-log plot of $P(E)$ at zero temperature for the Ohmic model with $g = 5$. Also shown are the low energy asymptote according to (111) (long-dashed) and the high energy asymptote according to (114) (short-dashed). Energy is taken in units of E_c.

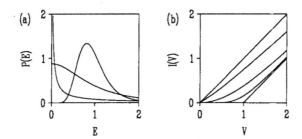

Figure 6. (a) $P(E)$ at zero temperature for the Ohmic model with lead conductances $g = 0.2$ (peaked around E_c), $g = 2$ (zero slope at $E = 0$), and $g = 20$ (diverging at $E = 0$). Energy is taken in units of E_c. (b) Zero temperature current-voltage characteristics for the Ohmic model with $g = \infty, 20, 2, 0.2$, and 0 from top to bottom.

$g = 5$ together with its low and high energy asymptotes. The data were obtained numerically by solving the integral equation (75). The dependence of $P(E)$ and of the corresponding current-voltage characteristics on the lead conductance is shown in Fig. 6 for three different values of g. According to (110) the $P(E)$ depicted in Fig. 6a has a singularity for the large conductivity $g = 20$, starts with zero slope for $g = 2$ and is peaked around E_c for the small conductivity $g = 0.2$. The current-voltage characteristics of Fig. 6b demonstrate that quantum fluctuations destroy the Coulomb blockade. Again, a clear Coulomb blockade is obtained for a high impedance environment. As a criterion for the occurrence of a Coulomb blockade one may require that for vanishing voltage the curvature of the current-voltage characteristic goes to zero. Since the curvature is given by $P(E)$ we find that this criterion is fulfilled if the dimensionless lead conductance is sufficiently small ($g < 2$). This is related to the fact that at this lead conductance $P(E)$ switches from a divergent behavior for $E \rightarrow 0$ to a regime where $P(E)$ vanishes in this limit. This singular behavior of $P(E)$ for $g > 2$ disappears for finite temperatures but then thermal fluctuations also contribute to the destruction of the Coulomb blockade.

4.3. A mode with a finite quality factor

We now combine the two models considered previously and discuss the case of finite temperatures. As in the first model we start with a single mode. But now we allow for a finite quality factor which means that the resonance is broadened. Technically

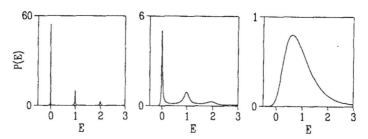

Figure 7. $P(E)$ at $k_B T = 0.05\ E_c$ for the total impedance (116). The quality factor decreases from left to right $Q = 50, 5, 0.25$ and $\hbar\omega_s = E_c$. Energy is taken in units of E_c.

this is achieved by putting a resistor in series with the inductor of the single mode model. We may keep the notation of the previous sections where we introduced the mode frequency $\omega_s = (LC)^{1/2}$, the inverse relaxation time $\omega_R = 1/RC$, and the lead conductance $g = R_K/R$. However, it is useful to introduce the quality factor $Q = \omega_R/\omega_s$ which measures the broadening of the resonance or equivalently how fast an oscillation decays with respect to the oscillation period. The single mode case then corresponds to $Q = \infty$ while the Ohmic case is approached for $Q \to 0$. By varying the quality factor, we are able to change qualitative features of the environment. For an environment with a resistor and an inductor in series we get for the total impedance

$$\frac{Z_t(\omega)}{R_K} = \frac{1}{g}\frac{1 + iQ^2(\omega/\omega_R)}{1 + i(\omega/\omega_R) - Q^2(\omega/\omega_R)^2}. \tag{116}$$

For the calculation of $P(E)$ and of finite temperature current-voltage characteristics one has to resort to numerical methods. The results presented in Fig. 7 were obtained by means of an inhomogeneous integral equation which is a generalization of the integral equation (75). An inhomogeneous term, which allows for a simple recursive algorithm, was obtained by splitting off the Ohmic long time behavior of the correlation function $J(t) \sim t$ discussed in the last section.[20]

In Fig. 7 we have chosen the mode energy $\hbar\omega_s$ equal to the charging energy E_c. The quality factors range from 50 which gives a very good resonance over $Q = 5$ showing a considerable broadening to the rather low value of 0.25. The temperature $k_B T = 0.05\ E_c$ is very low so that for negative energies $P(E)$ is strongly suppressed as can be seen very clearly from the figure. The $P(E)$ for the high quality factor reflects the sharp resonance in the environmental impedance and also describes the possibility of exciting more than one quantum according to (99). The broadening of the lines which is connected to additional bath modes is clearly seen for $Q = 5$. For $Q=0.25$ one finds a broad distribution for $P(E)$ resembling the one found for the pure Ohmic model with a broad frequency range of environmental oscillators. It is obvious from this discussion that $P(E)$ contains a lot of information about the environment to which the junction is coupled. According to (80) $P(E)$ for normal tunnel junctions is proportional to the second derivative of the current-voltage characteristic and therefore rather difficult to measure. However, we will show in Sec. 5 that the Josephson current in ultrasmall Josephson junctions at $T = 0$ is related directly to $P(E)$. For normal tunnel junctions one may measure the first derivative of the current-voltage characteristic. For a single bath mode we had already seen that differential current-voltage characteristics show

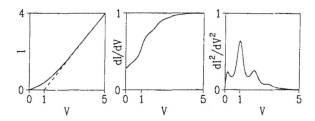

Figure 8. Current-voltage characteristic and its first and second derivatives as calculated from $P(E)$ for $Q = 5$ given in Fig. 7. The dashed line in the I-V characteristic indicates the ideal Coulomb blockade characteristic. Currents are taken in units of $e/2CR_T$ and voltages in units of $e/2C$.

more details than the I-V curve itself. As an example, Fig. 8 presents results for the case $Q = 5$. It is difficult to distinguish this current-voltage characteristic from the characteristic of the Ohmic model. However, in the first derivative with respect to voltage we find a step-like structure which we know is due to the resonance in the environment. It is smeared because of the finite quality factor and thermal fluctuations. In the second derivative we almost reproduce $P(E)$. According to (80) this would be exact at zero temperature. For finite temperatures the second derivative is roughly given by the antisymmetric part of $P(E)$. The antisymmetric part ensures that the current vanishes at zero voltage. This leads to noticeable deviations from $P(E)$ at low voltages as seen in Fig. 8.

4.4. Description of transmission lines

So far we have treated only impedances which can be described by at most two lumped circuit elements like a resistor and an inductor. To model a real experiment, however, this is often not sufficient. Thinking for example of wires attached to the junction, one has to model the environment by distributed resistors, inductors, and capacitors characterized by the three parameters R_0, L_0, and C_0 which are resistance, inductance, and capacitance per unit length, respectively. Before discussing two special cases of such transmission lines, let us first derive the impedance for a more general transmission line. We describe two wires by segments containing a resistor and an inductor in series with a capacitive coupling between the wires as shown in Fig. 3 of Chap. 1. We neglect a conductance between the wires which is sometimes also taken into account. The voltage drop along the line is connected with the current flowing through the wire and the impedance per unit length via the differential equation

$$\frac{\partial V}{\partial x} = -I(x)(R_0 + i\omega L_0) \tag{117}$$

where we assumed that the time dependence of the current and the voltage is given by $\exp(i\omega t)$. This equation is complemented by the continuity equation

$$i\omega q(x) + \frac{\partial I}{\partial x} = 0 \tag{118}$$

where $q(x) = C_0 V(x)$ is the charge sitting on the capacitor at position x. This charge can only change if current flowing through the wires charges the capacitor. Equations (117) and (118) describe the dynamics of the transmission line. Eliminating the current we obtain

$$\frac{\partial^2 V}{\partial x^2} = -k^2 V(x) \tag{119}$$

where we introduced the wave number

$$k = \sqrt{\omega(-i R_0 C_0 + \omega L_0 C_0)} \tag{120}$$

which indeed has the dimension of an inverse length since the parameters R_0, L_0, and C_0 are taking per unit length. We note that in general k is not real so that only for an LC transmission line ($R_0 = 0$) the propagation of undamped waves becomes possible. It is straightforward to solve (119) for the voltage yielding

$$V(x) = A e^{-ikx} + B e^{ikx}. \tag{121}$$

We make use of (117) to obtain the current

$$I(x) = \frac{ik}{R_0 + i\omega L_0}(A e^{-ikx} - B e^{ikx}). \tag{122}$$

If we attach a semi-infinite transmission line to the right of the point $x = 0$ we only have waves traveling to the right, i.e. $B = 0$. Then the impedance of the transmission line at $x = 0$ is

$$Z_\infty(\omega) = \sqrt{\frac{R_0 + i\omega L_0}{i\omega C_0}}. \tag{123}$$

In reality, a transmission line has a finite length ℓ. Let us determine the impedance Z at $x = 0$ for a transmission line terminated at $x = \ell$ by a load impedance Z_L. This leads to the boundary condition $V(\ell) = Z_L I(\ell)$ at the end of the line. From (121) and (122) we then get

$$V(x) = \frac{I(\ell)}{2}\left[(Z_\infty + Z_L)e^{-ik(x-\ell)} + (Z_L - Z_\infty)e^{ik(x-\ell)}\right] \tag{124}$$

and

$$I(x) = \frac{I(\ell)}{2Z_\infty}\left[(Z_\infty + Z_L)e^{-ik(x-\ell)} - (Z_L - Z_\infty)e^{ik(x-\ell)}\right]. \tag{125}$$

The impedance at $x = 0$ is given by $Z = V(0)/I(0)$ for which we find

$$Z = Z_\infty \frac{e^{2ik\ell} - \lambda}{e^{2ik\ell} + \lambda}. \tag{126}$$

Here, we introduced the reflection coefficient

$$\lambda = \frac{Z_\infty - Z_L}{Z_\infty + Z_L} \tag{127}$$

which is obtained from (124) as the negative of the ratio between the voltages at $x = \ell$ of the reflected and incident waves. For $Z_\infty \gg Z_L$ we have a short at the end of the line and the voltage vanishes there. In the opposite limit $Z_L \gg Z_\infty$ the line is open at its end and the voltage has a maximum.

According to the form of the impedance (123) we may distinguish between two cases of relevance as far as the effect on Coulomb blockade phenomena is concerned. If the relevant frequencies of order $E_c/\hbar$ are much larger than R_0/L_0 we may neglect the resistance in (123) and consider an LC transmission line. If, on the other hand, the relevant frequencies are much smaller than R_0/L_0 we may neglect the inductance and end up with an RC transmission line. Typical experimental values for the capacitance and inductance per unit length are of the order of $C_0 \simeq 10^{-16}\text{F}/\mu\text{m}$ and $L_0 \simeq 10^{-13}\text{H}/\mu\text{m}$. Therefore, the adequate model depends to a large extent on the specific resistance of the wire material. For a pure metal like aluminium the wire resistance is typically of the order of $R_0 \simeq 10^{-3}\Omega/\mu\text{m}$ and the crossover frequency is then $R_0/L_0 \simeq 10^{10}\text{Hz}$. For capacitances in the fF-range this frequency is below $E_c/\hbar$ and the LC transmission line model is applicable. On the other hand, for wires made of high resistive alloys, R_0 may be larger than $10\Omega/\mu\text{m}$ and the crossover frequency then exceeds 10^{14}Hz. In this case the RC line will render a reasonable description. In the following two sections we discuss the influence of these transmission lines on charging effects more specifically.

4.5. LC transmission line

The limit of an LC transmission line is obtained from the case considered in the previous section by setting $R_0 = 0$. The wave number of the solutions (121) and (122) becomes $k = \omega(L_0 C_0)^{1/2}$ and thus describes waves propagating along the line with velocity $u = 1/(L_0 C_0)^{1/2}$. From (123) it follows that the impedance of an infinite line is purely Ohmic, i.e. $Z_\infty = (L_0/C_0)^{1/2}$. The line impedance Z_∞ varies with geometry and typically ranges between 10 and a few 100 Ω. Hence, it is of the order of the free space impedance $(\mu_0/\varepsilon_0)^{1/2} = 377\Omega$, that is much smaller than the quantum resistance R_K. As discussed in the previous section we may terminate the line at $x = \ell$ with a load resistor Z_L and get for the external impedance

$$Z(\omega) = Z_\infty \frac{\exp(2i\omega\ell/u) - \lambda}{\exp(2i\omega\ell/u) + \lambda}. \tag{128}$$

This impedance exhibits resonances at $\omega_n = \pi n u/\ell$ for $Z_\infty \ll Z_L$ and at $\omega_n = \pi(n - 1/2)u/\ell$ for $Z_\infty \gg Z_L$. From our experience with the single mode model we expect these features to show up in the differential current-voltage characteristic as steplike increases of the dynamic conductance dI/dV at the voltages $\hbar\omega_n/e$. Every step corresponds to a new inelastic channel which is opened as the voltage increases.

To see this more explicitly we assume $Z_\infty, Z_L \ll R_K$ which is frequently the case. Under these conditions we may expand $\exp[J(t)]$ in the definition (50) of $P(E)$. Keeping

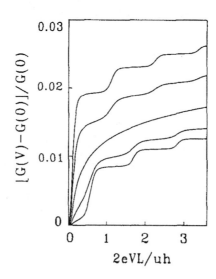

Figure 9. Results of numerical calculations of the differential conductance of a tunnel junction attached to an LC-line of finite length L. The Ohmic conductance $G(0) = V/R_T$ has been subtracted off. The line impedance $Z_\infty = 50\,\Omega$, and the ratio $Z_L/Z_\infty = 10, 3, 1, 1/3, 1/10$ from the upper to the lower curve. The temperature is 4.2 K and the voltage is taken in units of $uh/2eL = 12\,\text{mV}$. In this figure taken from Ref. [25] the length of the line is denoted by L instead of ℓ used in the text.

the first two terms we get by virtue of (78) for the current-voltage characteristic [25]

$$I(V) = \frac{1}{eR_T}\left[eV + \int_{-\infty}^{+\infty} \frac{\mathrm{d}E}{E} \frac{1}{1 - e^{-\beta E}} \frac{\mathrm{Re}Z_t(E/\hbar)}{R_K}\right.$$
$$\left. \times \left(\frac{(eV - E)(1 - e^{-\beta eV})}{1 - e^{-\beta(eV - E)}} - eV\right)\right]. \qquad (129)$$

As mentioned earlier, the environmental effect becomes more apparent in derivatives of the current-voltage characteristic. Fig. 9 shows numerical results for the differential conductance $G(V) = \mathrm{d}I/\mathrm{d}V$ for various ratios Z_L/Z_∞ and finite temperature $T = 4.2\,\text{K}$. For simplicity, the difference between $Z_t(\omega)$ and $Z(\omega)$ was neglected which is appropriate if the junction capacitance is very small. The expected steps can be seen very clearly except for the case $Z_L = Z_\infty$ where the terminating resistance matches the line impedance and thus no resonances are present.

4.6. RC transmission line

We now consider the RC transmission line which is obtained in the limit $L_0 = 0$ from the more general model discussed above. For the impedance of an infinite line we obtain from (123)

$$Z(\omega) = \sqrt{\frac{R_0}{i\omega C_0}} \qquad (130)$$

so that the impedance increases with decreasing frequency. For a finite line there will always be a cutoff and $Z(\omega)$ remains finite for $\omega \to 0$. From (130) the total impedance of an infinite line takes the form

$$Z_t(\omega) = \frac{1}{i\omega C + \sqrt{i\omega C_0/R_0}} \; .$$ (131)

Since the influence of the environment depends on the ratio between $Z_t(\omega)$ and R_K, the relevant dimensionless parameter is $\kappa = (R_0 C/C_0)/R_K$. This gives the resistance of a piece of wire whose capacitance equals the capacitance of the tunnel junction.

In the limit $\kappa \to \infty$ we approach the high impedance limit and a classical Coulomb blockade picture emerges. The tunneling is completely suppressed for $V < e/2C$. At higher voltages we find the shifted linear characteristic $I = V - e/2C$. If κ is finite but large the sharp curve is smoothed and an exponentially small tunneling current appears for voltages below $e/2C$. This is in accordance with our earlier findings.

It is surprising that there is also a substantial suppression of tunneling in the opposite case $\kappa \ll 1$. In this limit we see from (131) that at frequencies of order $E_c/\hbar$ the effective shunt resistance is much smaller than R_K. Hence, there is no blockade in this region. However, at lower frequencies the impedance increases and reaches R_K for frequencies of the order of $eV_c/\hbar$ where $V_c = 2eR_0/(C_0 R_K) = 2\kappa e/C$. Provided the line is sufficiently long, so that $Z_t(\omega)$ does not yet saturate at $\omega \simeq eV_c/\hbar$, tunneling is strongly suppressed for voltages smaller than V_c. It is worth noting that this phenomenon does not depend on the junction capacitance.

Thus, for a long RC line environment two types of Coulomb blockade exist.[7] Let us consider in more detail the second type of blockade occurring in the limit $\kappa \ll 1$ for voltages below V_c. For small frequencies the total impedance is approximately given by the line impedance (130). At zero temperature we then may calculate $P(E)$ for small energies and find

$$P(E) = \sqrt{\frac{eV_c}{4\pi E^3}} \exp(-eV_c/4E).$$ (132)

According to (80) the second derivative of the current-voltage characteristic with respect to the voltage is proportional to $P(E)$ and therefore its low voltage behavior is determined by (132). As a consequence the current is suppressed exponentially like $\exp(-V_c/4V)$ at very small voltages. The current-voltage characteristic of a junction coupled to an RC-line together with its second derivative is presented in Fig. 10. The current is noticeably suppressed for voltages $\simeq 2V_c$ whereas it becomes exponentially small only for voltages below about $0.05V_c$. This is quite unusual for a one-parameter behavior. Temperatures of the order of eV/k_B wash out the suppression of tunneling at voltages less than eV and Ohm's law is restored for $eV \ll k_B T$. Finally, we note that at large voltages $V \gg e/(\kappa C)$ the effect of the junction capacitance dominates resulting in an offset of the I-V curve by $e/2C$.

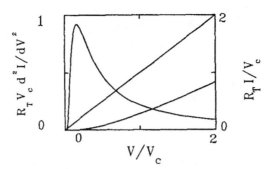

Figure 10. Coulomb blockade of the second type in a tunnel junction attached to an RC-line. The current-voltage characteristic shows a significant deviation from the straight line representing Ohm's law at voltages $V \simeq V_c$. On the other hand, the current is suppressed exponentially only for voltages up to about $0.05V_c$ as can be seen from the second derivative of the I-V characteristic which rises sharply above this voltage.

5. Tunneling rates in Josephson junctions

5.1. Introduction

So far we have studied the effect of an external circuit on electron tunneling rates in normal tunnel junctions. It is also interesting to consider Josephson junctions. In this case we have two kinds of charge carriers, namely Cooper pairs and quasiparticles. While the concepts developed for normal junctions are still valid, it will turn out that the influence of the environment on Cooper pair tunneling is even simpler to describe than its effect on electron tunneling in normal junctions considered so far. Furthermore, the supercurrent provides a more direct mean to measure the environmental influence. The experimental relevance of superconducting junctions is also due to the fact that most metals used to fabricate tunnel junctions become superconducting at sufficiently low temperatures. Often one even has to apply an external magnetic field to drive these junctions normal.

As in the previous sections, we will concentrate on the environmental influence on single charge tunneling. For other aspects of ultrasmall Josephson junctions we refer the reader to Chap. 4 and to the review provided by Ref. [5]. In the previous sections, the concept of a phase proved to be very useful to determine the current-voltage characteristics of normal tunnel junctions. It is clear that the phase will be even more important in the superconducting case where it has a non-vanishing expectation value due to the long-range order in the superconducting leads. In contrast to the phase which is usually introduced by means of the Josephson relation we keep the phase as defined in (13). For quasiparticle tunneling this is the adequate choice. The factor of two explicitly appearing in expressions for the supercurrent will always remind us of the fact that Cooper pairs carry twice the electron charge. According to the commutator (15), the operator $\exp(-2i\varphi)$ leads to a change of the junction charge by $2e$ associated with the tunneling of a Cooper pair.

For normal tunnel junctions we have seen that the relevant energy scales were the charging energy E_c and the thermal energy k_BT. In Josephson junctions an additional energy scale appears in form of the Josepson coupling energy E_J. One may distinguish the regime $E_c \gg E_J$, which means that the charge is well defined, from the regime

$E_J \gg E_c$ where the phase fluctuates only little. We will calculate the tunneling rates for weak Josephson coupling and then present a duality transformation which relates the weak coupling regime to the strong coupling regime. In the following we treat the tunneling of Cooper pairs and of quasiparticles separately thereby neglecting effects which couple the two processes.

5.2. Tunneling of Cooper pairs

In this section we consider the tunneling of Cooper pairs in an ultrasmall Josephson junction. We neglect quasiparticle excitations which is a good approximation at temperatures very low compared to the critical temperature and voltages below the gap voltage $2\Delta/e$, where 2Δ is the superconducting gap. Before we start calculating the tunneling rates, we need to discuss the main differences between Cooper pair tunneling and quasiparticle tunneling. In Sec. 3.1. we had decomposed the total Hamiltonian (31) into contributions of the quasiparticles and the environment, and both were coupled by the tunneling Hamiltonian. In contrast, for Cooper pair tunneling we only have a Hamiltonian acting in the Hilbert space of Q, φ and the environmental degrees of freedom. The Cooper pairs form a condensate and therefore do not lead to additional dynamical degrees of freedom. The only consequence of the coupling between the superconducting leads is the Josephson energy given by the second term in the total Hamiltonian

$$H = H_{\text{env}} + E_J \cos(2\varphi). \tag{133}$$

The environmental Hamiltonian was defined in (21) and remains unchanged. Rewriting the Josephson term as

$$E_J \cos(2\varphi) = \frac{E_J}{2} e^{-2i\varphi} + \text{H.c.} \tag{134}$$

we see that it replaces the electron tunneling Hamiltonian H_T defined in (25). The operator $e^{-2i\varphi}$ changes the charge Q on the junction by $2e$. This process is connected with the tunneling of a Cooper pair, although the Cooper pairs appear in the Hamiltonian only through the phase difference between the condensate wave functions on both sides of the barrier. The Hamiltonian (133) is similar to the total Hamiltonian (31) for electron tunneling except that there are no electronic degrees of freedom. This allows us to calculate tunneling rates for Cooper pairs in the spirit of Sec. 3.2. However, the steps performed in Sec. 3.2.1. are now obsolete and we can start the calculation by tracing out the environment. Considering forward tunneling, the expression analogous to (37) reads

$$\vec{\Gamma}(V) = \frac{\pi}{2\hbar} E_J^2 \sum_{R,R'} |\langle R|e^{-2i\varphi}|R'\rangle|^2 P_\beta(R)\delta(E_R - E_R'). \tag{135}$$

This is just the golden rule rate with (134) as perturbation averaged over an equilibrium distribution of initial states. So far we have kept the dependence on the external voltage in the phase. The trace over the environmental degrees of freedom is performed like in Sec. 3.2.2. We then arrive at

$$\vec{\Gamma}(V) = \frac{E_J^2}{\hbar^2} \int_{-\infty}^{+\infty} dt \, \exp\left(\frac{2i}{\hbar}eVt\right) \langle e^{2i\tilde{\varphi}(t)} e^{-2i\tilde{\varphi}(0)}\rangle \tag{136}$$

where we introduced $\tilde{\varphi}(t)$ according to (17). We may again exploit the generalized Wick theorem and express the correlation function in (136) in terms of the phase-phase correlation function $J(t)$ given by (64). In analogy to (50) we define

$$P'(E) = \frac{1}{2\pi\hbar} \int_{-\infty}^{\infty} dt \exp\left[4J(t) + \frac{i}{\hbar}Et\right] \tag{137}$$

and get for the forward tunneling rate for Cooper pairs

$$\vec{\Gamma}(V) = \frac{\pi}{2\hbar} E_J^2 P'(2eV). \tag{138}$$

The probability $P'(E)$ differs from the probability $P(E)$ introduced for normal junctions only by a factor 4 in front of the phase-phase correlation function $J(t)$ which arises from the fact that the charge of Cooper pairs is twice the electron charge. In view of the relation (64) for the correlation function $J(t)$ we may absorb this factor into the "superconducting resistance quantum" $R_Q = h/4e^2 = R_K/4$. Since the total impedance must now be compared with R_Q the influence of the environment on Cooper pair tunneling rates, and thus on the supercurrent, is more pronounced than for the current through a normal junction.

Before calculating the supercurrent let us briefly discuss the range of validity of the perturbative result (138). Since the Josephson coupling was considered as a perturbation, the Josephson energy E_J has to be small. From an analysis of higher order terms, one finds that our lowest order result is correct if $E_J P'(2eV) \ll 1$. Obviously, this condition depends on the voltage and on the environmental impedance. To be more specific let us choose an Ohmic environment and zero temperature. If the impedance Z is of the order of the resistance quantum R_Q, the probability $P'(2eV)$ will be a broad distribution with a maximum height of the order of the inverse charging energy E_c^{-1} (cf. Fig. 6a). Then our rate expression is correct if $E_J \ll E_c$. On the other hand, for a high impedance environment, $P'(2eV)$ is peaked around E_c. Now, $P'(E_c)$ is found to be of order $(1/E_c)(Z/R_Q)^{1/2}$ for $Z \gg R_Q$ and the rate formula (138) holds provided the condition $E_J \ll E_c(R_Q/Z)^{1/2}$ is satisfied. This latter condition is more restrictive. In the opposite case of a low impedance environment $P'(2eV)$ is sharply peaked at $V = 0$ and decreases with increasing voltage. The condition $E_J \ll 1/P'(2eV)$ is then always violated at sufficiently low voltages.

From the rate expression (138) together with the symmetry $\vec{\Gamma}(V) = \overleftarrow{\Gamma}(-V)$ we immediately get for the supercurrent [26]

$$I_S(V) = 2e\left(\vec{\Gamma}(V) - \overleftarrow{\Gamma}(V)\right) = \frac{\pi e E_J^2}{\hbar}\left(P'(2eV) - P'(-2eV)\right) \tag{139}$$

where we accounted for the charge $2e$ which each tunneling process transports. This result reflects the fact that a Cooper pair tunneling in the direction of the applied voltage carries an energy $2eV$. This energy has to be transferred to the environment since the Cooper pairs have no kinetic energy that could absorb a part of $2eV$. The probability for this transfer of energy is $P'(E)$. Since the supercurrent depends directly on the probability $P'(E)$, it enables one to measure properties of the environment more directly. For normal junctions it was necessary to measure the second derivative of the current-voltage characteristic which is more complicated. On the other hand, this

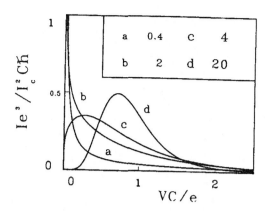

| a | 0.4 | c | 4 |
| b | 2 | d | 20 |

Figure 11. Zero-temperature supercurrent-voltage characteristics for a Josephson junction in the limit $E_J \ll E_c$. The junction is coupled to Ohmic resistors with four different values given in kΩ. $I_c = 2eE_J/\hbar$ is the critical current.

relation in principle provides a possibility to check the consistency of the theory. Of course, one always has to account for the relative factor of 4 in the definitions of $P(E)$ and $P'(E)$.

In Sec. 3.4. we had derived some general properties of the probability $P(E)$. The sum rules discussed there now become sum rules for supercurrent-voltage characteristics at zero temperature. Since at $T = 0$ the probability $P'(E)$ vanishes for negative energies, we may directly employ (65) and (66) yielding

$$\int_0^\infty dV\, I_S(V) = \frac{\pi E_J^2}{2\hbar} \tag{140}$$

and

$$\int_0^\infty dV\, V I_S(V) = \frac{\pi e E_J^2}{2\hbar C}. \tag{141}$$

For specific environments one may derive further properties of the supercurrent-voltage characteristics in accordance with our discussion of $P(E)$ for normal junctions. Here, we concentrate on an Ohmic impedance $Z(\omega) = R = R_K/g$ and consider first the case of zero temperature. As for normal junctions we find a zero-bias anomaly which is now given by $I_S \sim V^{2/g-1}$. This behavior is shown in Fig. 11 where supercurrent-voltage characteristics are shown for various values of the dimensionless conductance g. For $g > 2$ the supercurrent is peaked at $V = 0$ and decreases with increasing voltage. On the other hand, for $g < 2$ the supercurrent increases with voltage for small V, thereby leading to a peak at finite voltage. For rather small conductances a well marked gap is present at small voltages. Fig. 11 of course corresponds to Fig. 6a which shows $P(E)$ at zero temperature for a normal junction coupled to an Ohmic environment.

Let us now have a closer look at the peak developing at $V = e/C$ for low conductance g. Thermal and quantum fluctuations broaden this peak. Its shape close to the maximum is given by the Gaussian

$$I_S(V) = I_{\max} \exp\left[-\frac{(V - e/c)^2}{W}\right] \tag{142}$$

with the width [27]

$$W = \begin{cases} \dfrac{e^2}{2\pi^2 C^2}g & \text{for } k_B T \ll \dfrac{e}{C}g \\[3mm] \dfrac{2}{C}k_B T & \text{for } k_B T \gg \dfrac{e}{C}g. \end{cases} \tag{143}$$

In the first case, for very low temperatures, the peak is broadened by quantum fluctuations which decrease as the conductance is decreased. The second case describes the thermal broadening in analogy to the result (85) derived for normal tunnel junctions in the high impedance limit where the conductance $g \ll k_B T C/e$.

5.3. Charge-phase duality and incoherent tunneling of the phase

In the previous subsection we have discussed the case of weak Josephson coupling where the charge on the junction is well defined. For the following discussion it is convenient to express the Hamiltonian (133) in terms of charge states $|N\rangle$, for which $Q = 2eN$. As mentioned before, the operator $e^{-2i\varphi}$ changes the charge Q by $2e$. From (134) we then find the equivalence

$$E_J \cos(2\varphi) \leftrightarrow \frac{E_J}{2}\sum_N \left(|N+1\rangle\langle N| + |N\rangle\langle N+1|\right). \tag{144}$$

In the charge representation the Hamiltonian may be written as

$$H = \frac{E_J}{2}\sum_N \left(|N+1\rangle\langle N| + |N\rangle\langle N+1|\right) + 2e(V + \tilde{V})\sum_N N|N\rangle\langle N| \tag{145}$$

where the environment couples to the charge via the external voltage V and a voltage operator $\tilde{V}$ describing the voltage fluctuations at the junction induced by the environment. The Hamiltonian (145) could alternatively be used to derive the expressions for the tunneling rates.

In the limit of large Josephson coupling $E_J \gg E_c$ the phase is well defined and localized in one of the wells of the Josephson potential. We introduce phase states $|n\rangle$ where the phase is given by $\varphi = \pi n$. Using these states we may write the Hamiltonian as

$$H = \sum_n \Delta_0 \left(|n+1\rangle\langle n| + |n\rangle\langle n+1|\right) + \frac{\pi\hbar}{e}(I + \tilde{I})\sum_n n|n\rangle\langle n|. \tag{146}$$

The first term describes tunneling of the phase from one well to a neighboring one and Δ_0 is the tunnel matrix between adjacent ground states in the wells. The second term couples the phase to an external current I and an operator $\tilde{I}$ describing a fluctuating current through the Josephson junction caused by the environment. The tight-binding Hamiltonian (146) makes sense if only the ground states in the wells can be occupied. The excitation energy is related to the oscillation frequency in the well given by $(2E_J E_c)^{1/2}/\hbar$, and we thus find that this approach is valid if frequency, current, and temperature fulfill the requirements $\omega, I/e, k_B T/\hbar \ll (E_J E_c)^{1/2}/\hbar$.

Since phase and charge have to be exchanged when going from (145) to (146), i.e. from the weak coupling limit to the strong coupling limit, the influence of the environment is now described by the charge-charge correlation function $\langle [\tilde{Q}(t) - \tilde{Q}(0)]\tilde{Q}(0)\rangle$ replacing the phase-phase correlation function $J(t)$. The charge $\tilde{Q}$ is related to the fluctuating current $\tilde{I}$ by

$$\tilde{Q}(t) = \int_{-\infty}^{t} \mathrm{d}t' \, \tilde{I}(t'). \tag{147}$$

The correlation function of the fluctuating current is determined by the environmental admittance $Y(\omega)$ as

$$\langle \tilde{I}(t)\tilde{I}(0)\rangle = \int_0^\infty \frac{\mathrm{d}\omega}{\pi} \hbar\omega \mathrm{Re}[Y(\omega)] \left\{ \coth\left(\frac{1}{2}\beta\hbar\omega\right) \cos(\omega t) - i\sin(\omega t) \right\}. \tag{148}$$

From (147) we then get for the charge-charge correlation function

$$\langle [\tilde{Q}(t) - \tilde{Q}(0)]\tilde{Q}(0)\rangle = \frac{\hbar}{\pi} \int_0^\infty \frac{\mathrm{d}\omega}{\omega} \mathrm{Re}[Y(\omega)]$$

$$\times \left\{ \coth\left(\frac{1}{2}\beta\hbar\omega\right) [\cos(\omega t) - 1] - i\sin(\omega t) \right\}. \tag{149}$$

Now we are able to transform results obtained for weak coupling into results for the strong coupling case by means of simple replacements. Comparing (145) and (146) we see that we have to replace the Josephson coupling by the tunnel splitting, the voltage by the current, and the Cooper pair charge by the flux quantum. Furthermore, according to (13) and (147) the charge replaces the phase, and according to (64) and (149) we have to substitute the environmental admittance for the total impedance. Thus, we arrive at the well-known phase-charge duality transformations [5]

$$\frac{E_J}{2} \Leftrightarrow \Delta_0 \qquad V \Leftrightarrow I \qquad 2e \Leftrightarrow \frac{h}{2e} \qquad \varphi \Leftrightarrow \frac{\pi}{2e}Q \qquad Z_t(\omega) \Leftrightarrow Y(\omega). \tag{150}$$

The process dual to the tunneling of Cooper pairs in the weak coupling limit is incoherent tunneling of the phase in the strong coupling limit.

In Fig. 12 we give an example of the voltage as a function of the bias current of a Josephson junction in the strong coupling regime as calculated from the equation dual to (139). The environment is described by an LC transmission line of length ℓ as discussed in Sec. 4.5. For $Z_\infty \ll R_K \ll Z_L$ one finds peaks separated by current intervals $e\omega_0/\pi$ where $\omega_0 = \pi u/\ell$ and u is the wave propagation velocity. If one averages over the oscillations of the characteristic one finds the corresponding characteristic for an Ohmic resistor Z_∞, which for curve b is curve d. Such an averaging occurs when the temperature becomes of the order of $\hbar I/e$.

5.4. Tunneling of quasiparticles

In Josephson junctions apart from Cooper pairs also quasiparticles may tunnel. Basically, we have to treat quasiparticle tunneling in a superconducting junction like quasiparticle tunneling in a normal junction. Therefore, most of the calculations per-

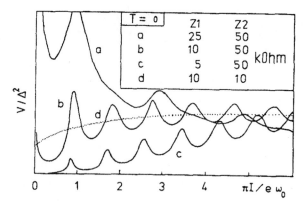

Figure 12. Voltage across a strongly coupled Josephson junction at zero temperature. The junction is connected to a finite LC transmission line of length $\pi u/\omega_0$ where u is the wave propagation velocity. The line impedance Z_1 and the load impedance Z_2 take different values given in the insert. In this figure taken from Ref. [26] Z_1 and Z_2 correspond to Z_∞ and Z_L as defined in Sec. 4.4., respectively.

formed in Sec. 3.2. can be taken over to the superconducting case. There is however one important difference. In Sec. 3.2.1. we had assumed that the density of states at the Fermi surface is constant. In a superconductor the quasiparticle density of states close to the gap depends very strongly on energy. Within the BCS-theory one finds for the reduced quasiparticle density of states [1]

$$\frac{N_S(E)}{N(0)} = \begin{cases} \dfrac{|E|}{(E^2 - \Delta^2)^{1/2}} & \text{for } |E| > \Delta \\[2ex] 0 & \text{for } |E| < \Delta. \end{cases} \tag{151}$$

The density of states is taken relative to the density of states in the normal metal at an energy in the middle of the gap. 2Δ is again the size of the superconducting gap within which the quasiparticle density of states vanishes. For the forward tunneling rate we then have as an extension of (51)

$$\vec{\Gamma}(V) = \frac{1}{e^2 R_T} \int_{-\infty}^{+\infty} \mathrm{d}E \mathrm{d}E' \frac{N_S(E)N_S(E' + eV)}{N(0)^2}$$

$$\times f(E)[1 - f(E' + eV)]P(E - E'). \tag{152}$$

Here, the probability to exchange energy with the environment is given by $P(E)$ since quasiparticles carry the charge e.

As for the normal tunnel junction we use the symmetry $\vec{\Gamma}(V) = \overleftarrow{\Gamma}(-V)$ to obtain from the rate expression (152) the quasiparticle current

$$I_{\mathrm{qp}}(V) = \frac{1}{eR_T} \int_{-\infty}^{+\infty} \mathrm{d}E \mathrm{d}E' \frac{N_S(E)N_S(E')}{N(0)^2} \Big[f(E)[1 - f(E')]P(E - E' + eV)$$

$$- f(E')[1 - f(E)]P(E' - E - eV) \Big]. \tag{153}$$

Using the detailed balance symmetry (72) of $P(E)$ and the relation (54) for Fermi functions, this equation may be rewritten in a more convenient way as

$$I_{\mathrm{qp}} = \frac{1}{eR_T} \int_{-\infty}^{+\infty} \mathrm{d}E\,\mathrm{d}E'\, \frac{N_S(E')N_S(E'+E)}{N(0)^2} \frac{1-e^{-\beta eV}}{1-e^{-\beta E}}$$

$$\times P(eV-E)[f(E')-f(E'+E)]. \qquad (154)$$

In the absence of an external impedance we recover the familiar quasiparticle current of a voltage biased Josephson junction

$$I_{\mathrm{qp,0}}(V) = \frac{1}{eR_T} \int_{-\infty}^{+\infty} \mathrm{d}E\, \frac{N_S(E)N_S(E+eV)}{N(0)^2}[f(E)-f(E+eV)]. \qquad (155)$$

For zero temperature the integral may be evaluated yielding [28]

$$I_{\mathrm{qp,0}}(V) = \frac{\Delta}{eR_T}\left[2xE(m)-\frac{1}{x}K(m)\right] \quad \text{for } x > 1 \qquad (156)$$

where $m = 1-1/x^2$ with $x = eV/2\Delta$. $K(m)$ and $E(m)$ are the complete elliptic integrals of the first and second kind, respectively.[23] For voltages below $2\Delta/e$ the quasiparticle current vanishes as a consequence of the energy gap 2Δ. We may use (155) to express the quasiparticle current in the presence of an environment as [21]

$$I_{\mathrm{qp}}(V) = \int_{-\infty}^{+\infty} \mathrm{d}E\, \frac{1-e^{-\beta eV}}{1-e^{-\beta E}} P(eV-E) I_{\mathrm{qp,0}}\!\left(\frac{E}{e}\right). \qquad (157)$$

This expression is rather general. For example, if we insert for $I_{\mathrm{qp,0}}(V)$ the Ohmic current-voltage characteristic of a normal tunnel junction we directly get our earlier result (78).

For an Ohmic environment and zero temperature we may calculate the current-voltage characteristic for voltages slightly above $2\Delta/e$ by inserting the low energy behavior (111) of $P(E)$ into (157). Expanding (156) we obtain to leading order

$$I_{\mathrm{qp}}(V) = \frac{\pi g\Delta}{4eR_T} \frac{e^{-2\gamma/g}}{\Gamma(2/g)} \left[\frac{\pi}{gE_{\mathrm{c}}}(eV-2\Delta)\right]^{2/g} \qquad (158)$$

where $g = R_K/R$ is the Ohmic lead conductance and γ is again the Euler constant. As for normal junctions and the supercurrent in Josephson junctions we find an anomaly $I_{\mathrm{qp}} \sim (eV-2\Delta)^{2/g}$ [21], which now is shifted in voltage by $2\Delta/e$. Fig. 13 shows the formation of a Coulomb gap with decreasing lead conductance g in accordance with the power law (158). For any nonvanishing lead conductance the current-voltage characteristic for large voltages approaches $I_{\mathrm{qp,0}}$ shifted in voltage by $e/2C$. As for normal junctions the high voltage behavior exhibits a Coulomb gap for $g \neq 0$ even though the gap might not be apparent at voltages close to $2\Delta/e$.

Finite temperature current-voltage characteristics for Ohmic environments with different conductances are shown in Fig. 14. Due to the finite temperature the gap is smeared. Interestingly, for voltages below $2\Delta/e$ one finds an increase of the current due to the environmental coupling. In contrast to the behavior at high voltages and the current in normal tunnel junctions, the quasiparticle tunnel current increases with decreasing lead conductance for low voltages.

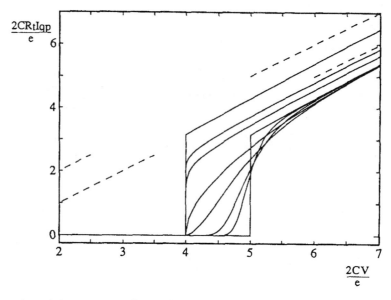

Figure 13. The quasiparticle current-voltage characteristic at zero temperature is shown for an Ohmic environment with $g = R_K/R = \infty, 40, 20, 4, 1.2, 0.2, 0.05$, and 0 from left to right. The superconducting gap is choosen as $\Delta = 2E_c$. The two dashed lines represent the large voltage asymptotes for $g = \infty$, i.e. for $I_{qp,0}$, and for the other values of g. Here, the tunneling resistance is denoted by R_t. Taken from [21] with permission.

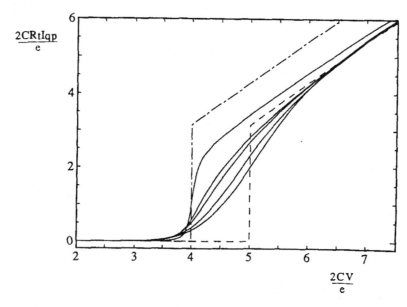

Figure 14. The quasiparticle current-voltage characteristic for finite temperature $k_BT/E_c = 0.25$ is shown for an Ohmic environment with $g = R_K/R = 20, 4.8, 3.2, 1.2$, and 0.2 from left to right. The superconducting gap is choosen as $\Delta = 2E_c$. The dash-dotted line and the dashed line are $I_{qp,0}$ and the same curve shifted by E_c/e, respectively, taken at the same temperature. Here, the tunneling resistance is denoted by R_t. Taken from [21] with permission.

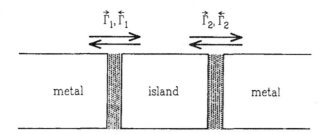

Figure 15. Schematic drawing of a metallic double junction system. The arrows indicate forward and backward tunneling through the two insulating barriers.

In the next section we consider multijunction systems and restrict the discussion to normal junctions. For superconducting double junction systems the combined tunneling of Cooper pairs and quasiparticles leads to new effects. For details we refer the reader to the literature.[29]

6. Double junction and single electron transistor

6.1. Island charge

In this section we discuss circuits of tunnel junctions. As we shall see, as far as the calculation of tunneling rates is concerned, most of the new features arising when several tunnel junctions are combined are already present in a double junction setup. Hence, we shall mainly discuss two-junction systems and briefly address more complicated circuits at the end of this section. Systems containing two tunnel junctions in series as shown in Fig. 15 differ significantly from a single junction because of the metallic island between the two junctions.[4, 30] While earlier work has entirely disregarded the influence of the electromagnetic environment we shall take it into account here following the line of reasoning in [10, 31]. The external circuit sees the two tunnel junctions with capacitance C_1 and C_2 as a capacitor of total capacitance

$$C = \frac{C_1 C_2}{C_1 + C_2}. \tag{159}$$

Since the voltage across the two junctions is $U = Q_1/C_1 + Q_2/C_2$ the total charge seen from the outside is

$$Q = CU = \frac{C_2 Q_1 + C_1 Q_2}{C_1 + C_2}. \tag{160}$$

As for a single junction this charge is to be considered as a continuous variable. The metallic island carries the charge

$$Q_1 - Q_2 = ne \tag{161}$$

which may change only by tunneling of electrons to or from the island. This leads to the quantization of the island charge which will turn out to be very important for the

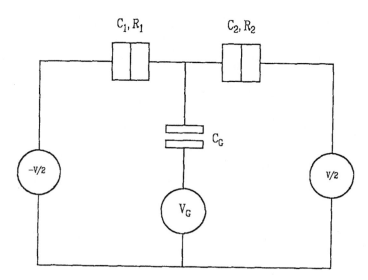

Figure 16. The single electron transistor setup consisting of a double junction with a control voltage source coupled capacitively to the island.

behavior of double junction systems. To describe the charges on the capacitors one may either use Q_1 and Q_2 or Q and ne. The corresponding charging energy reads

$$\frac{Q_1^2}{2C_1} + \frac{Q_2^2}{2C_2} = \frac{Q^2}{2C} + \frac{(ne)^2}{2(C_1 + C_2)}. \tag{162}$$

Compared with the single junction the charging energy now contains a contribution arising from the island charge.

In real double junction systems charged defects are frequently present in the vicinity of the junction. They lead to an effective island charge where the discrete set of island charges ne is shifted by an offset charge. To influence the effective island charge in a controlled way one frequently uses the single electron transistor setup presented in Fig. 16 where a gate voltage V_G is coupled capacitively to the island. We will show now that the voltage source V_G together with the capacitance C_G effectively leads to a shift of the island charge by $Q_0 = C_G V_G$. To this end we first determine the average charges on the capacitors in electrostatic equilibrium for given applied voltages V and V_G and given island charge

$$ne = Q_1 - Q_2 - Q_G. \tag{163}$$

Using Kirchhoff's law for two loops we find

$$Q_1 = \frac{C_1}{C_\Sigma} \left[(C_2 + \frac{C_G}{2})V + C_G V_G + ne \right] \tag{164}$$

$$Q_2 = -\frac{C_2}{C_\Sigma} \left[-(C_1 + \frac{C_G}{2})V + C_G V_G + ne \right] \tag{165}$$

$$Q_G = -\frac{C_G}{C_\Sigma} \left[\frac{1}{2}(C_2 - C_1)V - (C_1 + C_2)V_G + ne \right] \tag{166}$$

where we introduced the capacitance

$$C_\Sigma = C_1 + C_2 + C_G. \tag{167}$$

We suppose now that an electron has tunneled through the left junction onto the island thereby changing Q_1 into $Q_1 - e$ and ne into $(n - 1)e$. The new charges $Q_1 - e$, Q_2, and Q_G do no longer satisfy electrostatic equilibrium since the replacement of n by $n - 1$ in (164-166) does not result in a change of Q_1 by e. Equilibrium is reestablished by a transfer of charge through the voltage sources leading to the following difference of charges before and after the tunneling process

$$\delta Q_1 = -(C_1/C_\Sigma)e = -e + \delta Q_2 + \delta Q_G \tag{168}$$

$$\delta Q_2 = (C_2/C_\Sigma)e \tag{169}$$

$$\delta Q_G = (C_G/C_\Sigma)e. \tag{170}$$

Apart from the energy transfer to or from the environmental modes the energy determining the tunneling rates is the difference in electrostatic energy of the entire circuit. In contrast to the case of a single junction, this energy difference now not only consists of contributions from the work done by the various voltage sources. It also has to account for the change in charging energy. The total charging energy may be decomposed into a contribution depending on the voltages V and V_G which does not change and a contribution $(ne)^2/2C_\Sigma$ depending on the island charge. Thus the change in charging energy is entirely due to the change of the island charge. We finally obtain for the difference in electrostatic energy associated with the tunneling of an electron through the first junction onto the island

$$\frac{(ne)^2}{2C_\Sigma} - \frac{[(n-1)e]^2}{2C_\Sigma} - \frac{V}{2}(\delta Q_1 + e) + \frac{V}{2}\delta Q_2 + V_G \delta Q_G$$

$$= \frac{e}{C_\Sigma}\left[(C_2 + \frac{C_G}{2})V + C_G V_G + ne - \frac{e}{2}\right]. \tag{171}$$

The extra elementary charge added to δQ_1 is due to the fact that an electron has tunneled through the first junction and therefore the charge transferred by the voltage source is diminished by $-e$. From the right hand side of (171) it becomes clear now that the work done by the gate voltage source leads indeed to an effective island charge $q = ne + Q_0$ with $Q_0 = C_G V_G$.

If the gate capacitance C_G is small compared to the junction capacitances C_1 and C_2 and if no other impedance is present in the gate branch, the only effect of C_G is the shift of the effective island charge which we just discussed. To retain this shift we may let the gate capacitance go to zero. However, we have to keep the work done by the voltage source finite. In this limit the charge on the gate capacitor (166) is negligible as is the charge (170) transferred after tunneling. On the other hand the gate voltage is assumed to be sufficiently large to cause an offset charge $C_G V_G$.

In the following, we will restrict ourselves to this limit where C_G may be neglected. In the literature the reader will find a more complete discussion of the effect of gate and

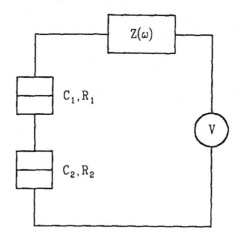

Figure 17. A double junction system with capacitances C_1, C_2 and tunneling resistances R_1, R_2 coupled to a voltage source V via the external impedance $Z(\omega)$.

stray capacitances and of gate impedances.[32, 33] In the limit we are considering here, the single electron transistor is reduced to a double junction with an offset of the island charge. We will therefore concentrate on the double junction and discuss the effects arising due to the offset whenever appropriate.

6.2. Network analysis

The system we are considering in the sequel is the double junction shown in Fig. 17 which is coupled to a voltage source via an external impedance $Z(\omega)$. It would be straightforward to carry out a golden rule calculation as we have done for the single junction. However, it turns out that this is not necessary because some simple network considerations yield the same result.[10] Furthermore, they will give us some additional insight. Before starting we would like to mention an important assumption underlying our approach. Second order perturbation theory or golden rule is only sufficient if tunneling through both junctions may be considered as uncorrelated. This means that when we are calculating tunneling rates for one junction the other junction may be viewed as a capacitor. Especially in the blockade region where our approach predicts no flow of current, higher order perturbation theory leads to important corrections. These are due to virtual transitions involving simultaneous tunneling through both junctions. This so-called co-tunneling which is not hindered by the Coulomb interaction is relevant if the tunneling resistances are no longer large compared to the resistance quantum R_K.[15] For a detailed discussion see Chap. 6.

The basic rule which will be needed for the network analysis is the transformation between the Thevenin and Norton configurations shown in Fig. 18. The two configurations form two-terminal devices through which a current $I_0(\omega)$ flows if a voltage $V_0(\omega)$ is applied. From the outside the two configurations appear as equivalent if the same voltage V_0 leads to the same current I_0. In the Thevenin configuration the voltage drop is given by $V_0(\omega) = I_0(\omega)Z(\omega) + V(\omega)$ where the current and the voltages may in general be frequency-dependent. On the other hand, the current flowing into the Norton configuration is given by $I_0(\omega) = -I(\omega) + V_0(\omega)/Z(\omega)$ if the current of the current source in

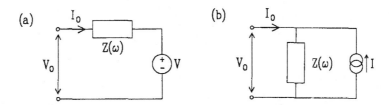

Figure 18. (a) Thevenin configuration: A voltage source V in series with the impedance $Z(\omega)$. (b) The equivalent Norton configuration: A current source $I(\omega) = V(\omega)/Z(\omega)$ in parallel with the impedance $Z(\omega)$.

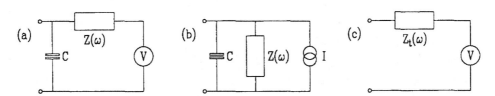

Figure 19. Transformation of a single junction circuit into an equivalent effective circuit. (a) Original circuit as seen from the junction. (b) Equivalent Norton configuration. (c) Effective single junction circuit.

Fig. 18b is flowing upwards. These two equations lead to the relation $V(\omega) = Z(\omega)I(\omega)$ between the voltage and current sources in the two configurations.

We introduce the method of network analysis by applying it to the single tunnel junction. In a first step we separate the tunneling junction into a tunneling element in parallel with a capacitor describing the junction capacitance. The tunneling element transfers electrons through the circuit which appears as the two-terminal device depicted in Fig. 19a. We simplify the circuit by transforming it into the Norton configuration shown in Fig. 19b. The current source is given by $I(\omega) = V(\omega)/Z(\omega)$. While transforming circuits we always keep the frequency dependence which is especially important when capacitors are involved as is the case for the double junction. Only at the end we account for the fact that we have a dc voltage source by taking the limit $\omega \to 0$. In Fig. 19b the capacitance C and the external impedance $Z(\omega)$ are seen to form the total impedance $Z_t(\omega)$ defined in (12). Returning to the Thevenin configuration of Fig. 19c we get a voltage source $V(\omega)Z_t(\omega)/Z(\omega)$ which reduces to the original voltage V in the limit $\omega \to 0$. Electrons are now transferred through the effective circuit. This leads to the work eV done by the voltage source. The effective impedance seen by the tunneling element is the total impedance $Z_t(\omega)$ containing the capacitance C and the impedance $Z(\omega)$ in parallel. For very low impedances the capacitor is thus shortened out and charging effects become unimportant. In contrast, for a very large impedance the capacitor remains and charging effects become apparent unless they are smeared out by thermal fluctuations. This picture fits our earlier considerations very well.

Having gained confidence in this method we apply it to the double junction system shown in Fig. 20. We consider tunneling through the first junction and therefore treat the second junction as a capacitor thereby disregarding the possibility of electron tunneling through the latter junction. We then arrive at the Thevenin configuration shown in Fig. 20a. For sake of simplicity we will not keep track of the charges during the transformations we are going to perform. While in principle this would be possible,

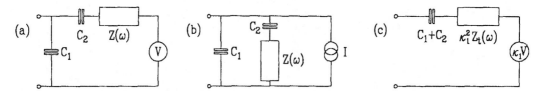

Figure 20. Transformation of a double junction circuit into an equivalent effective circuit. (a) Original circuit as seen from the first junction. The second junction is treated as a capacitor. (b) Equivalent Norton configuration. (c) Effective circuit for tunneling through the first junction.

it will turn out that we know the charges on the capacitors of the effective circuit from our considerations in the previous subsection. It is straightforward to apply the transformation to the Norton configuration (Fig. 20b) and back to the Thevenin configuration (Fig. 20c) as we did for the single junction. The new circuit contains a capacitance, an effective impedance, and an effective voltage source which can all be simply interpreted.

The voltage source has an effective voltage $\kappa_1 V$ where

$$\kappa_i = C/C_i \quad (i = 1, 2). \tag{172}$$

Since the total capacitance C is always smaller than the smallest capacitance C_i, the ratio κ_i is always less than one. The effective voltage source can easily be interpreted in terms of the work $\kappa_1 eV$ done by the source when an electron is transferred by the tunneling element. After an electron has tunneled, charge is transferred in the original circuit through the voltage source in order to reestablish electrostatic equilibrium according to (168)–(170). For $C_G = 0$ we indeed find that the transferred charge is $\kappa_1 e$. The charge which has to be transferred through the voltage source after an electron has tunneled through one junction is smaller than an elementary charge. Only after the electron has also tunneled through the other junction, the two charges transferred through the voltage source add up to an elementary charge. This is in agreement with $\kappa_1 + \kappa_2 = 1$ which follows directly from (159) and (172).

The effective impedance $\kappa_1^2 Z_t(\omega)$ with

$$Z_t(\omega) = \frac{1}{i\omega C + Z^{-1}(\omega)} \tag{173}$$

has the same structure as for the single junction if one replaces the single junction capacitance by the total capacitance (159) seen by the external circuit. In addition, there is again a reduction factor which for the impedance seen by the first junction is κ_1^2. As a consequence, the influence of the external circuit is reduced. For a system consisting of N junctions of about the same capacitance one finds as a generalization that the effective impedance is reduced by a factor of $1/N^2$. This means that one may apply the global rule for circuits containing many junctions. However, one should bear in mind that for sufficiently large voltages one will always find a crossover to the local rule due to the sum rules satisfied by $P(E)$ unless the external impedance vanishes. This crossover will occur at voltages which are about a factor $1/\kappa_i$ larger for a double junction system than for a single junction. How can one understand the reduced environmental influence? From a physical point of view the tunnel junction is to a certain extent decoupled from the external circuit by the other junction. More formally, one has two equivalent sets of charges $\{Q_1, Q_2\}$ and $\{Q, q\}$ for which one introduces the canonically conjugate phases $\{\varphi_1, \varphi_2\}$ and $\{\varphi, \psi\}$. Here, φ_1 and φ_2 are defined as straightforward generalizations of the single junction phase according to (13). Now, $\{\varphi, \psi\}$ are related to $\{\varphi_1, \varphi_2\}$ by

$$\psi = \kappa_2 \varphi_1 - \kappa_1 \varphi_2 \tag{174}$$

and

$$\varphi = \varphi_1 + \varphi_2. \tag{175}$$

The nonvanishing commutators between phases and charges are

$$[\varphi_1, Q_1] = ie, \quad [\varphi_2, Q_2] = ie \tag{176}$$

and

$$[\varphi, Q] = ie, \quad [\psi, q] = ie. \tag{177}$$

In the spirit of the tunneling Hamiltonian (25) for a single junction we write for the tunneling Hamiltonian of the first junction

$$H_{T,1} = \sum_{kq\sigma} T_{kq} c_{q\sigma}^{\dagger} c_{k\sigma} \exp(-i\varphi_1) + \text{H.c.} \tag{178}$$

where we may express the operator changing the charge on the first junction as

$$\exp(-i\varphi_1) = \exp(-i\kappa_1 \varphi - i\psi). \tag{179}$$

Since ψ is conjugate to q the operator $\exp(-i\psi)$ describes the change of the island charge by one elementary charge. The operator $\exp(-i\kappa_1 \varphi)$ couples the tunneling process to the environment. It is the factor κ_1 appearing there which leads to the reduction of the environmental coupling by κ_1^2.

Finally, in our effective circuit of Fig. 20c we have a capacitance $C_1 + C_2$ which is related to the charging energy of the island $(ne)^2/2(C_1 + C_2)$. The charging energy corresponding to the total charge Q may become irrelevant if the external impedance is small. In contrast, the capacitor in series with the total impedance is always affected by an electron which is transferred through the effective circuit and thus the charging energy of the island will affect the rate for any environment.

Having applied network analysis, we have now a rather clear picture of the relevant quantities governing the dynamics of double junctions. Hence, we are in a position to immediately write down the expressions for the double junction tunneling rates which will be discussed in the next subsection. We only mention that network considerations become especially useful when considering more complicated circuits. A straightforward extension of the double junction is the one-dimensional array of junctions which will be discussed briefly in Sec. 6.9. and in more detail in Chap. 7. Another application is the single electron transistor if gate and stray capacitances are taken into account.[32]

6.3. Tunneling rates in a double junction system

The changes in the expressions for the tunneling rates in double junction systems compared with those for single junctions can be motivated by taking into account the discussion in the previous subsection. We emphasize again that the results we are going to discuss could as well be obtained from an explicit calculation of the rates in second order perturbation theory.[10]

In the previous sections we have found that the environmental influence on tunneling rates in normal as well as superconducting single junctions may be described by means of the probability $P(E)$ of energy exchange between the tunneling electron and the external circuit. For double junction systems we have to account for the reduced coupling to the environment, and the probability to transfer energy to the environmental modes is given by

$$P(\kappa_i, E) = \frac{1}{2\pi\hbar} \int_{-\infty}^{+\infty} dt \exp\left[\kappa_i^2 J(t) + \frac{i}{\hbar} Et\right]. \tag{180}$$

The correlation function $J(t)$ is defined as for the single junction in (64) provided the capacitance C appearing in the total impedance is the total capacitance (159) of the double junction system.

The energy difference for elastic tunneling of an electron through the i-th junction onto the island is given by

$$\begin{aligned}
E_i(V, q) &= \kappa_i eV + \frac{q^2}{2(C_1 + C_2)} - \frac{(q-e)^2}{2(C_1 + C_2)} \\
&= \kappa_i eV + \frac{e(q - e/2)}{C_1 + C_2} \quad (i = 1, 2)
\end{aligned} \tag{181}$$

where the effective island charge $q = ne$ for a double junction and $q = ne + Q_0$ for a SET transistor in the limit $C_G \to 0$. For practical purposes it is often useful to express this energy difference in terms of quantities of the i-th junction only. Using (160) and (161) we may rewrite (181) to obtain

$$E_i(Q_i) = \frac{e}{C_i}(Q_i - Q_i^c). \tag{182}$$

Here, we have introduced a critical charge

$$Q_i^c = \frac{e}{2}(1 - \kappa_i). \tag{183}$$

Although (182) contains only quantities of the i-th junction it still describes the change of electrostatic energy for the entire circuit.

It is now straightforward to write down the forward tunneling rate through the first junction [10, 31]

$$\vec{\Gamma}_1(V, q) = \frac{1}{e^2 R_1} \int_{-\infty}^{+\infty} dE \frac{E}{1 - \exp(-\beta E)} P(\kappa_1, E_1(V, q) - E). \tag{184}$$

Here, R_1 is the tunneling resistance of the first junction. The forward and backward tunneling rates for the first junction are connected by

$$\overleftarrow{\Gamma}_1(V, q) = \overrightarrow{\Gamma}_1(-V, -q) \tag{185}$$

since in the backward tunneling process the electron is tunneling from the island opposite to the direction favored by the applied voltage. As for the single junction rates there exists a detailed balance symmetry which now connects rates for different island charges

$$\overleftarrow{\Gamma}_1(V, q - e) = \exp[-\beta E_1(V, q)]\overrightarrow{\Gamma}_1(V, q). \tag{186}$$

The forward and backward tunneling rates through the second junction are obtained from the respective rates for the first junction by exchanging the indices 1 and 2 and by changing q into $-q$. For the forward tunneling rate one thus finds

$$\overrightarrow{\Gamma}_2(V, q) = \frac{1}{e^2 R_2} \int_{-\infty}^{+\infty} dE \frac{E}{1 - \exp(-\beta E)} P(\kappa_2, E_2(V, -q) - E). \tag{187}$$

The relations corresponding to (185) and (186) now read

$$\overleftarrow{\Gamma}_2(V, q) = \overrightarrow{\Gamma}_2(-V, -q) \tag{188}$$

and

$$\overleftarrow{\Gamma}_2(V, q + e) = \exp[-\beta E_2(V, -q)]\overrightarrow{\Gamma}_1(V, q). \tag{189}$$

We end this general part on tunneling rates by noting that for a symmetric double junction system with equal capacitances $C_1 = C_2$ and equal tunnel resistances $R_1 = R_2$ all tunneling rates are related to each other by $\overrightarrow{\Gamma}_1(V, q) = \overleftarrow{\Gamma}_1(-V, -q) = \overrightarrow{\Gamma}_2(V, -q) = \overleftarrow{\Gamma}_2(-V, q)$.

6.4. Double junction in a low impedance environment

For explicit analytical results of the environmental influence on electron tunneling rates we restrict ourselves to the limits of very low and very high impedance environments. The first case is of relevance for most practical cases because of the reduced effective impedance. The high impedance case, on the other hand, determines the behavior at very large voltages.

In the limit of vanishing external impedance tunneling is elastic and we have like in the single junction case $P(\kappa_i, E) = \delta(E)$. Then, the tunneling rates may easily be evaluated and we get from (184) for the forward tunneling rate through the first junction

$$\overrightarrow{\Gamma}_1(V, q) = \frac{1}{e^2 R_1} \frac{E_1(V, q)}{1 - \exp[-\beta E_1(V, q)]}. \tag{190}$$

For zero temperature this reduces to

$$\overrightarrow{\Gamma}_1(V, q) = \frac{1}{e^2 R_1} E_1(V, q) \Theta(E_1(V, q)) \tag{191}$$

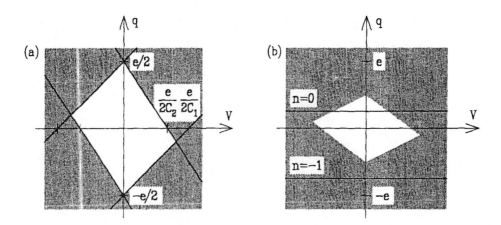

Figure 21. (a) Low impedance stability diagram for the effective island charge q in dependence on the transport voltage V. In the non-shaded area the state $n = 0$ is stable while in the shaded area one or more tunneling rates are non-vanishing. (b) Stability diagram for a transistor with $Q_0 = e/4$. The possible effective island charges are shifted accordingly.

where $\Theta(E)$ is the unit step function. Therefore, at zero temperature $\overrightarrow{\Gamma}_1(V, q)$ is only different from zero if $E_1(V, q) > 0$. This justifies that we call Q_i^c, which was defined in (182) and (183), a critical charge. The effective charge on the island through which electron tunneling is considered has to exceed the critical charge to allow for a finite tunneling rate. Together with (181) we now find the following conditions under which the rates are nonvanishing:

$$\overrightarrow{\Gamma}_1(V, q): \quad V + \frac{1}{C_2}\left(q - \frac{e}{2}\right) > 0 \tag{192}$$

$$\overleftarrow{\Gamma}_1(V, q): \quad V + \frac{1}{C_2}\left(q + \frac{e}{2}\right) < 0 \tag{193}$$

$$\overrightarrow{\Gamma}_2(V, q): \quad V - \frac{1}{C_1}\left(q + \frac{e}{2}\right) > 0 \tag{194}$$

$$\overleftarrow{\Gamma}_2(V, q): \quad V - \frac{1}{C_1}\left(q - \frac{e}{2}\right) < 0. \tag{195}$$

The heavy lines in Fig. 21a indicate the parameter region where one of the Eqs. (192)–(195) is fulfilled as an equality. The area inside these lines is the region where the island charge $n = 0$ is stable because all tunneling rates vanish. In the shaded areas one or more rates are different from zero. Suppose now that we have an ideal double junction system without offset charges so that the effective island charge is given by $q = ne$. If $|V| < \min(e/2C_1, e/2C_2)$ and $n \neq 0$ then the rates force the electrons to tunnel in such a way that after some time the island charge is zero. The state $n = 0$ is stable in this voltage regime since all rates vanish. As a consequence at zero temperature there is no current if the absolute value of the voltage is below $\min(e/2C_1, e/2C_2)$ and we find a Coulomb gap even in the low impedance case. This important difference as compared to the case of a single junction is due to the charging energy related to the island charge.

If we now apply an offset charge either by placing a charge near the island or by using a transistor setup according to Fig. 16 with $Q_0 = C_G V_G$, the Coulomb gap will be

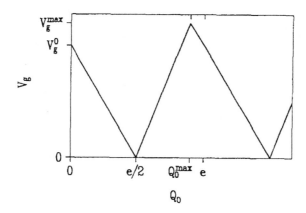

Figure 22. Low impedance Coulomb gap for a single electron transistor as function of the offset charge $Q_0 = C_G V_G$. If $C_1 < C_2$ one has $V_g^0 = e/2C_2$, $V_g^{max} = e/(C_1 + C_2)$, and $Q_0^{max} = e/2 + \kappa_2 e$.

affected. The possible effective island charges are then no longer given by $q = ne$ but by $q = ne + Q_0$. With this replacement we immediately get the tunneling rates for the single electron transistor from the double junction rates (184), (185), (187), and (188). The transistor rates then depend on V, n, and V_G. Note that going from forward to backward tunneling rates now not only involves a change in sign of V and n but also of V_G. To obtain the influence of the offset charge on the Coulomb gap we have to consider only charges in the range $-e/2 < Q_0 < e/2$ since an integer number of elementary charges can always be absorbed in n. This means that for a transistor the stable island charge is not necessarily given by $n = 0$ but depends on the gate voltage. However, if Q_0 is in the range just mentioned the stable state will be $n = 0$. The situation for $Q_0 = e/4$ is shown in Fig. 21b. It becomes clear from this figure that the range in which $n = 0$ is stable is decreased as compared to $Q_0 = 0$. For $Q_0 = e/2$ there is no voltage range for which an island charge is stable. This means that in the low impedance limit the Coulomb gap will vanish for $Q_0 = e/2$. The dependence of the Coulomb gap on the offset charge is shown in Fig. 22. For larger offset charges the picture is continued periodically with period e according to the argument given above. Finally, we mention that the results for a low impedance environment may easily be generalized for a transistor with finite gate capacitance. If the transport voltage is divided symmetrically as shown in Fig. 16 one finds from (171) that the replacement $C_1 \to C_1 + C_G/2$ and $C_2 \to C_2 + C_G/2$ will account for the gate capacitance.[32]

6.5. Double junction in a high impedance environment

The similarity of the tunneling rates for single and double junctions shows also in the rate expressions for a double junction in a high impedance environment. For finite temperatures we obtain

$$P(\kappa_i, E) = \frac{1}{\sqrt{4\pi \kappa_i^2 E_c k_B T}} \exp[-\frac{(E - \kappa_i^2 E_c)^2}{4 \kappa_i^2 E_c k_B T}] \tag{196}$$

which differs from the corresponding single junction result (85) only by the factor κ_i^2

in front of E_c which is due to the reduced coupling of the double junction to the environment. Accordingly, at zero temperature (196) reduces to $P(\kappa_i, E) = \delta(E - \kappa_i^2 E_c)$. Together with (184) we then find for the forward tunneling rate through the first junction

$$\vec{\Gamma}_1(V, q) = \frac{1}{e^2 R_1}[E_1(V, q) - \kappa_1^2 E_c]\Theta(E_1(V, q) - \kappa_1^2 E_c) \quad \text{for } T = 0. \tag{197}$$

By rewriting the energy difference as

$$E_1(V, q) - \kappa_1^2 E_c = \frac{Q_1^2}{2C_1} - \frac{(Q_1 - e)^2}{2C_1} \tag{198}$$

it becomes clear that the local rule determines the zero temperature tunneling rates for a high impedance environment as it is the case for single junctions. Note that one may rewrite (198) in the local form (182) with a critical charge $e/2$. This high impedance critical charge is unaffected by the reduced coupling to the environment since the local rule knows only about the capacitor of the junction through which the electron is tunneling. As for the low impedance environment we give the conditions under which the four rates are nonvanishing at zero temperature:

$$\vec{\Gamma}_1(V, q): \quad V + \frac{q}{C_2} - \frac{e}{2C} > 0 \tag{199}$$

$$\overleftarrow{\Gamma}_1(V, q): \quad V + \frac{q}{C_2} + \frac{e}{2C} < 0 \tag{200}$$

$$\vec{\Gamma}_2(V, q): \quad V - \frac{q}{C_1} - \frac{e}{2C} > 0 \tag{201}$$

$$\overleftarrow{\Gamma}_2(V, q): \quad V - \frac{q}{C_1} + \frac{e}{2C} < 0. \tag{202}$$

These conditions are of course equivalent to the requirement that the charge on the junction should be larger than the critical charge $e/2$. Again (199)–(202) define a region of stability for the state $n = 0$ which is shown in Fig. 23. In comparison with the low impedance case this state is stable here for a wider range of parameters. In the absence of an offset charge the Coulomb gap is given by $e/2C$ which always exceeds the low impedance Coulomb gap because $C < C_1, C_2$. If offset charges are present we may apply the same arguments as for the low impedance case. We observe that in q-direction the stable region extends over a range exceeding one elementary charge. As a consequence, one finds a Coulomb gap even for an offset charge $Q_0 = e/2$ where the gap vanishes for low impedance environments. The high impedance gap as a function of the offset charge is shown in Fig. 24. Another consequence of this wide stability region in q-direction is that for certain voltages two states with different island charges may be stable. Which one of the two states is realized depends on how the stability region is reached. Such a multistability may also result from a capacitor in series with the tunnel junction thereby producing a high impedance environment. This situation is discussed in more detail in Chap. 3 in connection with the single electron trap and related devices.

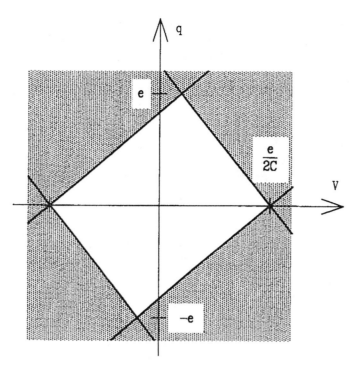

Figure 23. High impedance stability diagram for the effective island charge q in dependence on the transport voltage V. In the non-shaded area the island charge $q = 0$ is stable while in the shaded area one or more tunneling rates are non-vanishing.

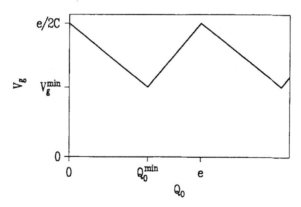

Figure 24. High impedance Coulomb gap for a single electron transistor as function of the offset charge $Q_0 = C_G V_G$. For $C_1 < C_2$ one has $V_g^{min} = e/2C - e/(C_1 + C_2)$ and $Q_0^{min} = \kappa_1 e$.

6.6. Current-voltage characteristics of a double junction

For a single junction the current could be calculated by subtracting the backward from the forward tunneling rate and multiplying this result by the elementary charge. This was possible since every tunneling process contributed to the current in the respective direction. For a double junction the situation is more complicated because the tunneling rates depend on previous tunneling processes which lead to a certain island charge. As we know from subsection 6.3. the tunneling rates depend on the external voltage V and the effective island charge q. The external voltage is taken to be constant and it is always assumed that electrostatic equilibrium at the double junction is established before the next electron tunnels. Then the state of the double junction is characterized by the number n of electrons on the island. Neglecting correlations between different tunneling processes we may write down a master equation which connects states with different island charge. The probability to find the double junction in the state n may change by leaving this state or by coming into this state from the states $n - 1$ or $n + 1$

$$\dot{p}_n = \Gamma_{n,n+1}p_{n+1} + \Gamma_{n,n-1}p_{n-1} - (\Gamma_{n+1,n} + \Gamma_{n-1,n})p_n. \tag{203}$$

Here, $\Gamma_{k,l}$ is the rate for a transition from state l to state k. Since each tunneling process changes the island charge by e the states l and k are neighbors on the island charge ladder with $|l - k| = 1$. There exist two independent possibilities to change the island charge, namely by tunneling through the first or through the second junction. Accordingly, the two rates have to be summed up yielding

$$\Gamma_{n+1,n} = \overleftarrow{\Gamma}_1(n) + \overrightarrow{\Gamma}_2(n) \tag{204}$$

$$\Gamma_{n-1,n} = \overrightarrow{\Gamma}_1(n) + \overleftarrow{\Gamma}_2(n) \tag{205}$$

where we suppressed the dependence on the external voltage. In verifying these two relations one should keep in mind that the island charge is defined as ne, and hence an additional electron on the island decreases the island charge and thereby n. Since we are not interested in the transient behavior we calculate the stationary probabilities p_n by requiring $\dot{p}_n = 0$. It is easy to see that probabilities satisfying the detailed balance condition

$$\Gamma_{n,n+1}p_{n+1} = \Gamma_{n+1,n}p_n \tag{206}$$

are a solution of the master equation (203). Since only nearest neighbor states are connected by nonvanishing rates it can be shown that this solution where the upward flow equals the downward flow is the only nontrivial solution. Starting from a neutral island one finds from (206) the stationary solution

$$p_n = p_0 \prod_{m=0}^{n-1} \frac{\Gamma_{m+1,m}}{\Gamma_{m,m+1}} \tag{207}$$

and

$$p_{-n} = p_0 \prod_{m=-n+1}^{0} \frac{\Gamma_{m-1,m}}{\Gamma_{m,m-1}} \tag{208}$$

with $n > 0$ in both formulas. The only free parameter left is p_0 which is determined by the normalization condition

$$\sum_{n=-\infty}^{+\infty} p_n = 1. \tag{209}$$

Knowing the stationary probability to find the island charge ne we may now calculate the current-voltage characteristics for a double junction from

$$I = e \sum_{n=-\infty}^{+\infty} p_n(\vec{\Gamma}_1(n) - \overleftarrow{\Gamma}_1(n)) = e \sum_{n=-\infty}^{+\infty} p_n(\vec{\Gamma}_2(n) - \overleftarrow{\Gamma}_2(n)). \qquad (210)$$

Because of current conservation it does not matter for which junction we calculate the current. The equality of the second and third expression in (210) is ensured by the detailed balance condition (206).

While the above considerations are valid also for finite temperatures we will restrict ourselves to zero temperature to further illustrate the calculation of current-voltage characteristics. In the simplest case the voltage is below the gap voltage. According to our earlier discussions the state $n = 0$ then is stable and only rates leading to a decrease of the absolute value of the island charge are nonvanishing. This means that the stationary solution of the master equation is given by $p_0 = 1$, $p_n = 0$ for $n \neq 0$. Since all rates vanish for $n = 0$ we find from (210) that indeed the current vanishes in the blockade region. Let us now increase the voltage beyond the gap voltage for a double junction with different capacitances $C_1 < C_2$ in a low impedance environment (cf. Fig. 21a). We begin by considering voltages satisfying $e/2C_2 < V < e/2C_1$. Setting $n = 0$, Fig. 21a tells us that tunneling of electrons through the first junction onto the island is allowed while tunneling through the second junction is forbidden. Being at $n = -1$ the rates only allow the transition back to $n = 0$ by tunneling through the second junction. Consequently, two states, namely $n = 0$ and $n = -1$, are involved. From (208) and (209) one readily gets

$$p_0 = \frac{\vec{\Gamma}_2(V, -e)}{\vec{\Gamma}_1(V, 0) + \vec{\Gamma}_2(V, -e)} \qquad (211)$$

and

$$p_{-1} = \frac{\vec{\Gamma}_1(V, 0)}{\vec{\Gamma}_1(V, 0) + \vec{\Gamma}_2(V, -e)}. \qquad (212)$$

The probabilities p_0 and p_{-1} together with (210) yield for the current

$$I = e\Gamma(V) \qquad (213)$$

where the effective rate $\Gamma(V)$ is given by

$$\frac{1}{\Gamma(V)} = \frac{1}{\vec{\Gamma}_1(V, 0)} + \frac{1}{\vec{\Gamma}_2(V, -e)}. \qquad (214)$$

Since the two tunneling processes occur one after the other, the rates are added inversely and the total rate is dominated by the slower rate. We note that $\Gamma(V)$ here is really only an effective rate because the two step tunneling process does not lead to a purely exponential time dependence. Still $1/\Gamma(V)$ is the average time between tunneling events across the double junction system.

Let us now increase the voltage to the regime where $e/2C_1 < V < 3e/2C_2$. Assuming $1 < C_2/C_1 < 3$ we are sure that it is not possible that the island is charged with two electrons. According to Fig. 21 we now have two possibilities to leave the state $n = 0$. Either an electron may tunnel through the first junction to the right or it may tunnel through the second junction to the right. Depending on what actually happens, the island charge is then either $-e$ or e. The island may not be charged further. Therefore, in the next step an electron has to tunnel through the other junction restoring the neutral island. We thus have two competing processes, namely $n = 0 \to e \to 0$ and $n = 0 \to -e \to 0$. These two mechanisms now allow for two subsequent tunneling processes occurring at the same junction. The sequence of tunneling through the two junctions is no longer fixed but contains a statistical element.

It is obvious that by increasing the voltage the situation will become more and more complicated. In the following section we will derive some properties of the current-voltage characteristic which also hold at higher voltages. In general, however, one has to resort to numerical techniques. It is rather straightforward to use (207)–(209) to determine the stationary probabilities p_n and calculate the current-voltage characteristics by means of (210).

6.7. Coulomb staircase

In the first part of this subsection we will calculate the current through a double junction at zero temperature for special values of the voltage for the limits of low and high impedance environments. We will assume that the capacitances C_1 and C_2 of the tunnel junctions are equal $C_1 = C_2 = C_J$ while the tunneling resistances R_1 and R_2 may take arbitrary values. For the positive voltages $V_m = (e/C_J)(m + 1/2), (m = 0, 1, 2, \ldots)$ in the low impedance case and $V_m = (e/C_J)(m+1), (m = 0, 1, 2, \ldots)$ in the high impedance case the tunneling rates are given by

$$\vec{\Gamma}_1(m, n) = \frac{1}{eC_J(R_1 + R_2)} \left(1 + \frac{R_2}{R_1}\right) \frac{m + n}{2} \Theta(m + n) \tag{215}$$

$$\overleftarrow{\Gamma}_1(m, n) = -\frac{1}{eC_J(R_1 + R_2)} \left(1 + \frac{R_2}{R_1}\right) \frac{m + n}{2} \Theta(-m - n) \tag{216}$$

$$\vec{\Gamma}_2(m, n) = \frac{1}{eC_J(R_1 + R_2)} \left(1 + \frac{R_1}{R_2}\right) \frac{m - n}{2} \Theta(m - n) \tag{217}$$

$$\overleftarrow{\Gamma}_2(m, n) = -\frac{1}{eC_J(R_1 + R_2)} \left(1 + \frac{R_1}{R_2}\right) \frac{m - n}{2} \Theta(-m + n). \tag{218}$$

Here, m and n correspond to voltages V_m and island charges ne, respectively. We calculate the occupation probabilities of the n-th state by starting from $n = 0$. For $n > 0$ one immediately finds from (216) that $\overleftarrow{\Gamma}_1(m, n)$ vanishes. Furthermore, we find from (217) that $\vec{\Gamma}_2(m, n)$ vanishes for $n \geq m$ and thus, according to (207), $p_{m+k}(m) = 0$ for $k > 0$. Together with (218) this means that $\overleftarrow{\Gamma}_2(m, n)$ vanishes for all n for which $p_n \neq 0$. The detailed balance condition (206) thus yields

$$p_{n+1}(m) = p_n(m) \frac{\vec{\Gamma}_2(m, n)}{\vec{\Gamma}_1(m, n + 1)}. \tag{219}$$

From similar considerations for $n < 0$ one finds

$$p_{n-1}(m) = p_n(m)\frac{\vec{\Gamma}_1(m,n)}{\overleftarrow{\Gamma}_2(m,n-1)}. \tag{220}$$

Making use of the rates (215) and (217), the two equations (219) and (220) may be recast into

$$p_n(m) = \left(\frac{R_1}{R_2}\right)^n \frac{(m!)^2}{(m-|n|)!(m+|n|)!}p_0(m). \tag{221}$$

Exploiting properties of binomial coefficients, we find for the normalization condition

$$\sum_{n=-m}^{m} p_n = p_0(m)\frac{(m!)^2}{(2m)!}\left(\frac{R_1}{R_2}\right)^m \left(1+\frac{R_2}{R_1}\right)^{2m} = 1. \tag{222}$$

This determines $p_0(m)$ and we finally get the probabilities

$$p_n(m) = \frac{(2m)!}{(m-|n|)!(m+|n|)!}\frac{(R_1/R_2)^{n+m}}{(1+R_1/R_2)^{2m}}. \tag{223}$$

When this is combined with (210), we obtain for the current

$$I = \frac{e}{C_J(R_1+R_2)}\frac{1}{2}\left(1+\frac{R_2}{R_1}\right)\sum_{n=-m}^{m}(m+n)p_n(m). \tag{224}$$

Here, we have evaluated the current through the first junction and taken into account that the backward rate does not contribute. The first term in the sum is obtained from the normalization condition (209) while the second term is given by

$$\sum_{n=-m}^{m} np_n(m) = m\frac{R_1-R_2}{R_1+R_2}. \tag{225}$$

The latter result may be derived by viewing $p_n(m)$ as a function of R_1/R_2 and applying the same trick used to derive (53). From (224) and (225) we get our final result for special points of the current-voltage characteristic [10]

$$I(V_m) = \frac{e}{C_J(R_1+R_2)}m \qquad (m=0,1,2,\ldots) \tag{226}$$

at voltages

$$V_m = \frac{e}{C_J}(m+\frac{1}{2}) \qquad \text{(low impedance environment)} \tag{227}$$

or

$$V_m = \frac{e}{C_J}(m+1) \qquad \text{(high impedance environment)}. \tag{228}$$

Thus, for certain voltages the current-voltage characteristic touches an Ohmic current-voltage characteristic with resistance R_1+R_2 which is shifted by the gap voltage $e/2C_J$ in the low impedance case and by e/C_J in the high impedance case.

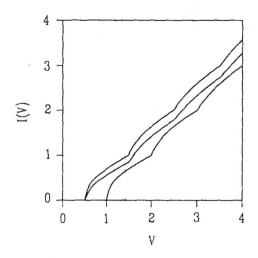

Figure 25. Zero temperature current-voltage characteristics of a double junction in a low impedance environment (left curve), an environment with Ohmic resistance $R_K/5$ (middle curve), and a high impedance environment (right curve). The junction parameters are $C_2 = C_1$ and $R_2 = 10R_1$. Voltage is given in units of $e/2C$ and current is given in units of $e/(2C(R_1 + R_2))$.

Let us now discuss the current-voltage characteristics between the voltage points for which we just calculated the current. To keep things as simple as possible we assume the tunneling resistance R_2 of the second junction to be very large compared to R_1. Then the island will be charged through the first junction up to a maximum charge. Occasionally an electron will tunnel through the second junction resulting in a current through the double junction. From the condition (192) for a non-vanishing rate through the first junction, one finds that the maximum island charge is given by $n_{max}e = -e[C_J V/e - 1/2]$, where $[\ldots]$ denotes the largest integer smaller or equal to the argument. In the limit $R_2 \gg R_1$ the current in the presence of a low impedance environment then reads

$$I(V) = e\vec{\Gamma}_2(V) = \frac{1}{2R_2}\left(V - \frac{e}{C_J}\left(n_{max} + \frac{1}{2}\right)\right). \tag{229}$$

At the voltages given by (227) we recover the current (226). Increasing the voltage we observe a jump in current by $e/(2R_2C_J)$ because the maximum island charge is increased by e. This jump is followed by a linear current-voltage characteristic with differential resistance $2R_2$. This ensures the validity of (226) and (227). One may apply the same arguments for the high impedance case where the whole picture is just shifted in voltage by $e/2C_J$.

The steplike structure which we have found is called the Coulomb staircase.[34]–[37] It is very distinct if the ratio of the tunnel resistances is very different from one. For $R_1 \approx R_2$ the steps are barely visible. For general parameters one has to evaluate (207)–(210) numerically. In Fig. 25 current-voltage characteristics are shown for low and high impedance environments as well as for the case of an Ohmic resistance $Z(\omega) = R_K/5$. As for the single junction one finds for such a low conductance a crossover from the low impedance characteristic to the high impedance characteristic. For $C_1 \neq C_2$, the staircase need not be as regular as it appears for a double junction with equal capacitances.

In this section on double junction systems we mentioned in several places how the results have to be generalized to account for an offset charge Q_0. It is now straightforward to calculate current-voltage characteristics for different offset charges. As an example we

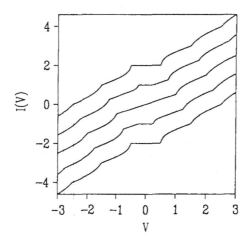

Figure 26. Zero temperature current-voltage characteristics of a single electron transistor in a low impedance environment. The junction parameters are $C_1 = C_2 = 2C$ and $R_2 = 10R_1$. The offset charge is increased from $Q_0 = 0$ for the lowest curve to $Q_0 = e$ in the highest curve in steps of $e/4$. The voltage is given in units of $e/2C$ and the current is given in units of $e/(2C(R_1 + R_2))$.

present in Fig. 26 the zero temperature current-voltage characteristics of a single electron transistor in a low impedance environment. This figure shows the Coulomb staircase as well as the dependence of the Coulomb gap on the offset charge. The junctions chosen here have equal capacitances but a ratio of tunneling resistances $R_2/R_1 = 10$ to produce a marked Coulomb staircase. The offset charge changes from $Q_0 = 0$ in the lowest curve to $Q_0 = e$ in the uppermost curve. As we have discussed already, the Coulomb gap depends on the offset charge. The middle curve with $Q_0 = e/2$ does not show a Coulomb gap as is expected from our earlier considerations. Furthermore, the characteristics for an offset charge different from 0 and $e/2$ exhibit an asymmetry. As we found earlier all orientations of tunneling processes are reversed if we make the replacement $V \to -V$ and $q \to -q$. As $-e/4$ and $3e/4$ are equivalent offset charges we find that the characteristics for $V > 0$ and $Q_0 = e/4$ should be identical to the characteristics for $V < 0$ and $Q_0 = 3e/4$ and vice versa. This can clearly be seen in Fig. 26.

6.8. SET-transistor and SET-electrometer

In a SET-transistor setup like the one shown in Fig. 16 the current through the junctions depends on both the transport voltage V and the gate voltage V_G. So far we have mainly concentrated on current-voltage characteristics $I(V)$. In this section we will keep V fixed and discuss how the current changes with the offset charge Q_0. The offset charge may be due to a gate voltage coupled capacitively to the island or due to some other mechanism. The dependence of the current on the offset charge can be exploited in two ways. By means of the gate voltage one may control the current thereby realizing a transistor.[4] On the other hand, one may use the current to measure the offset charge. In this case one uses the circuit as a very sensitive electrometer.[38] For the practical aspects of these devices we refer the reader to Chaps. 3 and 9. Here, we want to apply the results obtained above to the calculation of I-Q_0-characteristics. As in the previous sections it will not be possible to give a closed analytical expression for arbitrary transport voltages. We therefore restrict ourselves to the regime below the

Coulomb gap voltage. Since the performance of the transistor and electrometer reaches an optimum when biased at the gap voltage, this is the regime of practical interest.

To be more specific, let us choose a setup with equal junction capacitances $C_1 = C_2 = 2C$ but arbitrary ratio of tunneling resistances R_1/R_2. In practice, the assumption of a low impedance environment will be well satisfied. Furthermore, we restrict ourselves to the case of zero temperature. From the discussion of the current-voltage characteristics of a double junction it is clear how to generalize the calculation to finite temperatures. However, in this case one has to resort to numerical methods.

As already mentioned, we consider transport voltages below the low impedance gap $e/4C$. The offset charge is assumed to satisfy $0 \le Q_0 < e$. Then, once a stationary situation is reached, only two island charge states are occupied (cf. Fig. 21b). Taking the transport voltage to be positive we find from (193) and (195) that the backward tunneling rates $\overleftarrow{\Gamma}_1$ and $\overleftarrow{\Gamma}_2$ vanish if the island charge takes one of the two allowed values. When initially $n = 0$ and thus $q = Q_0$ the tunneling rate $\overrightarrow{\Gamma}_1$ through the first junction is nonvanishing while $\overrightarrow{\Gamma}_2$ is zero according to (192) and (194). After an electron has tunneled through the first junction q has changed to $q - e$. Now, $\overrightarrow{\Gamma}_1$ vanishes and $\overrightarrow{\Gamma}_2$ is different from zero allowing the electron to tunnel from the island. We now have the same situation as described in Sec. 6.6. The corresponding Eqs. (211) and (212) read

$$p_0 = \frac{\overrightarrow{\Gamma}_2(V, Q_0 - e)}{\overrightarrow{\Gamma}_1(V, Q_0) + \overrightarrow{\Gamma}_2(V, Q_0 - e)} \tag{230}$$

and

$$p_{-1} = \frac{\overrightarrow{\Gamma}_1(V, Q_0)}{\overrightarrow{\Gamma}_1(V, Q_0) + \overrightarrow{\Gamma}_2(V, Q_0 - e)}. \tag{231}$$

Inserting these probabilities for the two possible island charge states we find with (210) for the current

$$I(V, Q_0) = e\frac{\overrightarrow{\Gamma}_1(V, Q_0)\overrightarrow{\Gamma}_2(V, Q_0 - e)}{\overrightarrow{\Gamma}_1(V, Q_0) + \overrightarrow{\Gamma}_2(V, Q_0 - e)}. \tag{232}$$

According to (184) and (187), for equal capacitances $C_1 = C_2$ the rates through the first and second junction are related by

$$\overrightarrow{\Gamma}_2(V, q) = \frac{R_1}{R_2}\overrightarrow{\Gamma}_1(V, -q). \tag{233}$$

Together with (191) and (181) we finally obtain for the current at fixed transport voltage

$$I(Q_0) = \frac{1}{2}\frac{\left(\dfrac{Q_0 - e/2}{2C}\right)^2 - V^2}{(R_1 - R_2)\dfrac{Q_0 - e/2}{2C} - (R_1 + R_2)V}$$

$$\times\Theta(Q_0 - \frac{e}{2} + 2CV)\,\Theta(-Q_0 + \frac{e}{2} + 2CV). \tag{234}$$

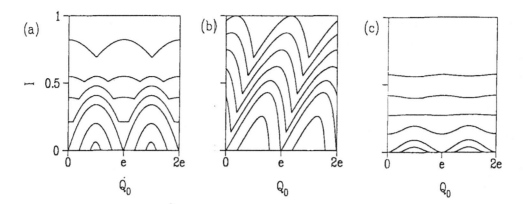

Figure 27. I-Q_0 characteristics at zero temperature for SET transistors with gate capacitance $C_G = 2C$ and Ohmic environment $Z_1(\omega) = Z_2(\omega) = R/2$. (a) Symmetric transistor with $R/R_K = 0.05$. The transport voltages in units of the gap voltage $V_g^0 = V_g(Q_0 = 0) = e/C_\Sigma$ are from bottom to top $V = 0.2, 0.6, 1.0, 1.2, 1.4, 1.6, 2.0$. (b) Asymmetric transistor with $R_1/R_2 = 10$ and external resistance $R/R_K = 0.05$. The transport voltages are $V = 0.6, 1.0, 1.2, 1.4, 1.6, 1.8, 2.0$. (c) Symmetric transistor with $R/R_K = 1$. The transport voltages are $V = 0.6, 1.0, 1.6, 2.0, 2.4, 2.8$. The current is given in units of $V_g^0/(R_1 + R_2)$.

The two step-functions $\Theta(x)$ are a consequence of the Coulomb gap. For very low transport voltages one needs an offset charge rather close to $e/2$ to obtain a current. As a function of the gate voltage the zero bias conductance thus displays a sequence of peaks at $Q_0 = C_G V_G = e(k + 1/2)$, k integer. For semiconductor nanostructures these Coulomb blockade oscillations are discussed in great detail in Chap. 5. With increasing transport voltage the range of offset charges leading to a nonvanishing current becomes larger. Finally, if the transport voltage equals the gap voltage, the current only vanishes for $Q_0 = 0$ as expected. Above this voltage the control of the current by the gate rapidly decreases and the performance of the transistor or electrometer is reduced as can be seen from Fig. 27.

For equal tunneling resistances $R_1 = R_2$ the I-Q_0-characteristics at transport voltages below the gap voltage are given by parabolas symmetric to $Q_0 = e/2$. This behavior is shown in Fig. 27a together with some curves at higher transport voltages. The environmental impedance is taken to be Ohmic and rather small ($R/R_K = 0.05$). If the tunneling resistance R_1 is much larger than R_2 the I-Q_0-characteristic (234) becomes

$$I(Q_0) = \frac{1}{2R_1} \left(\frac{Q_0 - e/2}{2C} + V \right) \Theta(Q_0 - \frac{e}{2} + 2CV) \Theta(-Q_0 + \frac{e}{2} + 2CV) \qquad (235)$$

which at the gap voltage reduces to

$$I(Q_0) = \frac{Q_0}{4R_1C}. \qquad (236)$$

In contrast to the parabolic characteristic obtained for the symmetric transistor we now have a sawtooth-like characteristic with the slope determined by the larger tunneling resistance. For the opposite case, $R_2 \gg R_1$, one finds from (234) at the gap voltage

$$I(Q_0) = \frac{e - Q_0}{4R_2C}, \qquad (237)$$

i.e. a reversed sawtooth characteristic. Numerical results for a ratio of tunneling resistances $R_1/R_2 = 10$ and low Ohmic damping are presented in Fig. 27b. Note that by choosing an asymmetric transistor one may obtain a very high sensitivity for a certain range of offset charges. Fig. 27c shows numerical results for a rather large Ohmic impedance $R = R_K$ of the environment. Obviously the sensitivity on the offset charge is drastically reduced as compared with an electrometer embedded in a low impedance environment.

6.9. Other multijunction circuits

The methods we have discussed so far in this section are not only applicable to double junction systems but also to circuits containing more than two junctions. Such multijunction systems can exhibit interesting physical behavior and are therefore discussed extensively in other chapters of this book. We mention systems containing few tunnel junctions (Chaps. 3 and 9), one-dimensional arrays (Chap. 7), and two-dimensional arrays (Chap. 8).

The network analytical approach introduced in Sec. 6.2. may be applied to a general multijunction circuit. To calculate tunneling rates across a given junction the circuit can be reduced to an effective single junction circuit containing a tunneling element, an environmental impedance and an effective voltage source. Such a reduction becomes possible by applying the Norton-Thevenin transformation discussed earlier. To disentangle complex circuits one usually will also need the transformation between a star-shaped and a triangle-shaped network, the so-called T-π-transformation.[39] Generally, the effective impedance will contain a contribution diverging like ω^{-1} for small frequencies. By splitting off this pole one separates the effective impedance into a capacitance related to the charging energy and an impedance describing the environmental influence. In this way, one directly finds expressions for the tunneling rates through the individual junctions as long as simultaneous tunneling through more than one junction is neglected. For a discussion of phenomena arising if co-tunneling is taken into account, we refer the reader to Chap. 6.

For the remainder of this section, we concentrate on the influence of the environment and choose as an example a one-dimensional array of tunnel junctions as shown in Fig. 28. Although the network analysis for an array is straightforward, we shall first present here a more microscopic approach which brings out the underlying physics. For simplicity, we neglect a capacitive coupling to ground which may be present in a real setup and which is of importance for the description of charge solitons in one-dimensional arrays (cf. Chap. 7). For our purposes it is sufficient to consider the capacitances associated with the N tunnel junctions carrying the charges Q_k ($k = 1, \ldots, N$). As for the single and double junction systems we introduce phases φ_k ($k = 1, \ldots, N$) satisfying the commutation relations

$$[\varphi_j, Q_k] = ie\delta_{jk}. \tag{238}$$

To describe the environmental influence it is convenient to introduce another set of charges and phases. From the surrounding circuit the array of tunnel capacitors may be

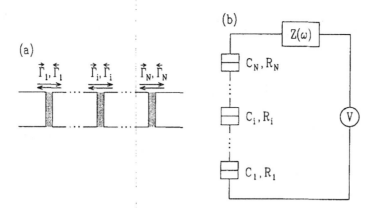

Figure 28. (a) Schematic drawing of a one-dimensional N-junction array. The arrows indicate forward and backward tunneling through the barriers. (b) Circuit containing a one-dimensional N-junction array with capacitances C_i, $(i = 1, \ldots, N)$ and tunneling resistances R_i coupled to a voltage source V via the external impedance $Z(\omega)$.

viewed as a single capacitor with total capacitance

$$C = \left(\sum_{k=1}^{N} \frac{1}{C_k} \right)^{-1} \tag{239}$$

carrying the total charge

$$Q = C \sum_{k=1}^{N} \frac{Q_k}{C_k}. \tag{240}$$

In addition there are $N - 1$ island charges

$$q_k = Q_k - Q_{k+1} \quad (k = 1, \ldots, N - 1). \tag{241}$$

This set of charges $\{Q, q_k\}$ is associated with a set of phases $\{\varphi, \psi_k\}$ satisfying the commutation relations

$$[\varphi, Q] = ie, \quad [\psi_k, q_k] = ie \tag{242}$$

with all other commutators vanishing. The two sets of phases are related by

$$\varphi_1 = \psi_1 + \frac{C}{C_1} \varphi \tag{243}$$

$$\varphi_k = \psi_k - \psi_{k-1} + \frac{C}{C_k} \varphi \quad (k = 2, \ldots, N - 1) \tag{244}$$

$$\varphi_N = -\psi_{N-1} + \frac{C}{C_N} \varphi \tag{245}$$

as for the double junction system in Sec. 6.2. The operator $\exp(-i\varphi_k)$ describes tunneling through the k-th junction. According to (244) this may be decomposed into an electron leaving the $k{-}1$-th island $(-\psi_{k-1})$ and an electron entering the k-th island (ψ_k). In (243) and (245) only one ψ-operator occurs since tunneling through the first or last junction affects only one island charge. The operators $\kappa_k \varphi$ are associated with the change of the total charge Q seen by the environment. Here, $\kappa_k = C/C_k$ is the obvious generalization of (172) to the multijunction case. The relations (243)–(245) essentially contain all information we need to know about tunneling in a one-dimensional array. The change in electrostatic energy connected with a tunneling process consists of contributions arising from the change of island charges and a contribution from the work done by the voltage source to restore the total charge. The latter is given by $\kappa_k eV$ since according to (243)–(245) the charge transferred to the voltage source after an electron has tunneled through the k-th junction is $\kappa_k e$. As in (180) for the double junction, the factor κ_k leads to a reduced influence of the environment due to a decoupling of the k-th tunnel junction from the environment by the other junctions. The total impedance is thus effectively reduced by a factor κ_k^2. Assuming that the capacitances C_k of the tunnel junctions are all of the same order, we find from (239) that the total capacitance is smaller by a factor $1/N$. As a consequence, the influence of the environment on a N-junction system is reduced by a factor $1/N^2$ as compared to a single junction. The assumption of a low impedance environment is therefore usually well satisfied.

To determine the tunneling rates we first derive an explicit expression for the internal charging energy due to the island charges. Solving (240) and (241) for the charges on the junctions we find

$$Q_1 = Q + C \sum_{i=1}^{N} \sum_{k=1}^{i-1} \frac{q_k}{C_i}$$

$$Q_n = Q + C \sum_{i=1}^{N} \sum_{k=1}^{i-1} \frac{q_k}{C_i} - \sum_{k=1}^{n-1} q_k \quad (n = 2, \ldots, N). \tag{246}$$

After some algebra one obtains for the internal charging energy of the array

$$\varepsilon(q_1, \ldots, q_{N-1}) = \sum_{i=1}^{N} \frac{Q_i^2}{2C_i} - \frac{Q^2}{2C} = \sum_{k,l=1}^{N-1} \frac{1}{2}(C^{-1})_{kl} q_k q_l \tag{247}$$

with

$$(C^{-1})_{kl} = C \sum_{m=1}^{\min(k,l)} \frac{1}{C_m} \sum_{n=\max(k,l)+1}^{N} \frac{1}{C_n}. \tag{248}$$

Here, $(C^{-1})_{kl}$ is the inverse of the capacitance matrix

$$C_{kl} = \begin{cases} C_k + C_{k+1} & \text{for } l = k \\ -C_{k+1} & \text{for } l = k+1 \\ -C_k & \text{for } l = k-1 \\ 0 & \text{otherwise.} \end{cases} \tag{249}$$

We note that the structure of (247) is also valid for more complicated multijunction circuits (cf. Chap. 3). Only the explicit form of the capacitance matrix will differ.

As a generalization of the energy difference (181) for electron tunneling in a double junction system we have for the one-dimensional array the changes in electrostatic energy

$$E_1(V, q_1, \ldots, q_{N-1}) = \kappa_1 eV + \varepsilon(q_1, \ldots, q_{N-1}) - \varepsilon(q_1 - e, q_2, \ldots, q_{N-1}),$$

$$E_i(V, q_1, \ldots, q_{N-1}) = \kappa_i eV + \varepsilon(q_1, \ldots, q_{N-1})$$

$$-\varepsilon(q_1, \ldots, q_{i-2}, q_{i-1} + e, q_i - e, q_{i+1}, \ldots, q_{N-1}) \qquad (250)$$

$$(i = 2, \ldots, N - 1),$$

$$E_N(V, q_1, \ldots, q_{N-1}) = \kappa_N eV + \varepsilon(q_1, \ldots, q_{N-1}) - \varepsilon(q_1, \ldots, q_{N-2}, q_{N-1} + e)$$

containing contributions from the change in the internal charging energy as well as from the work done by the voltage source. While the general structure of E_i becomes very apparent if expressed in the set of variables $\{V, q_1, \ldots, q_{N-1}\}$ the explicit form is rather complicated. On the other hand, we may as well choose the set of charges $\{Q_1, \ldots, Q_N\}$. Making use of (240) and (241) we then find from (250) together with (247) the very simple result

$$E_i(Q_i) = \frac{e}{C_i}(Q_i - Q_i^c) \qquad (251)$$

with the critical charge

$$Q_i^c = \frac{e}{2}(1 - \kappa_i). \qquad (252)$$

This is a straightforward generalization of the double junction result (182) and (183).

Before turning to the tunneling rates we will shortly outline how the result (251) can be obtained by network analysis. Considering electron tunneling through the i-th junction, we may combine the other junctions into one capacitor of capacitance

$$\tilde{C} = \left(\sum_{j \neq i} \frac{1}{C_j}\right)^{-1} = \frac{CC_i}{C_i - C} \qquad (253)$$

carrying the charge

$$\tilde{Q} = \frac{C_i C}{C_i - C} \sum_{j \neq i} \frac{Q_j}{C_j}. \qquad (254)$$

We thus have reduced the one-dimensional array to an effective double junction system. We now view the capacitances C_i and $\tilde{C}$ as the two capacitances in a double junction system. Accordingly, Q_i and $\tilde{Q}$ are the charges sitting on the two junction capacitors. Using our results from Sec. 6.2. we find an effective single junction circuit like the one depicted in Fig. 20c. The capacitor with effective capacitance

$$C_{\text{eff}} = C_i + \tilde{C} = \frac{C_i^2}{C_i - C} \qquad (255)$$

carries the charge

$$q_{\text{eff}} = Q_i - \tilde{Q} = \frac{C_i}{C_i - C}(Q_i - CV) \tag{256}$$

which corresponds to the island charge of a double junction. As already discussed above the effective impedance is given by $\kappa_i^2 Z_i(\omega)$ and the effective voltage is $\kappa_i V$. The difference in electrostatic energy E_i is now readily obtained as

$$E_i(Q_i) = \frac{q_{\text{eff}}^2}{2C_{\text{eff}}} - \frac{(q_{\text{eff}} - e)^2}{2C_{\text{eff}}} + \kappa_i eV = \frac{e}{C_i}(Q_i - Q_i^c) \tag{257}$$

in agreement with (251) and (252).

We are now in a position to write down the tunneling rates for a one-dimensional array. Using the same kind of reasoning as for the single junction or double junction we obtain for the forward tunneling rate through the i-th junction [10]

$$\vec{\Gamma}_i(V, q_1, \ldots, q_{N-1}) = \frac{1}{e^2 R_i} \int_{-\infty}^{+\infty} dE \frac{E}{1 - \exp(-\beta E)}$$
$$\times P(\kappa_i, E_i(V, q_1, \ldots, q_{N-1}) - E). \tag{258}$$

The backward tunneling rate is related to the forward tunneling rate by

$$\overleftarrow{\Gamma}_i(V, q_1, \ldots, q_{N-1}) = \vec{\Gamma}_i(-V, -q_1, \ldots, -q_{N-1}) \tag{259}$$

and the detailed balance symmetry

$$\overleftarrow{\Gamma}_1(V, q_1 - e, q_2, \ldots, q_{N-1}) = \exp[-\beta E_1(V, q_1, \ldots, q_{N-1})]\vec{\Gamma}_1(V, q_1, \ldots, q_{N-1}),$$

$$\overleftarrow{\Gamma}_i(V, q_1, \ldots, q_{i-2}, q_{i-1} + e, q_i - e, q_{i+1}, \ldots, q_{N-1}) \tag{260}$$

$$= \exp[-\beta E_i(V, q_1, \ldots, q_{N-1})]\vec{\Gamma}_i(V, q_1, \ldots, q_{N-1}) \quad (i = 2, \ldots, N-1),$$

$$\overleftarrow{\Gamma}_N(V, q_1, \ldots, q_{N-2}, q_{N-1} + e)$$

$$= \exp[-\beta E_N(V, q_1, \ldots, q_{N-1})]\vec{\Gamma}_N(V, q_1, \ldots, q_{N-1}).$$

It was pointed out above that the influence of the environment on the tunneling of electrons in a N-junction array is reduced by a factor of the order of $1/N^2$. This means that for most applications the low impedance limit will be correct. The rate expression (258) then reduces to the global rule result

$$\vec{\Gamma}_i(V, q_1, \ldots, q_{N-1}) = \frac{1}{e^2 R_i} \frac{E_i(V, q_1, \ldots, q_{N-1})}{1 - \exp[-\beta E_i(V, q_1, \ldots, q_{N-1})]} \tag{261}$$

which at zero temperature yields

$$\vec{\Gamma}_i(V, q_1, \ldots, q_{N-1}) = \frac{1}{e^2 R_i} E_i(V, q_1, \ldots, q_{N-1}) \Theta\big(E_i(V, q_1, \ldots, q_{N-1})\big). \tag{262}$$

As for the simpler circuits, the zero temperature rate is only different from zero if $E_i(V, q_1, \ldots, q_{N-1}) > 0$, i.e., if the charge Q_i on the i-th junction exceeds the critical charge Q_i^c given by (252). This local formulation of the blockade criterion involving only the charge Q_i, the capacitance C_i of the i-th junction, and the capacitance of the surrounding junctions is of great importance for the understanding of few-junction systems (cf. Chap. 3).

Let us finally discuss the size of the Coulomb gap. As for the double junction system the stable state at voltages below the gap voltage is characterized by vanishing island charges $q_i \equiv 0$. According to (246) the charges on the junctions are then all given by $Q_i = CV$. Tunneling at the i-th junction can therefore occur if $|V| > (e/2C)(1 - \kappa_i)$. No current will flow if the voltage across the array is smaller than $(e/2C) \min_i(1 - \kappa_i)$. Since $\kappa_i = C/C_i$, the gap voltage is determined by the junction with the largest capacitance. For an array with equal capacitances, i.e. $C_i = C_J = NC$, the low impedance Coulomb gap is given by $(1 - 1/N)(e/2C) = (N-1)(e/2C_J)$. However, this gap and the Coulomb offset observed at large voltages where the high impedance gap $e/2C = N(e/2C_J)$ appears, differ by $e/2C_J$.

ACKNOWLEDGEMENTS. We have benefitted from the interaction with numerous colleagues of whom we can name only few here. One of us (G.-L. I.) has enjoyed a fruitful collaboration with H. Grabert and would like to acknowledge discussions with P. Wyrowski and the members of the Quantronics Group at Saclay. The other author (Yu. V. N.) is grateful to T. Claeson, D. Haviland, and L. S. Kuzmin for discussions on experimental aspects of single charge tunneling and to A. A. Odintsov and M. Jonson for critically reading the manuscript on the microscopic foundation. He enjoyed the hospitality of the Fachbereich Physik at Essen where this chapter was written. We thank G. Falci for granting us permission to reproduce Figs. 13 and 14.

A. Microscopic foundation

A.1. Introduction

One may note that in the previous sections the tunnel junction was treated as the primary object. The whole electrodynamics was reduced to the network theory and the only trace of solid state physics was the one-electron Fermi distribution function. Therefore, one may call it 'phenomenological approach'. It is difficult to underestimate its importance for applications. Nevertheless, there are some reasons to go deeper and to discuss the microscopic foundation of the method applied. First, the microscopic derivation is always a good way to test and probably to confirm the phenomenology. Second, the range of the applicability becomes clearly visible. Third, the interesting links between different approaches and different phenomena can be comprehended and some new effects can be described.

In this appendix we move from the microscopic description of the metals on both sides of the tunnel junction towards the phenomenology step by step. We encounter interesting physics on every step and have a general look on it.

We are starting with the formulation of the problem in a general way: how does the electromagnetic interaction affect the tunneling rate? We develop a semiclassical approach to electron motion and derive the basic formula which allows us to evaluate the effect in terms of Maxwell electrodynamics and a Boltzmann master equation description for the electrons. Then we discuss the fundamental relation between the voltage applied to the junction and the effective scales at which the tunneling electron feels the electrodynamic environment.

After that we apply our approach to Altshuler's diffusive anomalies in the density of states.[40] Within this framework these anomalies and the Coulomb blockade appear to be two sides of one coin. While proceeding we consider the most important case when the field induced by the electron moves faster than the electron itself. It allows us to forget about this electron. Moving along this way we remind the reader of some simple electrodynamics in a metal-thin insulator-metal system and consider fingerprints of this electrodynamics on the tunnel junction I-V curve. In the very end we show how one can move from the continuous electrodynamics to the network theory finishing the consequent microscopic derivation of the phenomenological approach.

A.2. General problem

As it is widely known, it is easier to answer general questions than specific ones so that we try to formulate the problem in a most general way. Let us answer the question: How does the electromagnetic interaction affect the electron tunneling rate?

First, of course, this interaction forms the crystal lattice of metals and of the insulator layer through which the electrons tunnel. It determines the energy spectrum of elementary excitations in these condensed media and thus provides every decoration of the scene for this solid state physics performance. Everything is made here by the electromagnetic fields acting on the length, energy, and time scales of the order of atomic values. The question how it makes these things is in fact the basic problem of condensed matter physics and we are not going to solve it just here. We come on the scene formed and we are interested in the part of the electromagnetic field which is:

1. slow enough not to change the electron energy on an atomic scale
2. weak enough not to turn electrons from its trajectories
3. described by linear electrodynamic equations
4. basically uniform on the atomic scale.

We call it the low-frequency part of the electromagnetic interaction and will deal only with this part. Also we assume the voltage applied to the junction to be much less than the typical atomic energy or the Fermi energy for electrons. So we will operate on scales which are much larger than the atomic ones. One may call this formulation a mesoscopic problem but we prefer not to do it. The evaluation of the effect under consideration has been done in Ref. [41] by using the standard formalism of field trajectory integrals. This approach is inconvenient due to the use of an imaginary time representation and the relative complexity. Here we use another way to derive it which is based mostly on physical reasoning. This way can be compared with the one used in Ref. [42].

First let us note that on this scale a semiclassical approach to the electron motion is valid or, in simple terms, we can consider electrons as classical particles that are scattered and jump through the tunnel barrier. Switching off the low-frequency part of the electromagnetic interaction we introduce a probability $w(y, k_{in}, k_f)$ to jump through the barrier at the point y on the barrier surface with initial electron wave vector k_{in} and wave vector k_f after tunneling. Only the electrons near the Fermi surface can tunnel so we need to know this probability only on this surface. The tunneling is completely elastic because the interaction is switched off. The total tunneling rate then is given by

$$\Gamma(V) = \int d^2y\, d^2k_{in} d^2k_f\, \nu_1\nu_2 w(y, k_{in}, k_f) \int d\epsilon\, d\epsilon'\, f(\epsilon)[1 - f(\epsilon' - eV)]\delta(\epsilon - \epsilon'). \quad (263)$$

Here we integrate over y belonging to the junction area, the two-dimensional wave vectors k_{in} and k_f parametrize the Fermi surface, the δ-function ensures the tunneling to be elastic, $\nu_{1,2}$ denote the densities of states per energy interval at the Fermi surface for the two metal banks, respectively, and $f(\epsilon)$ is the Fermi distribution function.

Let us consider now the problem from the quantum mechanical point of view. We introduce the electron propagation amplitude K connecting the electron wave functions ψ at different times by [42]

$$\psi(x, t) = \int d^3x'\, K(x, t; x', t')\psi(x', t'). \quad (264)$$

Within the semiclassical approach the different classical trajectories contribute to $K(x, t; x', t')$ without interfering with each other. If we are interested in evaluating the tunneling rate at a given point y with given k_{in} and k_f there is a unique trajectory determined by these parameters which contributes. The phase of the propagation amplitude is proportional to the classical action along this trajectory:

$$K(x, t; x', t') \sim \exp(iS/\hbar). \quad (265)$$

This is the way to take the interaction into account. Let us note that electromagnetic interaction can be treated as the exchange of photons and as a first step consider the electron motion in the photon field. In accordance with the correspondence principle one should add to the total action the interaction term and thus obtain the propagation amplitude in the presence of a field:

$$K_{field}(x', t'; x'', t'') = K(x', t'; x'', t'')\exp(iS_{int}/\hbar) \quad (266)$$

with

$$S_{int} = e\int_{t''}^{t'} dt\left[\frac{v^\alpha(k(t))A^\alpha(x(t), t)}{c} + \phi(x(t), t)\right]. \quad (267)$$

Here, e is the electron charge, $A^\alpha(x, t)$ and $\phi(x, t)$ are the vector and scalar potential of the electromagnetic field, respectively, and $x(t)$ and $k(t)$ are trajectory parameters. On a large time scale the electron is scattered many times by impurities and the metal surface and the trajectory is extremely complicated. This is a reason to characterize this trajectory by the probability $p(x, k, t)$ for the electron to be in the point x with wave vector k at time t. It is worth to emphasize that the probability is conditional:

the electron must jump at the point y with certain k_{in} and k_f at time 0. Therefore, it is more convenient to rewrite (267) in the form

$$S_{\text{int}} = e \int_{t''}^{t'} dt \left[\frac{j^\alpha(x,t) A^\alpha\big(x(t),t\big)}{c} + \rho(x,t)\phi(x,t) \right] \tag{268}$$

where

$$j^\alpha(x,t) = \int d^2k\, v^\alpha(k) p(x,k,t-T_1-t'') \tag{269}$$

and

$$\rho(x,t) = \int d^2k\, p(x,k,t-T_1-t''). \tag{270}$$

Here, T_1 is the time needed to move from x'' to y along the trajectory. A careful analysis shows that there is also a probability flow

$$j^\alpha(x,t) = v^\alpha(k_{\text{in}}) N^\alpha p(y,k_{\text{in}},t-T_1-t'') = v^\alpha(k_f) N^\alpha p(y,k_f,t-T_1-t'') \tag{271}$$

at an x belonging to the oxide barrier volume, where N is a vector normal to the junction surface. The necessity of this term results from the conservation of probability flow. The barrier thickness is of the order of a few atomic sizes and at first look the contribution of this term to the action is much smaller than that of the probability flow in the metal. However, the electromagnetic field in the oxide barrier may be much larger than in the surrounding metals and therefore one has to keep this term.

Performing the field quantization we replace classical potentials by appropriate Heisenberg time-dependent operators. The propagation amplitude also becomes an operator. The tunneling rate at given energies E_{in}, E_f in initial and final states is proportional to the square of the Fourier transform of the propagation amplitude

$$\Gamma(E_{\text{in}}, E_f) \sim \langle \hat{K}(E_{\text{in}}, E_f) \hat{K}^+(E_{\text{in}}, E_f) \rangle \tag{272}$$

where $\langle \ldots \rangle$ denotes the average over the equilibrium density matrix of the electromagnetic field and

$$\hat{K}(E_{\text{in}}, E_f) = \int dt' dt'' \exp\left[\frac{i}{\hbar}(E_f t'' - E_{\text{in}} t') \right] \hat{K}(t',x';t'',x''). \tag{273}$$

In this equation x'' and x' are chosen to be very far from the jump point y in the one bank and in the other one, respectively. They can be characterized by the two traversal times T_1 needed to move from x'' to y and T_2 needed to move from y to x'. For $T_{1,2} \to \infty$ the square of the propagation amplitude reaches a certain limiting value related to the rate.

If we are not interested in relativistic effects concerning the electromagnetic field propagation we are able to omit the vector potential in (267) and (268). Using (268) we then obtain

$$\hat{K}(E_{\text{in}}, E_f) \sim \int dt'' \exp\left[\frac{i}{\hbar}(E_f - E_{\text{in}})t''\right]$$

$$\times \exp\left[\frac{i}{\hbar}\int_{t''-T_1-T_2}^{t''} dt \int d^3x\, \rho(t - t'' - T_1, x)\hat{\phi}(t, x)\right]. \quad (274)$$

According to the previous considerations we can consider $\hat{\phi}(t, x)$ as the sum of a large number of boson operators as it has been considered within the phenomenological approach in Sec. 2.3. It allows us to use the simple rules for operator multiplication when calculating the square. For example,

$$\left\langle \exp\left(i\int dt\, C_1(t)\hat{\phi}(t, x)\right) \exp\left(-i\int dt\, C_2(t)\hat{\phi}(t, x)\right)\right\rangle$$

$$= \exp\left(-\int dt_1 dt_2\, F(t_1, t_2)\langle\hat{\phi}(t_1)\hat{\phi}(t_2)\rangle\right) \quad (275)$$

with

$$F(t_1, t_2) = \frac{C_1(t_1)C_1(t_2) + C_2(t_1)C_2(t_2)}{2} - C_1(t_1)C_2(t_2). \quad (276)$$

After some algebra one gets

$$\Gamma(E_{\text{in}}, E_f) \sim P(E_f - E_{\text{in}}) \quad (277)$$

where we introduced

$$P(E) = \int \frac{d\tau}{2\pi} \exp\left[\frac{i}{\hbar}E\tau\right] \exp(-Y(\tau)) \quad (278)$$

with

$$Y(\tau) = \frac{e^2}{\hbar^2} \int \frac{d\omega}{2\pi}(1 - e^{-i\omega\tau}) \int d^3x d^3x'\, \rho_\omega(x)\rho_{-\omega}(x')D^{+-}(\omega; x, x'). \quad (279)$$

The latter can be expressed in terms of the Fourier transforms of ρ and the field correlation function $\langle\hat{\phi}(0)\hat{\phi}(t)\rangle$:

$$\rho_\omega(x) = \int dt\, \rho(t, x)e^{i\omega t} \quad (280)$$

$$D^{+-}(\omega; x, x') = \int dt\, \langle\hat{\phi}(0, x)\hat{\phi}(t, x')\rangle e^{i\omega t}. \quad (281)$$

The coefficient of proportionality in (277) can be determined if we switch off the electromagnetic interaction. Thus we obtain for the total tunneling rate as a nice generalization of (263)

$$\Gamma(V) = \int d^2y d^2k_{\text{in}} d^2k_f\, \nu_1\nu_2 w(y, k_{\text{in}}, k_f)$$

$$\times \int d\epsilon d\epsilon'\, f(\epsilon)[1 - f(\epsilon' - eV)]P(\epsilon' - \epsilon; y, k_{\text{in}}, k_f). \quad (282)$$

Now we should discuss what we have done. The physical picture in fact is rather simple. In (263) we replace the δ-function, which reflects the fact that in the absence of the interaction the tunneling is elastic, by the probability $P(E)$ to have a certain energy change when tunneling in the presence of the interaction. This change is provided by emission or absorption of photons. Due to the semiclassical nature of electron motion and due to the small energy transfer for every emission/absorption act all of these acts happen independently. Thus for every photon energy the probability to emit a certain number of photons obeys the Poissonian statistics and (278) and (279) are simply a generalized mathematical formulation of this fact. Within the phenomenological approach this was discussed in Sec. 4.1. Within the microscopic theory we were able to express these photon emission/absorption probabilities in terms of the electron motion along the classical trajectory and the field correlation function. Thus we managed to simplify the problem considerably. Now, we can calculate the tunneling rate if we know all about classical electron motion and the electrodynamics in the region around the junction. Let us now express our results in a form appropriate for practical use.

First we express the field correlation function in terms of the response function by means of the fluctuation-dissipation theorem

$$D^{+-}(\omega; x, x') = -\left(\coth(\frac{\hbar\beta\omega}{2}) - 1\right) \operatorname{Im}D^A_\omega(x, x'). \tag{283}$$

Here $D^A_\omega(x, x')$ is the advanced response function

$$\Phi_\omega(x) = -\frac{e}{\hbar} \int \mathrm{d}^3x' D^A_\omega(x, x')\rho_\omega(x') \tag{284}$$

which determines the field response $\Phi_\omega(x)$ to an external charge placed at point x'. It is convenient to consider $e\rho_\omega(x)$ as this external charge. Φ is determined by the electrodynamic equations in a medium with a given source. It is easier to solve these equations than to completely calculate the response function D^A_ω. The result (279) may be rewritten in terms of Φ as

$$Y(\tau) = \frac{e}{\hbar} \int \frac{\mathrm{d}\omega}{2\pi}(1 - e^{i\omega\tau}) \int \mathrm{d}^3x \operatorname{Im}[\Phi_\omega(x)\rho_{-\omega}(x)] \left[\coth(\frac{\hbar\beta\omega}{2}) + 1\right]. \tag{285}$$

To calculate the probability $p(x, t, k)$ which characterizes the electron motion one may use the standard Boltzmann master equation approach. For example, for the probability to be on one of the banks after tunneling the Boltzmann equation reads

$$\frac{\partial p(k)}{\partial t} = v^\alpha(k)\frac{\partial p(k)}{\partial x_\alpha} + \int \mathrm{d}^2k' W(k, k', x)\big(p(k') - p(k)\big)$$

$$+\delta(t)\delta^3(x - y)\delta^2(k - k_f). \tag{286}$$

Here, $v^\alpha(k)$ is the electron velocity in the state with a wave vector k, $W(k, k', x)$ is the scattering rate from the state with k to the state with k' due to the impurities and the metal surface, and the source term describes the electron arrival from the other bank at $t = 0$. Often we need to describe the electron motion only on a time scale which is much larger than the time between the scattering events. In this case one may use the diffusion equation for the electrons:

$$\frac{\partial \rho}{\partial t} = D\Delta\rho + \delta(t)\delta^3(x - y) \tag{287}$$

with appropriate boundary conditions.

For the derivation of (282) we assumed a non-relativistic field. Strictly speaking this means that we are not able, for example, to describe the inductance in electric circuits. This is why it is worth to emphasize that all the field relativistic effects can be treated in the same manner. The most convenient gauge choice is $\phi(x,t) = 0$ and the field is described by the vector potential only. Acting along the same lines we express the answer

$$Y(\tau) = \frac{e^2}{\hbar^2} \int \frac{d\omega}{2\pi}(1 - e^{-i\omega\tau}) \int d^3x d^3x'\, j_\omega^\alpha(x) j_{-\omega}^\beta(x') D_{\alpha\beta}^{+-}(\omega; x, x') \tag{288}$$

in terms of the vector potential correlation function

$$D_{\alpha\beta}^{+-}(\omega; x, x') = \int dt e^{i\omega t} \frac{\langle \hat{A}^\alpha(0, x)\hat{A}^\beta(t, x')\rangle}{c^2}. \tag{289}$$

Here $\vec{j}$ is the probability flow introduced earlier. It is also possible to simplify this form by introducing the vector potential response $A^\alpha(x, t)$ on the external current $e\vec{j}(x, t)$ yielding

$$Y(\tau) = \frac{e}{\hbar c} \int \frac{d\omega}{2\pi}(1 - e^{i\omega\tau}) \left[\coth(\frac{\hbar\beta\omega}{2}) + 1\right] \int d^3x\, \mathrm{Im}[A_\omega^\alpha(x) j_{-\omega}^\alpha(x)]. \tag{290}$$

Now we are in a position to apply these general results to some illustrative examples.

A.3. Time of tunneling

Before making these applications it is important to discuss the relevant time scale on which we are going to operate. As far as we consider tunneling rates this relevant time should be the time of tunneling. The problem is not transparent and sometimes it leads to misunderstanding. To feel that let us consider the noninteracting electrons on a scale much larger then the atomic one. The electrons rush along the metal and sometimes jump through the tunnel barrier. The thickness of this barrier is of the atomic order and a jump takes no time. On the other hand, the tunneling is elastic and the energy loss equals zero. According to quantum mechanics, the energy uncertainty ΔE and the time for tunneling t obey the relation $\Delta E \cdot t \simeq \hbar$. It means that the time for elastic tunneling is infinitely long.

So we have a real choice: from zero to infinity. This is natural since quantum mechanics always gives rise to duality. The answer depends on the way how the time of tunneling is introduced or, in practical terms, it depends on the quantity measured.

If we are sure that our tunneling is elastic there are many ways to introduce the traversal time (see [43] for a review) and sometimes to measure it [44]. In this case the time is determined by the properties of the electron's motion under the barrier and there is some interesting physics due to the virtual nature of this motion. In contrast, for inelastic tunneling the voltage applied to the tunnel barrier and/or the temperature impose strict restrictions on the frequency ω of the radiation emitted/absorbed: $\hbar\omega \leq \max\{eV, k_BT\}$. This frequency determines the time scale as can be seen from the previous equations. If the most part of the tunneling events are inelastic it also determines the time of tunneling. If it is not so this argument is applicable only for inelastic events.

Thus the tunneling may be characterized by two time scales. The case when these scales are of the same order is described in Refs. [42, 45]. Note that in this case the tunnel barrier is suppressed by the applied voltage.

The frequency scale specifies also the length scale since we are able to estimate the distance which the electromagnetic field or electrons propagate for a given time. It allows us to find the effective geometry of the junction or, in other terms, to determine whether it can be considered as a point or as an interface of two semi-infinite metal banks. The length scale increases with decreasing frequency so that the lower the voltage and temperature the further away is the horizon which the junction sees. However, we should emphasize that the length scale can not be determined unambiguously. The electrons and the field propagate with different velocity. Moreover, different types of electromagnetic excitations differ in its velocities. This makes the length scale and effective geometry dependent on the inelastic process under consideration.

A.4. One-photon processes: anomalies and fingerprints

Now we try to find the scale for the strength of the effect considered. To characterize this strength it is convenient to introduce the effective frequency dependent impedance which the electron feels when tunneling:

$$Z_{\text{eff}}(\omega) = \frac{i\omega}{e} \int \mathrm{d}^3x\, \Phi_\omega(x)\rho_{-\omega}(x) \tag{291}$$

or for the other gauge choice (290):

$$Z_{\text{eff}}(\omega) = \frac{i\omega}{ec} \int \mathrm{d}^3x\, A^\alpha_\omega(x)j^\alpha_{-\omega}(x). \tag{292}$$

The value of this impedance in the frequency region considered governs the deviations from Ohm's law. If this value exceeds the quantum unit of resistance $R_K = h/e^2$ the probability for many-photon processes is significant, the deviations are large, and the tunneling rate is strongly suppressed in comparison to Ohm's law. If not, the most part of the tunneling events are elastic and there are only small deviations from an Ohmic behavior which are due to one-photon processes. Very roughly the effective impedance can be estimated as the resistance of a metallic piece on an appropriate length scale. Usually, on the microscopic scale this resistance is small in comparison with R_K and the deviations from Ohm's law are small. For our illustrative applications we use this fact and we will consider mostly one-photon processes.

Let us express the probability $P(E)$ in terms of the impedance. To do this we expand the exponent in (278) with respect to Z_{eff}. The first order dependence of $P(E)$ on Z_{eff} will be

$$\delta P(E) = P_1(E) - \delta(E) \int \mathrm{d}E' P_1(E') \tag{293}$$

where $P_1(E)$ is the probability to emit one photon in a unit energy interval

$$P_1(E) = \frac{2\mathrm{Re}Z_{\mathrm{eff}}(E/\hbar)}{ER_K} \frac{1}{1 - \exp(-\beta E)}. \tag{294}$$

The simple formula

$$R_T \frac{\partial^2 I}{\partial V^2} = \frac{2Z_{\mathrm{eff}}(eV/\hbar)}{R_K V} \tag{295}$$

valid at zero temperature reflects the influence of one-photon processes on the I-V curve. By the way there is an answer how one can observe this small deviations on the background of the main effect. These deviations are clearly visible on the differential conductance-voltage curve or on the second derivative of the current in the region of small voltages because Ohm's law is linear. It is convenient to divide the observable deviations into two classes:

1. anomalies: the anomalous power law is displayed in some voltage region,
2. fingerprints: the deviation is localized near a certain voltage.

A.5. Diffusive anomalies

So-called zero-voltage anomalies were observed in tunnel junction experiments from the early sixties. As it is comprehended now they were caused by different mechanisms. The early explanation ascribed the whole effect to the scattering by paramagnetic impurities.[46] It was confirmed that this can sometimes produce such anomalies [47], but there was also an effect in the absence of these impurities.

In 1975, Altshuler and Aronov proposed a more fundamental mechanism to be responsible for zero-voltage anomalies.[48] They calculated the interelectron Coulomb interaction effect on the carrier density of states near the Fermi level. Although the contribution to the density of states was found to be small it influences the observed anomalies because it depends nonanalytically on the distance from the Fermi level. A relation of the same kind as Eq. (263) was applied to calculate the tunnel current. The densities of states $\nu_{1,2}$ in this relation were allowed to have a small energy-dependent part. This results in a square-root contribution to the conductance of a junction and in the transition from the square-root to the logarithmic dependence as the thickness of the electrodes is reduced. These dependencies were perfectly confirmed by the experiments of Refs. [49] and [50].

In our opinion, these results are correct but there are two points of criticism concerning the link between the tunneling rate and the density of states. A discussion of these points will probably allow to better understand the physics involved.

First, the relation used assumes the tunneling rate to be proportional to the density of states in the banks of the junction. This assumption is undoubtedly correct within a one-particle theory where only elastic tunneling is possible, but it is not satisfied when the interelectron interaction is included. If the interaction affects the density of states, it is not consistent to neglect it when calculating the tunnel current.

Second, we are quite pessimistic about the principal possibility to measure the electron density of states in the presence of interaction excluding only few cases. The density of states in the presence of interaction is defined and calculated as being proportional to the probability to annihilate an electron with a given energy at a given point. Under realistic circumstances the number of electrons is conserved. We therefore can not annihilate an electron but only pull it out at a point and then measure the energy of this electron. In fact, this is the way how the experimental methods work, from X-ray to tunneling methods. But if there is an interaction by which the electron may loose or gain energy when being pulled out, there is a fundamental restriction on the resolution of these measurements.

This is why we prefer to discuss the effect of the electromagnetic interaction on the tunneling rate but not on the density of states. Now, we obtain anomalies involved in the framework of the method presented above and we will find them to be due to inelastic tunneling.

Let us first consider two semi-infinite metals separated by an insulating layer. We assume the frequency scale related to the voltage and temperature to be less then the inverse electron momentum relaxation time $1/\tau_{\text{imp}}$. This assumption allows us to use a diffusion equation for describing the electron motion. In order to evaluate the effective impedance we first calculate the Fourier transform of the conditional probability ρ_ω. Solving Eq. (287) we obtain

$$\rho_\omega(r) = \pm \frac{1}{2\pi D r} \exp(-\sqrt{\frac{i(\omega \pm i0)}{D}} r).$$

(296)

Here, different signs refer to the different banks and r is the distance from the point of tunneling. Now we should evaluate $\Phi_\omega(r)$. To do this, we solve the electrostatic equation with external charge $e\rho_\omega(r)$:

$$\Delta \Phi_\omega(r) = 4\pi \Big(q(r) + e\rho_\omega(r) \Big).$$

(297)

Here, $q(r)$ is the charge density formed by the metal electrons. As it is known metals are electroneutral and the sum on the right hand side of (297) equals zero in the metal. The only point where it is not zero is the mere point of the tunneling. The potential difference produces a current in the metal with density $\vec{j} = -\sigma \vec{\nabla} \Phi$, σ being the metal conductivity. Due to electroneutrality the total current through the point of tunneling must be equal to the flow of external charge through this point with the inverted sign. It allows us to obtain $\Phi_\omega(r)$:

$$\Phi_\omega(r) = \pm \frac{e}{2\pi\sigma r}.$$

(298)

After integration over space we find for the effective impedance

$$\text{Re} Z_{\text{eff}}(\omega) = \frac{1}{\pi\sigma} \sqrt{\frac{\omega}{2D}}.$$

(299)

With the aid of (295) we obtain for the anomaly of the tunneling current

$$R_T \delta I = \frac{4e^2}{3\pi^2 \sigma \hbar} \sqrt{\frac{eV}{2D\hbar}} V.$$

(300)

The result is the same as in [48] if taking into account the surface effect [51] but now we can ascribe it to inelastic tunneling. The length scale is of the order $(eV/D\hbar)^{-1/2}$ and the impedance can be estimated as the resistance of a metal piece of that size. The resistance increases with decreasing size provided the size is less than the electron mean free path l. The maximum value of resistance would be of the order of $R_K(k_F l) \ll R_K$ for a reasonable metal (k_F is the electron wave vector at the Fermi surface) and it ensures that the deviation is never comparable with the main current.

Now we change the effective geometry and let the electrodes be films of thickness d. The above consideration is valid if $(eV/D\hbar)^{-1/2} \gg d$ and we now investigate the opposite limiting case $(eV/D\hbar)^{-1/2} \ll d$. In this case we can treat the probability and voltage as to be approximately constant across the film. It is convenient to use Fourier transformation here with the wave vectors $\vec{q}$ along the film plane. For $\rho_\omega(q)$ we obtain

$$\rho_\omega(q) = \pm \frac{1}{\pm i\omega + Dq^2}. \tag{301}$$

The calculation of Φ_ω is a little bit more complex. The films separated by the insulating layer can be considered as a large capacitor with C_0 being the capacitance per unit area. The voltage difference between the electrodes produces the finite density of charge per unit area $\tilde{q} = C_0\Delta\phi$. Due to the symmetry of the system, $\Phi_\omega(x)$ has different signs but equal magnitude on the different electrodes. This allows us to write down the conservation law for the charge density in the following form:

$$\frac{\partial\tilde{q}}{\partial t} - \mathrm{div}(\sigma\nabla\phi) = e\delta(t)\delta^2(x). \tag{302}$$

As the next step we obtain

$$\Phi_\omega(q) = \pm \frac{e}{i\omega + D^*q^2}. \tag{303}$$

$D^* = \sigma d/2C_0$ may be interpreted as the diffusivity of the electric field. As a rule $D^* \gg D$ and the field propagates faster then the electron. To obtain the impedance we integrate over $\vec{q}$ and note that the dominant contribution to the integral comes from a wide region of q: $(\omega/D)^{1/2} \gg q \gg (\omega/D^*)^{1/2}$. We can ascribe the effect neither to field nor to electron propagation: there is something in between. It is convenient to express the answer introducing the film sheet resistance $R_\square = (\sigma d)^{-1}$

$$R_T \frac{\partial^2 I}{\partial V^2} = \frac{2R_\square}{\pi R_K V} \ln(D^*/D). \tag{304}$$

It differs from the one derived from the density of states by a logarithmic factor. This is natural because the approach of Refs. [40, 48, 51] does not take into account the field propagation induced by the tunneling electron.

As far as we know we considered here all observable anomalies related to electron diffusion. Considering the variety of other anomalies and the effect of the external circuit we needed not to take into account electron motion at all. The reason for that is the following:

A.6. Field moves faster than the electrons

Indeed this is the usual case as we already have seen when we considered field propagation along resistive films: the field diffuses faster than the electrons. If the resistivity of the electrodes is lower we encounter different types of electromagnetic excitations which can be as fast as the light and they effectively overtake electrons. Due to this fact the length scale for the electromagnetic field is much larger than that one for electron propagation. Thus the electromagnetic field is constant on the length scale for electrons. It means we can use the simplest expression for $\rho_\omega(x)$:

$$\rho_\omega(x) = \pm \frac{\delta(x)}{i\omega \pm 0}. \tag{305}$$

Different signs correspond to different banks. Integrating over x in Eq. (291) we obtain the simple but promising result:

$$Z(\omega) = \Phi_\omega^{(1)}(0) - \Phi_\omega^{(2)}(0). \tag{306}$$

Here, the superscripts (1) and (2) refer to different banks. Now, we are allowed to omit the subscript 'eff' because we have the honest electrodynamic impedance defined as voltage difference between banks at point 0 provided that this is the voltage response to the current produced by the electron jumping over the tunnel barrier. We have made one more step towards the phenomenology.

As an application of Eq. (306) we consider the influence of the undamped electromagnetic excitations which can propagate along the junction interface. We will assume that the junction is large enough so that the typical time scale defined by voltage/temperature is much smaller than the time needed for the electromagnetic excitation to cross the junction. The existence of these undamped excitations is provided by the insulating layer which separates the metallic banks and which can be considered as an infinite capacitor characterized by the capacitance $\tilde{C}$ per unit area. To calculate the impedance we write down the balance equation for the charge density of this capacitor

$$\frac{\partial \tilde{q}}{\partial t} = J_z + \delta(x)\delta(t) \tag{307}$$

where J_z is the volume density of the electrical current taken on the metal surface (z is normal to the junction interface). Performing the Fourier transform in time and in space coordinates along the interface and expressing all in terms of the voltage difference we can rewrite the previous equation as

$$\left(i\omega\tilde{C} - iB(\omega, q) \right) \Phi_\omega(q) = 1. \tag{308}$$

Here, we introduced the non-local link between the normal current and the voltage on the junction: $J_z(q) = iB(\omega, q)\Phi_\omega(q)$. From the previous equation we have for the impedance

$$Z(\omega) = -i \int \frac{\mathrm{d}^2 q}{(2\pi)^2} \frac{1}{\tilde{C}\omega - B(\omega, q)}. \tag{309}$$

We need only the real part of the impedance to evaluate its effect on the I-V curve. While the excitations are undamped the poles of the impedance at every q lie on the real axis and only these poles contribute to the real part. Thus for the real part of the impedance we obtain

$$\mathrm{Re}Z(\omega) = \frac{\pi}{\tilde{C}} \int \frac{\mathrm{d}^2 q}{(2\pi)^2} \delta\big(\omega - \Omega(q)\big) \tag{310}$$

where $\Omega(q)$ is the spectrum of electromagnetic excitations. To evaluate the effect we need to know only the capacitance and this spectrum.

Let us first consider plasma excitations. For a bulk metal the plasmons can not have an energy less than the plasma frequency ω_p. Nevertheless there are low energy plasma excitations localized on the junction interface. We consider them in some detail. The electrical field in the metal produces the current of electron plasma

$$\vec{J} = \frac{\omega_p^2}{4\pi i \omega} \vec{E}. \tag{311}$$

The electrostatic potential in the metal obeys Coulomb's law $\Delta\phi = 0$. It means that

$$\Phi(z, q) = \Phi(0, q) \exp(-qz) \tag{312}$$

and the electrical field on the metal surface is $E_z(0, q) = q\Phi(0, q)$. Combining these relations together with (308) we obtain for the spectrum

$$\Omega(q) = \omega_p \sqrt{q/8\pi\tilde{C}} \tag{313}$$

and for the deviation of the tunnel current

$$V \frac{\partial^2 I}{\partial V^2} = \mathrm{sign}(V) \frac{16\pi\tilde{C}}{R_K \omega_p} (V/\omega_p)^3. \tag{314}$$

We write 'sign' here in order to emphasize the non-analytical behavior of this deviation. As a rule the dimensionless factor on the right hand side of (314) is less than unity. For a reasonable thickness of the tunnel barrier of a few atomic lengths this coefficient is of the order E_F/ω_p. This ratio is less than unity for most metals.

One may note that the velocity of the plasma excitations increases with decreasing frequency so that at lower frequencies we should take into account relativistic effects omitted in the previous consideration. Actually, an electrical current in the plasma produces a magnetic field and an alternating magnetic field induces an electric one. Due to this fact the field penetration depth is restricted by the value of c/ω_p. By taking this fact into account the spectrum of electromagnetic excitations is given by

$$\Omega(q) = \omega_p \sqrt{\frac{q^2}{8\pi\tilde{C}\sqrt{(\omega_p/\mathrm{c})^2 + q^2}}} \tag{315}$$

which describes the crossover between high frequency plasma excitations and low frequency Swihart waves. In terms of voltage this crossover occurs at

$$eV \sim \omega_p/\hbar\sqrt{\omega_p/c\tilde{C}} \tag{316}$$

which is about 100 mV for aluminium. At lower voltages, we obtain for the deviation

$$R_T\delta(\frac{\partial I}{\partial V}) = 16\pi V \frac{e^3}{\hbar^2 v_{\rm sw}^2 \tilde{C}} \tag{317}$$

where the velocity of the Swihart waves is $v_{\rm sw} = \sqrt{\omega_p/8\pi c\tilde{C}}$.

Usually, the electrodes are thin metallic films. For low voltages when for a typical q we have $qd \simeq 1$, where d is the film thickness, the previous results should be modified. The velocity of electromagnetic waves in this system can be compared with the speed of light, so that Eq. (310) should also be modified. We present the result for the case $qd \ll 1$ [25]

$$v_1 = \omega_p^2 d/8\pi\tilde{C}, \qquad v = v_1/\sqrt{1+(v_1/c^*)^2},$$

$$\mathrm{Re}Z(\omega) = \frac{\pi}{\tilde{C}} \int \frac{{\rm d}^2 q}{(2\pi)^2} \delta(\omega - vq) \frac{1}{1+(v_1/c^*)^2} \tag{318}$$

$$R_T\delta(\frac{\partial I}{\partial V}) = 16\pi V \frac{e^3}{\hbar^2 v_1^2 \tilde{C}}.$$

Here, v_1 is the speed of electromagnetic excitations without taking into account relativistic effects, v is the real speed, c^* is the speed of light in the insulating layer and the Lorentz factor describes the relativistic effects.

There is a variety of different regimes of field propagation in this large area junctions and we will not review this matter here. Some of the cases were described in Refs. [25] and [52]. Now we return to small-area junctions which are mostly discussed in this book and consider the influence of finite size on the whole picture described above.

A.7. Junction-localized oscillations

We now concentrate our attention on the 100-10000 angstroms size junction that are under experimental investigation now. The time of electromagnetic excitation propagation along the whole junction corresponds to a voltage in the region from ten to several hundred microvolts. The excitations are practically not damped in this frequency region.

There are basically two ways to connect the junction with the contact wires: first, to interrupt the thin wire by the tunnel barrier; second, to form this barrier by overlapping of the two films. The junction area corresponds to the wire cross-section in the first case and may be much larger than the latter in the second case.

Consider now the electromagnetic excitation propagation along the junction. The first and the second case correspond to two-dimensional and three-dimensional geometries of the previous subsection, respectively. In the simplest case of a rectangular junction the boundary conditions permit only discrete values of $\vec{q}$. The modes with $\vec{q} \neq 0$ are junction-localized oscillations that can be excited by electron tunneling. The frequencies of these oscillations are simply $\omega_m = \Omega(\vec{q}_m)$, where $\vec{q}_m$ are permitted values of the wave vector. These frequencies evidently remain discrete regardless of the actual junction form.

We can make use of the formulae in the previous subsection for the effect of excitations on the junction I-V curve if we replace the integration over $\vec{q}$ by the summation over discrete values $\vec{q}_m$:

$$\int \frac{\mathrm{d}^2 q}{(2\pi)^2} \;\rightarrow\; \frac{1}{S} \sum_{\vec{q}_m} \tag{319}$$

where S is the junction area. This yields the very simple expression for the effect at $eV \gg k_B T$

$$\frac{\partial^2 I}{\partial V^2} R_T = E_c \sum_{\vec{q}_m} \frac{1}{V} \delta\big(eV - \hbar\omega(\vec{q}_n)\big). \tag{320}$$

Here, $E_c = e^2/2C$ is the charging energy of the whole tunnel capacitor having the capacitance C. Eq. (320) is valid only if the speed of excitations is much smaller than the speed of light. Otherwise it should be multiplied by the Lorentz factor of Eq. (319). Thus, every oscillation makes a fingerprint on the I-V curve at the corresponding voltage.

It is worth to compare these results with the predictions of the phenomenological theory. According to that one at voltages larger than the inverse time of discharge through the leads attached to the junction we have a linear I-V curve with offset E_c/e. From (320) one may see that it is valid only if the voltage does not exceed the lowest energy of the oscillation spectrum. At the voltages which are corresponding to the oscillation energies the junction conductance is jumping. The magnitude of the jump can be determined from the fact that the I-V curve is gaining the additional offset E_c/e at this point. If this frequency is n-fold degenerate, the additional offset is nE_c/e. At high voltages a large number of oscillations can be excited by the tunneling electron and the asymptotic law is determined by the appropriate expression for the infinite-area junction.

If we take into account the oscillation damping and/or the finite temperature, the jumps gain a finite width. The appropriate expressions one may find in Ref. [25]. Therefore the offset obviously increases with increasing voltage. It shows how the applicability of the phenomenology is restricted.

Now we finish by opening

A.8. A gateway into networks

In the previous subsection we have said nothing about the mode with $\vec{q} = 0$. The existence of this mode is a straightforward consequence of charge conservation: if the junction is included into some electrical circuit there should be a possibility for charge to go out of the junction, and modes with $\vec{q}_m \neq 0$ do not provide this possibility. In this zero mode the voltage difference is constant along the whole junction. Therefore, the mode dynamics does not depend upon the nearest junction environment but is determined by the external circuit. The phenomenological equations derived in the previous sections are completely valid for this dynamics, so that there is a gateway from the microscopic world to the networks created by human beings.

References

[1] See e.g. M. Tinkham, *Introduction to Superconductivity* (McGraw-Hill, New York, 1975).

[2] K. K. Likharev, IBM J. Res. Dev. **32**, 144 (1988).

[3] D. V. Averin and K. K. Likharev, J. Low Temp. Phys. **62**, 345 (1986).

[4] D. V. Averin and K. K. Likharev, in: *Mesoscopic Phenomena in Solids*, ed. by B. L. Altshuler, P. A. Lee, and R. A. Webb (Elsevier, Amsterdam, 1991), Chap. 6.

[5] G. Schön and A. D. Zaikin, Phys. Rep. **198**, 237 (1990).

[6] U. Geigenmüller and G. Schön, Europhys. Lett. **10**, 765 (1989).

[7] Yu. V. Nazarov, Pis'ma Zh. Eksp. Teor. Fiz. **49**, 105 (1989) [JETP Lett. **49**, 126 (1989)].

[8] M. H. Devoret, D. Esteve, H. Grabert, G.-L. Ingold, H. Pothier, and C. Urbina, Phys. Rev. Lett. **64**, 1824 (1990).

[9] S. M. Girvin, L. I. Glazman, M. Jonson, D. R. Penn, and M. D. Stiles, Phys. Rev. Lett. **64**, 3183 (1990).

[10] H. Grabert, G.-L. Ingold, M. H. Devoret, D. Esteve, H. Pothier, and C. Urbina, Z. Phys. B **84**, 143 (1991).

[11] G. Schön, Phys. Rev. B **32**, 4469 (1985).

[12] H. Haken, Rev. Mod. Phys. **47**, 67 (1975).

[13] A. O. Caldeira and A. J. Leggett, Ann. Phys. (N.Y.) **149**, 374 (1983).

[14] A. A. Odintsov, Zh. Eksp. Teor. Fiz. **94**, 312 (1988) [Sov. Phys. JETP **67**, 1265 (1988)].

[15] D. V. Averin and A. A. Odintsov, Phys. Lett. A **140**, 251 (1989).

[16] For a proof see e.g. W. H. Louisell, *Quantum Statistical Properties of Radiation* (Wiley, New York, 1973).

[17] R. Kubo, Rep. Prog. Phys. **29**, 255 (1966).

[18] H. Grabert, P. Schramm, and G.-L. Ingold, Phys. Rep. **168**, 115 (1988).

[19] P. Minnhagen, Phys. Lett. A **56**, 327 (1976).

[20] G.-L. Ingold and H. Grabert, Europhys. Lett. **14**, 371 (1991).

[21] G. Falci, V. Bubanja, and G. Schön, Europhys. Lett. **16**, 109 (1991); Z. Phys. B **85**, 451 (1991).

[22] M. H. Devoret, D. Esteve, H. Grabert, G.-L. Ingold, H. Pothier, and C. Urbina, Physica B **165&166**, 977 (1990).

[23] M. Abramowitz and I. A. Stegun, *Handbook of Mathematical Functions* (Dover, New York, 1972).

[24] S. V. Panyukov and A. D. Zaikin, J. Low Temp. Phys. **73**, 1 (1988).

[25] D. V. Averin and Yu. V. Nazarov, Physica B **162**, 309 (1990).

[26] D. V. Averin, Yu. V. Nazarov, and A. A. Odintsov, Physica B **165&166**, 945 (1990).

[27] Yu. V. Nazarov, preprint (1991).

[28] N. R. Werthamer, Phys. Rev. **147**, 255 (1966).

[29] A. Maassen van den Brink, A. A. Odintsov, P. A. Bobbert, and G. Schön, Z. Phys. B **85**, 459 (1991).

[30] I. O. Kulik and R. I. Shekhter, Zh. Eksp. Teor. Fiz. **68**, 623 (1975) [Sov. Phys. JETP **41**, 308 (1975)].

[31] H. Grabert, G.-L. Ingold, M. H. Devoret, D. Esteve, H. Pothier, and C. Urbina,

in: *Proceedings of the Adriatico Research Conference on Quantum Fluctuations in Mesoscopic and Macroscopic Systems*, ed. by H. A. Cerdeira, F. Guinea Lopez, and U. Weiss (World Scientific, Singapore, 1991).

[32] G.-L. Ingold, P. Wyrowski, H. Grabert, Z. Phys. B **85**, 443 (1991).

[33] A. A. Odintsov, G. Falci, and G. Schön, Phys. Rev. B **44**, 13089 (1991).

[34] K. K. Likharev, IEEE Trans. Magn. **23**, 1142 (1987).

[35] K. Mullen, E. Ben-Jacob, R. C. Jaklevic, and Z. Schuss, Phys. Rev. B **37**, 98 (1988).

[36] B. Laikhtman, Phys. Rev. B **43**, 2731 (1991).

[37] J.-C. Wan, K. A. McGreer, L. I. Glazman, A. M. Goldman, and R. I. Shekhter, Phys. Rev. B **43**, 9381 (1991).

[38] T. A. Fulton and G. J. Dolan, Phys. Rev. Lett. **59**, 109 (1987).

[39] See e.g. J. L. Potter and S. Fich, *Theory of Networks and Lines* (Prentice-Hall, London, 1963).

[40] B. L. Al'tshuler and A. G. Aronov, in: *Electron-Electron Interaction in Disordered Systems*, ed. by A. L. Efros and M. Pollak (Elsevier, Amsterdam, 1985).

[41] Yu. V. Nazarov, Zh. Eksp. Teor. Fiz. **95**, 975 (1989) [Sov. Phys. JETP **68**, 561 (1990)].

[42] Yu. V. Nazarov, Phys. Rev. B **43** 6220 (1991).

[43] E. H. Hauge and J. A. Støvneng, Rev. Mod. Phys. **61**, 917 (1989).

[44] E. Turlot, D. Esteve, C. Urbina, J. M. Martinis, M. H. Devoret, S. Linkwitz, and H. Grabert, Phys. Rev. Lett. **62**, 1788 (1989).

[45] Yu. V. Nazarov, Sol. St. Comm. **75**, 669 (1990).

[46] J. Appelbaum, Phys. Rev. Lett. **17**, 91 (1966).

[47] J. M. Rowell, in: *Tunneling Phenomena in Solids*, ed. by E. Burstein and S. Lundquist (Plenum, New York, 1969).

[48] B. L. Al'tshuler and A. G. Aronov, Zh. Eksp. Teor. Fiz. **77**, 2028 (1979) [Sov. Phys. JETP **50**, 968 (1979)].

[49] R. C. Dynes and J. P. Garno, Phys. Rev. Lett. **46**, 137 (1981).

[50] Y. Imry and Z. Ovadyahu, Phys. Rev. Lett. **49**, 841 (1982).

[51] B. L. Al'tshuler, A. G. Aronov, and A. Yu. Zyuzin, Zh. Eksp. Teor. Fiz. **86**, 709 (1984) [Sov. Phys. JETP **59**, 415 (1984)].

[52] Yu. V. Nazarov, Fiz. Tverd. Tela **31**, 188 (1989) [Sov. Phys. Solid State **31**, 1581 (1989)].

Chapter 3

Transferring Electrons One By One

D. ESTEVE

Groupe Quantronique, Service de Physique de l'Etat Condensé,
Centre d'Etudes de Saclay, 91191 Gif-sur-Yvette, France

1. Introduction

This chapter describes experiments in which the transfer of charge through small capacitance tunnel junction devices is controlled at the single electron level. From a fundamental point of view, these devices provide an insight into the basic concepts behind single charge tunneling such as global rules discussed in Chap. 2. More particularly, the detailed understanding of the transfer accuracy involves recently discovered effects, such as co-tunneling through several junctions discussed in Chap. 6. From a practical point of view, these devices could lead to metrological applications such as the accurate measurement of the fine structure constant α and the realisation of a current standard. We describe below three devices:

1) The single electron box [1], which provides the simplest example of the macroscopic quantization of the charge on an "island" of a small junction circuit on an integer number of electron charges. (We call an island a metallic electrode that electrons can enter or leave only by tunneling through a junction.) The charge on this island can thus be controlled at the single electron level with a gate voltage. This circuit is the basic building block for more elaborate designs.

2) The single electron turnstile [2], which was the first device to produce a current clocked electron by electron with an external periodic signal. This device requires a bias voltage and a gate voltage upon which the rf clock signal is superimposed. The direction of the current is determined by the sign of the bias voltage, and one electron is transferred through the circuit for each cycle of the rf signal. The transfer accuracy is limited mainly by electron heating and by co-tunneling. Electron heating results from the conversion of electrostatic energy into electron kinetic energy in the device, thus causing unwanted thermally activated tunneling events which degrade the regularity of the charge transfer. Co-tunneling across two or more junctions is also the source of some unwanted transitions, further degrading the control of electron transfer.

Single Charge Tunneling. Edited by H. Grabert and
M.H. Devoret, Plenum Press, New York, 1992

3) The single electron pump [3], which also produces a current clocked by an external periodic source. In contrast with the turnstile, this device requires two gate voltages upon which properly phase-shifted rf signals of the same frequency are superimposed. Moreover, the direction of the current is not determined by the sign of the bias voltage but by the phase-shift between the rf gate voltages. At low frequencies, energy is exchanged between the gate sources and the bias source, without any energy transfer to the electrons: the pump is therefore a reversible device. As a consequence, there is no heating at low frequencies, so the charge transfer accuracy of the pump is mainly limited by co-tunneling.

Following a detailed description of these devices, we shall address the more general question of the metrological applications of such charge transferring devices.

2. Basic concepts of small junction circuits

Before entering into the detailed description of these experiments, we first recall the main results important to the understanding of charge transferring devices.

2.1. Electrons and electronics

In standard electronic devices, the discreteness of charge carriers does not usually show up at the macroscopic level. Why is it so? Let us examine what happens in the charging of a capacitor with a battery. The charge on the surface of a metal plate results from the very small displacement of the electrons with respect to the fixed ions of the plate. Therefore, the charge transferred from the battery to the plate is a continuous variable: it is not constrained to be an integer number of electrons. Electrons do not show up at the macroscopic level not only because their charge is small but principally because they form a long range correlated liquid, which behaves as a continuous fluid. A charge corresponding to an integer number of electrons can however be placed on the capacitor plate if one interrupts the connection to the battery by opening a switch placed between the plate and the battery. Just as an isolated ion carries a well defined number of electrons, the charge of a piece of metal disconnected from any charge reservoir is an integer multiple of the electron charge and remains constant. Now, there is an intermediate situation in which a piece of metal contains at a given instant an integer number n of electrons which can be varied: If the switch is replaced by a tunnel junction of capacitance C and tunnel resistance R_T, such that $R_T \gg R_K = h/e^2 \simeq 25.8$ kΩ, then the charge passed through the junction will always be an integer number of electron charges. The discreteness of the charge passed through a junction is at the origin of the junction's shot noise, a phenomenon which does not occur in a metallic resistor. The electrode connecting the junction to the capacitor then forms an almost isolated island and contains, neglecting effects of order R_K/R_T, an integer number of electrons: it constitutes a "box" for electrons. We have called the complete circuit, shown in Fig. 1a, a single electron box. The gate capacitance C_S is connected to a bias voltage U. At zero temperature, the number n of extra electrons in the single electron box, (i.e. electrons whose charge is not canceled by the background of the positive ions), is such that the electrostatic energy of the complete circuit, which depends on the gate voltage

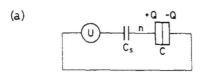

single electron box

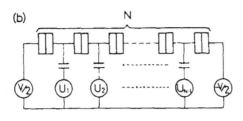

linear array

Figure 1. (a) "Electron box" circuit consisting of a junction of capacitance C in series with a true capacitor C_S and a voltage source U. The intermediate electrode between the junction and the capacitor forms an "island" which contains an integer number n of extra electrons. (b) Linear array consisting of N junctions in series. The array is voltage biased and the intermediate islands are coupled to gate voltages.

U, is at a minimum. At non-zero temperatures, n will have negligible fluctuations if the charging energy $e^2/2(C + C_S)$ of the island with one electron is much larger than the characteristic energy $k_B T$ of thermal fluctuations. This quantization of the charge on an isolated island is at the root of all the charge transferring devices described here. These more elaborate circuits contain more than one island and their states are described by more than one integer, so that their evolution allows richer behaviour than the simple addition and removal of single electrons to and from an island through one junction.

2.2. Charge configurations

Consider the linear array shown in Fig. 1b. It consists of N tunnel junctions and $(N - 1)$ islands. The islands are connected to gate voltages through capacitors and the end junctions are connected to a bias voltage. For example, the $N = 2$ case is simply the SET transistor with which Fulton and Dolan clearly demonstrated the existence of single electron charging effects [4]. The state of the array at a given instant of time can be described by its charge configuration, given by the set $(n_1, n_2, \ldots, n_{N-1})$ of the numbers of extra electrons on the islands and by the sum n_0 of electrons having passed through the end junctions. Two configurations that can be transformed from one into the other by a single tunnel event on one junction are called neighbouring configurations.

Using basic electrostatics, one readily shows that the island charges $(-n_i e)$ are related to the island potentials V_j by the following set of linear relations

$$\tilde{Q}_i - n_i e = \sum_{j=1}^{N-1} C_{ij} V_j. \tag{1}$$

For the islands ($i = 2, 3, \ldots, N - 2$), the bias charges $\tilde{Q}_i$ are simply equal to $\tilde{Q}_i = C_i U_i$ where C_i is the gate capacitance of island i. For the first island with an outer junction of

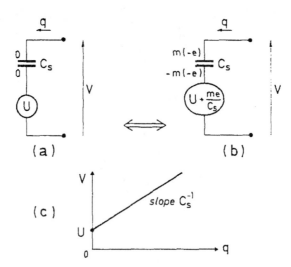

Figure 2. (a) Capacitor C_S in series with a voltage source U. (b) Same capacitor C_S, precharged with m electrons, in series with a voltage source $(U + me/C_S)$. (c) Charge-voltage characteristic common to both circuits. This electrical equivalence proves that the junction charge in the single electron box can only be a function of $(C_S U - ne)$.

capacitance C_1^J, one has $\tilde{Q}_1 = C_1 U_1 + C_1^J(V/2)$. Similarly, one has $\tilde{Q}_{N-1} = C_{N-1} U_{N-1} + C_N^J(-V/2)$. The elements C_{ij} of the capacitance matrix are simple linear combinations of junction and gate capacitances. The electrostatic energy of the whole circuit, including the voltage sources, is given by the following expression

$$E(n_0, n_1, n_2, \ldots, n_{N-1}) = \frac{1}{2} \sum_{i,j=1}^{N-1} C_{ij}^{-1}(\tilde{Q}_i - n_i e)(\tilde{Q}_j - n_j e) \ - \frac{n_0 eV}{2} \qquad (2)$$

where the C_{ij}^{-1} are the elements of the inverse capacitance matrix. In the absence of voltage sources, this expression is the usual expression for the energy of a set of charged conductors. At zero temperature, a configuration can only evolve to a configuration of lower energy. The energy spectrum of the configurations is thus particularly important. At zero bias voltage V, the energy spectrum has a lower bound, and there is a minimum energy configuration (possibly degenerate). At finite bias voltages, there is no minimum energy configuration, since one always can decrease the energy by changing n_0. A finite current can thus pass through the device. Let us first consider the simple example of the single electron box, for which there is a minimum energy configuration.

The minimum energy configuration for the single electron box as a function of the biasing charge $\tilde{Q} = C_S U$ can be obtained without any calculation. The integer function $n_{\min}(\tilde{Q})$ for which the energy is a minimum obeys the symmetry relations

$$n_{\min}(\tilde{Q}) = -n_{\min}(-\tilde{Q}), \qquad (3)$$

$$n_{\min}(\tilde{Q} + pe) = n_{\min}(\tilde{Q}) + p, \qquad (4)$$

where p is an integer. Equation (11) is trivial while Eq. (2) results from the complete electrical equivalence of the two circuits shown in Fig. 2. Using these two relations and the fact that $n_{\min}(\tilde{Q})$ is a monotonic function of $\tilde{Q}$, one shows that $n_{\min}(\tilde{Q})$ is the

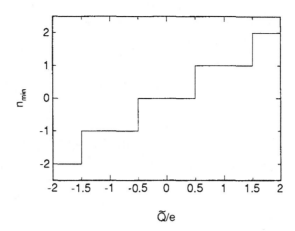

Figure 3. Variations of the integer n_{min} for which the single electron box energy is minimal, as a function of the bias charge $\tilde{Q} = C_S U$.

staircase function shown in Fig. 3. This result is readily recovered using the general expression (2) in the case of a single island.

Although there is no absolute minimum energy configuration for a linear array at finite bias voltages, two configurations differing only in n_0 are not neighbours, and a transition between them can only occur by co-tunneling through all N junctions, a process with a very low rate if N is large enough and if the junction tunnel resistances are much larger than R_K. A configuration can thus be locally stable in the sense that all its neighbouring configurations have a higher energy. If not even locally stable configurations exist, a current due to single tunneling events will flow through the device (the threshold bias voltage above which this type of conduction takes place is called the Coulomb gap). Hence, in order to transfer precisely one electron through a device, one has to force the state of the device to go through a particular sequence of neighbouring configurations called a transfer sequence. This motion in configuration space is achieved by applying gate voltages which vary with time so that each configuration in the transfer sequence becomes in turn more stable than the preceeding one. We will see in later sections the transfer sequences used for the turnstile and the pump.

2.3. Tunneling out of an unstable configuration

The tunneling transition from a configuration to a neighbouring one occurs at a rate which depends on the electromagnetic environment of the circuit. For a device which does not contain any large series resistance located very close to the junctions, the few millimiters of leads on the chip which connect the device to the bias circuitry provide an electromagnetic environment equivalent to that of a transmission line with a characteristic impedance of the order of the vacuum impedance $Z_V \simeq 377\ \Omega$, which is much smaller than R_K. Moreover, the connection pads provide a voltage bias because their capacitance is much larger than the junction capacitances. The theory of the influence of the environment [5] shows that under these circumstances the global rules for calculating transition rates almost perfectly apply (see Chap. 2). According to these rules, tunneling occurs without any excitation of the modes of the electromagnetic environment, just as a nucleus imbedded in a solid emits a photon without exciting the crystal phonons

(Mössbauer effect). The tunneling rate Γ_j of electrons across the junction j only depends on the electrostatic energy difference ΔE_j between the initial and final configurations [6]

$$\Gamma_j = \frac{\Delta E_j}{e^2 R_{Tj}(1 - \exp(-\Delta E_j/k_B T))},$$ (5)

where R_{Tj} is the tunnel resistance of junction j and T is the temperature. At zero temperature, only transitions which decrease the electrostatic energy are possible and the rate is

$$\Gamma_j = \frac{\Delta E_j}{e^2 R_{Tj}}.$$ (6)

The energy difference ΔE_j depends on the whole circuit configuration (see equation (2)). It can be expressed as a linear function of the charge Q_j on the junction capacitance that one calculates using classical electrostatics taking into account all voltage sources and the island charges:

$$\Delta E_j = \frac{e}{C_j}\left(Q_j - Q_j^c\right)$$ (7)

where C_j denotes here the capacitance of junction j and where the critical charge Q_j^c involves the total circuit capacitance C_j^{ext} seen from the junction j:

$$Q_j^c = \frac{e}{2}\left(\frac{C_j}{C_j + C_j^{\text{ext}}}\right).$$ (8)

Eqs. (6) and (7) show that, at zero temperature, an electron will tunnel only if the junction charge Q_j exceeds the critical charge Q_j^c. Note that the critical charge of the junction is always less than half an electron charge.

This calculation does not imply that the capacitor C_j really carries a charge Q_j. The charge on the capacitor is not simply a c-number, but an operator whose quantum average is equal to Q_j in the ground state of the complete electromagnetic circuit imbedding the junction. The quantum fluctuations around this average depend on the impedance of the biasing circuit: they decrease when the impedance increases. These fluctuations are much larger than e in the case of the low impedance environment found in the typical set-ups for multijunction circuits, but single charge effects can still occur because the integer charges on the islands can still have negligible quantum fluctuations.

When an electron tunnels across junction j, the junction charge Q_j is not simply changed by e because the modification of the charge configuration induces a rearrangement of the charge in the whole circuit. After the tunnel event, the junction charge is $Q_j^{\text{after}} = Q_j^{\text{before}} \mp 2Q_j^c$, while the charge displaced through the rest of the circuit is $\pm(e - 2Q_j^c)$.

In a charge transferring device, the charge of a junction evolves, between tunneling events, under the influence of the time-dependent control voltages. If at time t_0 the junction j reaches its critical charge and if no other transitions are possible, the probability

$P(t)$ that an electron has tunneled across this junction before time t is given at zero temperature by

$$P(t) = 1 - \exp\left[\int_{t_0}^{t} \frac{(Q_j - Q_j^c(\tau))}{eR_{Tj}C_j} d\tau\right] . \tag{9}$$

When the junction charge $Q_j(\tau)$ increases at a constant rate $\dot{Q}_j$, this expression reduces to

$$P(t) = 1 - \exp\left[-\left(\frac{\Delta t}{2\tau_j}\right)^2\right] \tag{10}$$

where $\Delta t = t - t_0$ and $\tau_j = (eR_{Tj}C_j/\dot{Q}_j)^{1/2}$ is a characteristic transition time. The stochastic nature of tunneling thus limits the operating frequency f of any single charge transferring device to the upper value $f^{\max} \approx 10^{-2}/(R_T C)$ if a transfer accuracy of the order of 10^{-8} is desired, independently of the detailed design of the device.

2.4. Co-tunneling out of a locally stable configuration

The phenomenon of co-tunneling, discovered by Averin and Odintsov [7] and discussed in Chap. 6, is a coherent tunneling of p electrons through p junctions, involving $(p-1)$ intermediate states. Co-tunneling thus enables a locally stable configuration to decay to any lower energy configuration. The co-tunneling sequence can be time ordered in $p!$ different ways whose corresponding transition amplitudes add coherently. The co-tunneling rate has been calculated perturbatively, at zero temperature, in the case of the complete transfer of a single electron through a linear array voltage biased below the Coulomb gap, where, in the absence of co-tunneling, no current would pass through the array because the configuration would find a locally stable configuration. In an array of two junctions in series, there are only two possible sequences, and the calculation can be carried out easily. The resulting expression diverges when the bias voltage approaches the gap voltage because the energy of one intermediate configuration becomes equal to the initial configuration energy. This divergence is not physical and is a consequence of the perturbative approach. In the case of longer arrays, the calculation has only been performed in the case of identical junctions with negligible gate capacitances, biased at voltages much smaller than the Coulomb gap, and with all the gate voltages set to zero. With these restrictions, the energy of an intermediate configuration depends only on the number of steps needed to reach it, and the sum of all the amplitudes of the various tunneling paths can be calculated. The co-tunneling rate $\Gamma_c(N, eV)$ is a function of the dissipated energy eV given by the following expression, valid for $eV \ll e^2/2C$

$$\Gamma_c(N, eV) = \frac{2\pi}{\hbar} \frac{e^2}{2C} \frac{N^{2N}}{(2N-1)!(N-1)!^2} \left(\frac{R_K}{4\pi^2 R_T}\right)^N \left(\frac{eV}{e^2/2C}\right)^{2N-1} . \tag{11}$$

The exponent N of the R_K/R_T factor is equal to the number of tunneling events necessary to transfer an electron through the array. The power $(eV)^{2N-1}$ which enters in the rate reflects the increase in the number of accessible electronic states when the voltage is increased. In order to be more quantitative, let us consider an array of $N = 4$ 1fF junctions. Expression (9) yields the following result

$$\Gamma_c \, [\text{Hz}] \, (4, eV \, [\mu\text{V}]) = (2.5 \times 10^{-3}) \frac{V^7 \, [\mu\text{V}]}{R_T^4 \, [\text{k}\Omega]} \ . \tag{12}$$

The calculation of the co-tunneling rate has recently been extended to finite temperatures. The result can be cast in a form similar to expression (11) with $(eV)^{2N-1}$ replaced by a polynomial $P(eV, k_B T)$ of the same total degree. However, this calculation does not include the possibility of combined thermally activated single tunnel events and co-tunneling transitions across a restricted number of junctions, but one can show that at low enough temperatures all these combined transitions are negligible in comparison with co-tunneling through the entire array.

The calculation can also be extended to the more general case of a co-tunneling event for which the charge configuration of the intermediate islands is not the same after the co-tunneling event. This case is particularly important for single charge transferring devices, and one example will be discussed later.

2.5. Coexistence of tunneling and co-tunneling

Co-tunneling has been considered only in the case where all the intermediate configuration energies lie above the initial state energy, taken as the zero for energy, and the perturbative approach used for the calculation gives a divergent result when any of these energies tends to zero. When the energy of one intermediate configuration is negative, a "real" transition from the initial configuration to this intermediate configuration becomes possible. This transition can occur by a usual tunnel event if this configuration is a neighbour of the initial one, or by co-tunneling if it is not. One then has to consider more than one decay channel for the initial configuration. This complex regime, which involves competing decay channels, is not completely understood. In fact, the concept of different decay paths is itself not valid: in quantum mechanics, one has to compute the transition amplitudes to all possible final states; whether an intermediate configuration can be reached by a real process or not is a meaningless question! The SET transistor [4] provides the simplest example of such a situation. The SET transistor, discussed in Chap. 2, consists of two junctions in series and has one gate capacitor C_0 to which a voltage U_0 is applied. It provides a very sensitive electrometer because the current I through the device is modulated by the charge $\tilde{Q} = C_0 U_0$ on the gate with periodicity e. When the device is biased at or above the Coulomb gap, an electron can be transferred by a sequence of two tunnel events or by co-tunneling across both junctions: co-tunneling and single events "coexist". Co-tunneling is of special importance here because it puts a limit on the maximal transfer coefficient $dI/d\tilde{Q}$ of the device. As R_T decreases, this coefficient increases, and for $R_T \approx R_K$ passes through a maximum whose value is not known. At that maximum, the sensitivity of the electrometer is optimal. If the gate capacitance is close to the junction capacitance C, the current noise through the electrometer is equivalent to a charge noise on the measured charge $\tilde{Q}$ with a spectral density $S_Q \approx \hbar C$. Well above the Coulomb gap, we have shown that the probability that a single electron has passed through the SET transistor is correctly given by the classical model of two incoherent transitions in sequence: the co-tunneling rate therefore

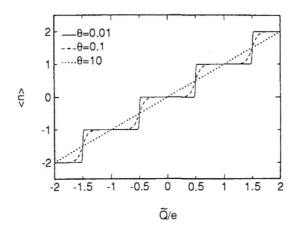

Figure 4. Variations of the island average electron number $\langle n \rangle$ as a function of the bias charge $\tilde{Q} = C_s U$ for $\theta = k_B T(C_s + C)/e^2 = 0.01$ (solid line), 0.1 (dashed line) and 10 (dotted line).

decreases above the gap and finally vanishes. This result suggests that one can perhaps neglect co-tunneling when a transition to any intermediate configuration is possible.

3. Single electron box

3.1. Average charge in the single electron box

At zero temperature, the charge $q = -ne$ of the single electron box island corresponds to the integer number n of electrons for which the electrostatic energy is minimal. At finite temperatures, this number can fluctuate, and one has to compute the thermodynamic average $\langle n \rangle$ by performing the Boltzmann weighting of the possible configurations according to their energies. The electrostatic energy $E(n)$ of a configuration is for this simple one island circuit:

$$E(n) = \frac{\left(ne - \tilde{Q} \right)^2}{2(C_S + C)} \tag{13}$$

where $\tilde{Q} = C_S U$. The energy $E(n)$ is minimum for the integer $n_{\min}$ closest to $\tilde{Q}/e$, as shown in Fig. 3. The average value $\langle n \rangle$ is given by

$$\langle n \rangle = \frac{\sum\limits_{n=-\infty}^{\infty} n \, \exp\left(-\frac{E(n)}{k_B T}\right)}{\sum\limits_{n=-\infty}^{\infty} \exp\left(-\frac{E(n)}{k_B T}\right)} \tag{14}$$

and is a function of $\tilde{Q}$ which depends on the reduced parameter $\theta = k_B T(C_S + C)/e^2$. The dependence of $\langle n \rangle$ on $\tilde{Q}$ for different values of θ are shown in Fig. 4.

The average junction charge $\langle Q \rangle$ is simply related to $\langle n \rangle$ by

$$\langle Q \rangle = \frac{C}{C + C_S}(\tilde{Q} - \langle n \rangle e). \tag{15}$$

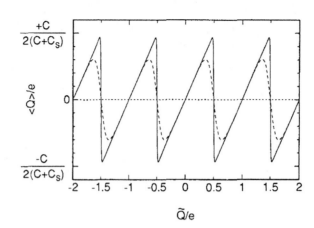

Figure 5. Variations of the average junction charge $\langle Q \rangle$ as a function of the bias charge $\tilde{Q}$ for the same set of values of the reduced temperature θ as in Fig. 4.

The variations of $\langle Q \rangle$ with $\tilde{Q}$ are shown in Fig. 5 for the same set of values of θ as in Fig. 4. One predicts a sawtooth pattern which is progressively more rounded as the temperature increases.

The sawtooth pattern of Fig. 5 is reminiscent of the single electron tunneling oscillations [8] predicted for a junction connected to a perfect current source I which completely suppresses the quantum fluctuations of charge. In this case, local rules apply and tunneling is blocked if the junction charge is less than $e/2$. After a tunnel event, the charge increases at a rate I and another tunnel event occurs only after it has exceeded $e/2$. During the event, the charge is reduced by e and starts to increase again immediately afterwards. This charging-discharging cycle repeats itself at a frequency $f = I/e$. The SET oscillations are a dynamic phenomenon which results from the combined effect of the blockade of tunneling and of the slow recharging of the junction.

In the single electron box, the oscillations of $\langle Q \rangle$ versus $\tilde{Q}$ are an equilibrium phenomenon, which is not based on the blockade of tunneling by the environment but on the quantization of the island charge. The junction charge can have large quantum fluctuations, but the integer island charge cannot. When $\tilde{Q}$ increases, $\langle Q \rangle$ increases until it becomes energetically favorable to let an electron enter the island. However, the steep side of the sawtooth pattern does not result from a single tunnel event. These sawtooth oscillations are equilibrium SET oscillations.

3.2. Measurement of the junction charge $\langle Q \rangle$

The measurement of the junction charge cannot be performed using a conventional capacitively coupled electrometer because its input capacitance would be much larger than the junction capacitance and is at any rate not sensitive enough. We have used a SET transistor electrometer [4] which has a sensitivity well below one electron, as well as a very small input capacitance. Of course, one has to place the electrometer close to the single electron box in order to avoid too much stray capacitance. We have thus fabricated a single electron box and a SET electrometer on the same chip, using e-beam nanolithography and shadow evaporation. The schematic representation of the complete circuit is shown in Fig. 6a. Note that the box junction consists of two junctions

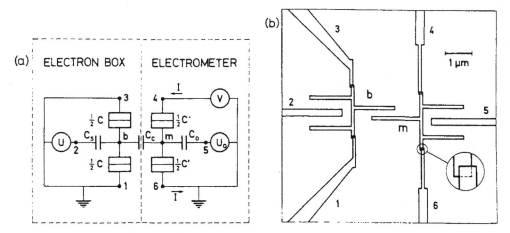

Figure 6. (a) Schematic representation of experimental set-up. An electron box with two junctions in parallel is coupled to a SET transistor used as an electrometer. (b) Electron beam lithography implementation of circuit shown in (a). Superfluous electrodes resulting from the use of the shadow mask evaporation technique have been omitted for clarity.

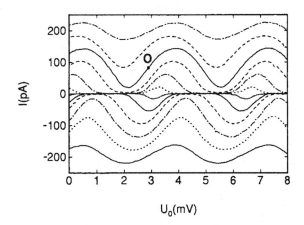

Figure 7. Electrometer current I versus electrometer gate votage U_0 for a set of values of bias voltage V separated by 25 μV. Temperature is 20 mK.

which are connected in parallel after fabrication; this is done to allow for a check of the junction tunnel resistance. The actual lithographic pattern is shown in Fig. 6b, where the numbers and letters labeling the aluminum electrodes refer to the corresponding nodes of the circuit schematic shown in Fig. 6a. The large-scale shape of the electrodes and of their guards were designed to minimize cross-talk capacitance. The coupling capacitance C_c between the box and the electrometer has been chosen to get an appreciable change of the current I through the electrometer when the single electron box charge varies, without disturbing the behaviour of the single electron box too much when electrons randomly enter and leave the electrometer island.

We first studied the electrometer itself in order to optimize its sensitivity. The device was thermally anchored to a dilution refrigerator, with all lines carefully filtered, and a 0.5 T magnetic field was applied to keep the aluminum of the junctions in the normal state. A series of $I - U_0$ characteristics of the electrometer is shown in Fig. 7 for different bias voltages V. These characteristics have a period e/C_0 in the gate voltage U_0; the maximum modulation depth is obtained for a bias V close to the Coulomb gap. At the

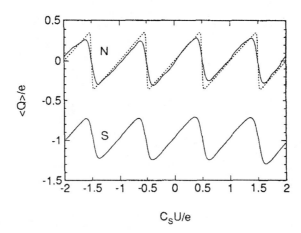

Figure 8. Solid lines: variations of the average junction charge in the normal state (curve labeled "N") and in the superconducting state (curve labeled "S") at 20 mK. Dashed line: theoretical prediction [Eqs. (14) and (15)].

optimal working point O, the transfer coefficient $dI/d(C_0U_0)$ is equal to 600 pA/e, and we have measured an equivalent charge noise of $10^{-4}e/\sqrt{\text{Hz}}$ at frequencies around 1 kHz. We attribute the lack of symmetry about the horizontal axis as due to an asymmetry in the junction capacitances of the electrometer, and we have inferred their values (0.15 and 0.45 fF) from an analysis of the set of $I(U_0, V)$ measurements.

The charge induced on the coupling capacitance C_c by the single electron box is $Q_c = \langle Q \rangle C_c/C$. Thus, the capacitances were chosen so that Q_c changes by $\simeq 0.1e$ when Q changes by e. It follows that the variation ΔI of the electrometer current is to a good approximation proportional to $\langle Q \rangle$ if the electrometer is placed in a linear region of its $I - U_0$ characteristic

$$\Delta I = \frac{C_c}{C} \frac{dI}{d(C_0U_0)} \langle Q \rangle. \tag{16}$$

We have recorded the variations ΔI of the current in the electrometer (placed at the optimal working point O) versus the single electron box control voltage U. We got rid of the low frequency noise by recording the derivative $d(\Delta I)/dU$ with a lock-in amplifier and integrating it afterwards. In order to compare measurements at different temperatures and at different working points, we have measured the coefficient dI/dU_0 prior to (and after) each measurement. The resulting sawtooth pattern of $\langle Q \rangle$ versus $\tilde{Q}$ at $T = 20$ mK, obtained with this procedure, is shown in Fig. 8 (curve labeled "N"). The horizontal axis has been calibrated assuming the period is e. The vertical axis has been calibrated assuming that the island charge increases by e during a period and that the junction capacitance of the box is equal to $C = 0.6\,$fF, which is the sum of the electrometer junction capacitances. We deduced the capacitances $C_S = 85\,$aF and $C_c = 75\,$aF, which are close to the expected values. The sawtooth pattern was reproducible, although sudden shifts along the horizontal axis occured on the time scale of one hour. We attribute these shifts to the sudden evolution of charges located close to the single electron box, perhaps within the substrate oxide.

We have measured $\langle Q \rangle$ versus $\tilde{Q}$ at different temperatures ranging from 20 to 312 mK. One period at each temperature is shown in Fig. 9 (solid line) together with the prediction of Eq. (14) (dashed line). Although the experimental results are in agreement

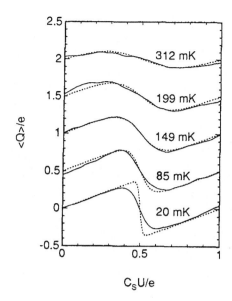

Figure 9. Solid lines: charge variations in the normal state at different temperatures. Dashed lines: theoretical predictions [Eqs. (14) and (15)].

with the theoretical predictions above 100 mK, there is a discrepancy at lower temperatures between the thermometer temperature and the temperature that would fit the data. At 20 mK, this discrepancy is 40 mK and is only partially understood. Parasitic rf signals on the U_0 and U lines could induce a broadening of the charge variations, although checks were performed to ensure that the direct influence of the background noise in the laboratory had no effect. The back action noise induced by the electrometer on the single electron box accounts for 25% of the excess temperature, as deduced from numerical simulations of the complete device. The heating of the single electron box electrons by the electrometer can be estimated and is negligible.

A more fundamental contribution to the rounding of the $\langle Q \rangle$ versus U curve at the lowest temperatures arises from the quantum fluctuations of n. It has been assumed so far that n is a classical variable, which is only true if the junction tunnel resistance is much larger than R_K. When this condition is not satisfied, the ground state is no longer a pure charge state n, but contains admixtures from other charge states $n \pm m$, because the tunneling Hamiltonian favors the delocalisation of the electrons on both sides of the junction. The ground state is dressed with virtual tunnel excitations, just as an atomic state is dressed with virtual photons. We have calculated to the first order in R_K/R_T the quantum corrections to the average junction charge

$$\langle Q \rangle = \frac{C}{C + C_S} \left[\tilde{Q} + e \frac{R_K}{4\pi^2 R_T} \ln \left(\frac{e/2 - \tilde{Q}}{\tilde{Q} + e/2} \right) \right] . \tag{17}$$

This expression diverges for $\tilde{Q} \to \pm e/2$ and is not valid close to these limits, where it should be zero by symmetry. Quantitatively, this correction is too weak to explain the observed rounding of the sawtooth pattern when evaluated for the estimated box effective tunnel resistance $R_T = 300$ kΩ. However, a complete non-perturbative theory is clearly needed [9].

3.3. Superconducting case

Since the SET transistor electrometer can be operated in the superconducting case by reducing the magnetic field [10], the charge of the junction can also be measured in the superconducting regime, with the electrometer biased at the edge of the superconducting gap. Strikingly, the observed sawtooth pattern is still e-periodic and is almost identical to the normal case sawtooth pattern, as can be seen in Fig. 8 (curve labeled "S"). One can infer from this observation that a quasiparticle enters the island on the time scale of the measurement ($> 10^{-2}$ s) as soon as the threshold voltage is slightly exceeded. This corresponds to a leakage resistance of less than 10^{11} Ω, a value well below the expected BCS leakage resistance due to thermally excited quasiparticles $R_{qp} = R_T \exp(\Delta/k_B T - \Delta/k_B T_C)$, calculated assuming that the gap Δ and the critical temperature T_c are the same as in bulk aluminum. In any case, the presence of these quasiparticles prevents the observation of single Cooper pair tunneling in this experiment.

One should not deduce from this experiment that a quasiparticle always enters an island before a Cooper pair, independent of the nature of the device. In particular, co-tunneling of a Cooper pair could occur prior to co-tunneling of a quasiparticle. A simple calculation shows that this could be the case in a modified box where the junction is replaced by a series of junctions, provided all of the parameters are properly chosen.

This modified box presents another interesting feature: in the normal state, co-tunneling of electrons can be slow enough to make the direct observation of a single co-tunneling event possible. At the time of writing, we have observed individual co-tunneling events, and we have found unstable configurations with lifetimes longer than 1 s, probably limited by the residual noise. This experiment should provide an accurate test of the co-tunneling theory at very low rates.

4. Single electron turnstile

4.1. Basic principles of the controlled transfer of single electrons

We have demonstrated with the single electron box that one can control a charge at the single electron level. This circuit is a basic building block in the design of more elaborate circuits which can transfer single charges.

In the single electron box, if the gate voltage U varies slowly enough, the configuration n adiabatically follows the minimum energy configuration. This evolution is fully reversible: if one reverses the time evolution of U, the evolution of the configuration is also reversed. Thus, any reversible device like the single electron box with a single control voltage cannot transfer electrons in a controlled way and satisfy the condition that all control voltages return to the same values after the transfer of a single electron, an essential condition for repeated transfer. A charge transferring device should thus either be irreversible, or have at least two control voltages. Both choices are in fact possible, and the simplest devices in each category respectively are the turnstile discussed in this section and the pump presented in the next section.

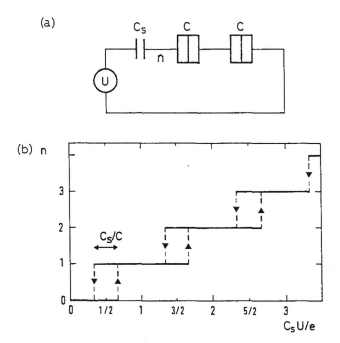

Figure 10. (a) Electron trap circuit consisting of two junctions in series with a capacitor and a voltage source. The main island contains n extra electrons. The island between the junctions cannot be charged in a locally stable configuration. (b) Local stability domains of the different configurations labelled by n. Two configurations can be locally stable for the same value of the voltage U. The evolution of the configuration is not reversible when U is varied, and thus allows hysteretic cycles.

4.2. The trap: an irreversible single electron box

Let us consider two identical junctions on either side of an island without a gate, imbedded inside a circuit. When the island contains no extra electron, both junctions have the same charge. If this charge just exceeds their common critical charge Q^c, a tunneling event will occur on one junction. After the event, the configuration is not yet locally stable, because the charge of the second junction is now equal to $(e - Q^c) > Q^c$. A tunneling event will thus occur in this junction, with an energy $(e/C)(e - 2Q^c)$ transformed into electron kinetic energy. The evolution of the charge configuration does not conserve the total electrostatic energy of the circuit, and cannot therefore be reversible with the control voltages (this would violate energy conservation). The trap, shown in Fig. 10a, is the simplest irreversible circuit. Although it has two islands, the island between both junctions cannot be charged in a locally stable configuration. A diagram of the locally stable configurations is given in Fig. 10b. The locally stable configuration is not unique for all values of the the control voltage U, as opposed to the case of the single electron box. When U varies, the evolution of the charge configuration is hysteretic. The entrance and the exit of an electron do not occur at the same value of U. "Trapping" electrons in the main island is thus possible, and the number of locally stable configurations existing for the same value of U grows with the ratio C_S/C. However, this trapping is only metastable, since co-tunneling processes are possible. As one configuration among the different possible locally stable configurations has the lowest energy and is absolutely stable, this configuration will eventually be reached by co-tunneling.

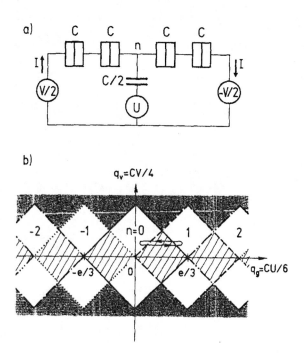

Figure 11. (a) Single electron turnstile schematic. The device is a trap with two arms connected to voltages $\pm V/2$. As in the trap, a locally stable configuration can only contain extra electrons in the central island. (b) Turnstile local stability diagram of the configurations with different n in the (U, V) plane. For example, the $n = 1$ configuration is locally stable inside the dashed line square. In the hatched regions, two different configurations are locally stable. In the grey regions, conduction takes place through the array. An example of a cycle which transfers a single electron per period across the device is shown (trajectory with arrows).

4.3. From the trap to the turnstile

The turnstile, shown in Fig. 11a, is a trap with two arms. At zero bias voltage V, it is equivalent to the trap. The gate capacitance is nominally equal to one half of a single junction capacitance, so that two locally stable configurations are possible over a range of U. At finite (but small) V, the turnstile differs from the trap only in the sense that when U is varied, electrons enter the main island only through the arm connected to the negative end of the bias voltage source, and leave the island only through the arm connected to the positive end. A single electron can thus be transferred through the device in a direction imposed by the biasing voltage V.

In order to determine how to use the turnstile to transfer single electrons, it is useful to know the stability domains in the (U, V) plane of the different island configurations. The borders of these domains are determined by the condition that the charge of the junctions on one of the two arms is equal to $\pm Q^c = \pm e/3$. As can be seen in Fig. 11b, these domains are diamond shaped regions and have an overlap (hatched areas) with their neighbours. The size of this overlap is determined by the gate capacitance and vanishes for zero capacitance. For example, inside the domain delimited by the dashed line, configuration $n = 1$ is locally stable. The positive (resp. negative) slope borders correspond to transitions on the positively (resp. negatively) voltage biased arm. Outside these domains, a current flows through the device (grey areas).

In order to transfer a single electron, a cycle of the gate voltage U has to cross two borders of different types without passing inside a conduction zone. An example of such

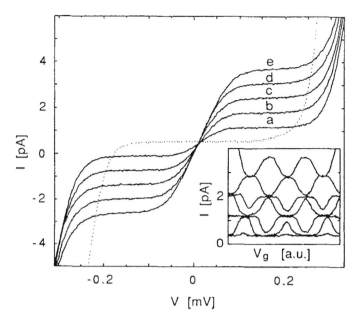

Figure 12. Turnstile current-voltage characteristics without rf gate voltage (dotted curve) and with applied rf gate voltage at frequencies f from 4 to 20 MHz in 4 MHz steps (a-e, full curves). Current plateaus are seen at $I = \pm ef$. Inset: current versus dc gate voltage for various amplitudes of the rf gate voltage at a frequency $f = 5$ MHz. Bias voltage is 0.15 mV and the rf amplitude increases from bottom to top. The curves tend to be confined between levels at $I = nef$ and $(n + 1)ef$, with n an integer. This can be understood by displacing a cycle of large amplitude along the horizontal axis in Fig. 14: An integer number n of electrons is transferred each period when the cycle encloses n hatched regions, and this number can at most vary by one unit when the mean value of the gate voltage is changed.

a cycle is shown in Fig. 11b. During the cycle, the kinetic energy of the electrons in the device increases by an amount at least equal to $2e^2/3C$. A given cycle works properly over a range of V which depends on the amplitude and on the average value of U. In practise, one records the current-voltage $I-V$ characteristic in presence of an rf signal of frequency f superimposed with dc voltage on the gate. The experimental manifestation of the single electron transfer consists in a plateau on the $I-V$ characteristic located at the current $I = ef$.

4.4. Experimental results

Experiments [2] have been carried on two different devices, and plateaus have been obtained at frequencies from 1 to 30 MHz. Plateaus obtained at different frequencies ranging from 4 to 20 MHz are shown in Fig. 12. This particular sample consisted of four Al junctions with $R_T = 340$ kΩ and $C = 0.5$ fF $((R_T C)^{-1} = 5.9$ GHz) in the normal state and at a temperature of 15 mK. These plateaus are not perfectly flat, and we define the current at the plateau I_{pl} as the current at the inflexion point of the $I-V$ characteristic. I_{pl} so obtained follows the expected law $I_{pl} = ef$ within the experimental accuracy of 3×10^{-3}.

The rounding of the plateaus results from random errors occuring during the transfer cycle. These errors can originate from missed transitions, thermally activated unwanted tunnel events and co-tunneling.

4.5. Transfer accuracy

4.5.1. Missed transitions

The probability that a transition has been missed because of the Poisson nature of the tunneling process can be estimated as in the case of a single transition [see Eq. (8)]. The resulting relative error $\epsilon = (I - ef)/ef$ in the determination of the current at the plateau is given by

$$\epsilon \simeq -\exp\left(-\frac{a}{fR_TC}\right) \tag{18}$$

where $a \simeq 5 \times 10^{-2}$ is a numerical coefficient whose precise value depends on the details of the transfer cycle. This error producing mechanism only gives a maximum error in the current $\delta I = -2 \times 10^{-6}$ pA for the plateaus reported here and is not therefore the main source of error.

4.5.2. Thermally activated unwanted tunnel events

Thermally activated tunneling events tend to wash out the current plateaus, as demonstrated by experiments at higher temperatures and by numerical simulations. The probability of such events is proportional to $\exp(-\Delta E/k_BT)$, where $\Delta E \approx 0.1e^2/C$ is an activation energy whose precise value depends on the transfer cycle.

Numerical simulations indicate that an electronic temperature of $T = 60$ mK would explain the observed shape of the plateaus, even without invoking co-tunneling, as shown in Fig. 12. Is this temperature plausible? Electrons are heated by the repeated energy release of more than $2e^2/3C$ produced by each transfer, and are in thermal contact with the phonon heat-bath through the electron-phonon coupling. One can estimate the electronic temperature T_e of the turnstile using the electron-phonon coupling theory [11]

$$T_e \simeq \left(\frac{e^2 f}{\Sigma \Omega C} + T_{\rm ph}^5\right)^{1/5}, \tag{19}$$

where Ω is the volume of the device islands, Σ a material-dependent coupling constant ($\Sigma_{\rm Al} \simeq 4 \times 10^9$ W/K^{5}m^3) and $T_{\rm ph}$ the phonon temperature. T_e calculated in this way is very weakly frequency dependent and is close to 60 mK. In this estimate, we have neglected the fact that the electron heating increases when the cycle amplitude and the frequency increase. If one increases the amplitude too much, T_e rises, and the plateaus are washed out even further. This picture is consistent with the observed shape of the plateaus at large cycle amplitudes for which a higher temperature is required to obtain a satisfactory fit, as shown in Fig. 13.

4.5.3. Co-tunneling

Co-tunneling should also contribute to the rounding of the plateaus. The analytical calculation of the plateau shape in presence of co-tunneling is tedious and one can only estimate the effect of co-tunneling. The relative correction $(I - ef)/ef$ is a sum of terms proportional to f^α, where $\alpha = -1/2, -1, -2, \ldots$ is always a negative exponent, with amplitudes depending on the position on the plateau current. The relative error decreases

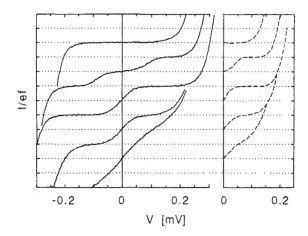

Figure 13. Left: turnstile current-voltage characteristics at $f = 5$ MHz for different levels of applied rf gate voltage. The dotted horizontal lines are at intervals $ef = 0.8$ pA. The curves have been offset in the y direction to display them more clearly. From top to bottom the estimated rf voltage amplitudes are 0, 0.60, 0.95, 1.50, and 1.89, expressed in units of $e/C \simeq 0.30$ mV. Right: corresponding simulated current-voltage characteristics. The fit temperature is 50 mK for the upper three curves and 75 mK for the lower two curves. This increase indicates that electron heating increases with the rf amplitude, as expected. Co-tunneling has not been taken into account in these simulations, but other simulations at zero temperature including co-tunneling have shown that the plateaus become washed out in the same way.

when the frequency increases, because less time is spent in unstable configurations with respect to co-tunneling. Numerical simulations show that co-tunneling modifies the shape of the plateaus in a way very similar to the effect of higher temperatures. However, in the case of the experiments reported here, the tunnel resistance of the junctions is large, and co-tunneling contributes less to the plateau rounding than electron heating.

5. Single electron pump

5.1. The principle of the pump: two coupled electron boxes

The single electron pump, shown schematically in Fig. 14a, is the simplest possible device with two gates. It can be seen as two SEB connected through a junction. It consists of three junctions and two gates whose capacitances C_1, C_2 are much smaller than the junction capacitances C, C', C''. Gate voltages U_1 and U_2 are the control voltages and V is a constant bias voltage. The configuration of the pump is given by the island integer charges (n_1, n_2) and by n_0, the number of electrons having passed through the pump.

At zero bias voltage V, one can prove that any locally stable configuration is also absolutely stable. Consequently, the local stability domains associated with each (n_1, n_2)–couple tile the $(\tilde{Q}_1 = C_1 U_1, \tilde{Q}_2 = C_2 U_2)$ plane. Although each domain consists of an elongated hexagon, the translational symmetry of this tiling is that of a square lattice, as shown in Fig. 14b. Neighbouring domains correspond to neighbouring configurations in configuration space. Three neighbouring domains share a triple point: for example, point P of Fig. 14b, is shared by domains (0,0), (1,0) and (0,1). The corresponding configurations have the same energy at that point. There are two triple points per unit

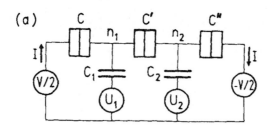

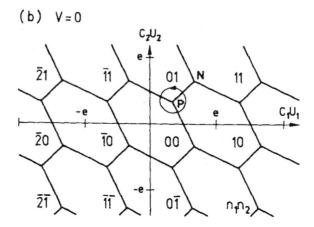

Figure 14. (a) Single electron pump circuit schematic. The pump consists of two single electron boxes coupled by a junction. n_1 and n_2 are the numbers of extra electrons in the two islands. (b) Stability domains of the different configurations at zero bias voltage. The hexagonal domains tile the plane with quadratic symmetry. One turn around the point P transfers one electron across the circuit, in a direction determined by the sense of rotation.

cell of the lattice, respectively congruent to point P and to point N in Fig. 14b. Every neighbouring point of a "P type" triple point is a "N type" triple point and vice versa. The existence of this triple point lattice is a generic topological consequence of the bidimensional character of the control space, and does not depend on the precise values of the junction capacitances.

At finite bias voltage V, the different configurations are no longer absolutely stable and their local stability domains do not completely tile the $(\tilde{Q}_1, \tilde{Q}_2)$ plane. The honeycomb pattern of Fig. 14b is distorted: the triple points are replaced by triangle-shaped regions inside which no stable configurations can exist and where conduction thus takes place. When V increases, these regions increase in size while the stability domains shrink and eventually disappear.

5.2. A pumping cycle

The pump is operated by first applying dc voltages to the gates so as to place the circuit in parameter space in the vicinity of a triple point, the bias voltage being much lower than the Coulomb gap voltage (equal to e/C if the junctions have the same capacitance C). Two periodic signals with the same frequency f but dephased by $\Phi \sim \pi/2$ are then superimposed on the gate voltages. The circuit thus follows a closed trajectory like the circle shown around point P in Fig. 14b. The frequency is chosen low enough $[f \ll (R_T C)^{-1}]$ to let the system adiabatically follow its ground state, which changes along the trajectory. Suppose that the initial island configuration is $(0, 0)$ and

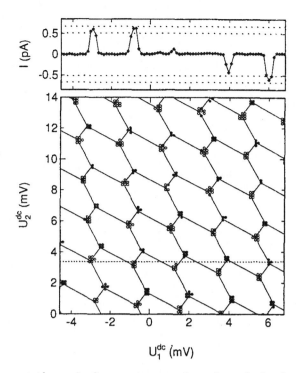

Figure 15. Zero bias current through the pump as a function of the dc gate voltage. Amplitudes and frequency of rf gate modulation superimposed on the dc gate voltage were 0.3 mV and 4 MHz, respectively. In the bottom panel white (black) dots denote current measured within the upper (lower) pair of dashed lines in the top panel. The small regions where the current is close to $\pm ef$ are centered on the vertices of the honeycomb lattice of Fig. 14. The calculated stability domain diagram that best fits the results is shown by the solid lines.

that the trajectory is followed counterclockwise. The circuit goes first from $(0,0)$ to $(1,0)$ by letting one electron tunnel through the leftmost junction. Then the island configuration changes to $(0,1)$ when one electron goes through the central junction. Finally, the system returns to its initial island configuration $(0,0)$ by letting one electron out through the rightmost junction. In a complete cycle one electron is transferred from left to right. If the sense of rotation in parameter space is reversed, in practise by adding π to the phase shift Φ, the electron transfer will take place in the opposite direction. Note also that the same original positive rotation around a "N type" triple point produces a negative current. In summary, these geometrical considerations show that for zero bias voltage V, two rf gate voltages induce a current $I = ef$ to flow through the circuit, provided the dc gate voltages are set in the vicinity of a triple point. The direction of current is solely determined by the phase shift and the type of the triple point.

As the voltage V is increased, electrons can still be pumped, even if V and I have opposite signs, provided that the trajectory followed in parameter space encloses the conduction regions. Numerical simulations have shown that regular electron transfers can persist up to one-fifth of the Coulomb gap voltage for an optimal rf amplitude.

5.3. Experimental observation of single electron pumping

Experiments have been carried on two samples [3]. We report here measurements on the second one, with improved gate capacitor design and with a total normal resistance of 380 kΩ. Measurements were performed at 20 mK. The average junction capacitance

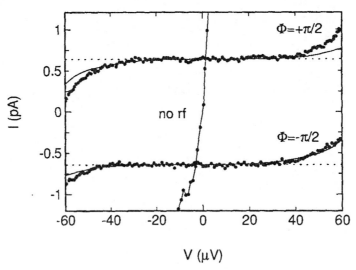

Figure 16. Current-voltage characteristic of the pump with and without a $f = 4$ MHz gate voltage modulation around a "P-type" triple point. The U_1 and U_2 rf amplitudes were respectively 1 mV and 0.6 mV. Dashed lines mark $I = \pm ef$. Full lines are the result of numerical simulations taking co-tunneling into account.

was estimated from the Coulomb gap to be 0.4 fF. In Fig. 15 we show the results of an experiment in which the current through the array was recorded as the dc gate voltages U_1^{dc} and U_2^{dc} were swept over the (U_1, U_2) plane. The bias voltage V was set to zero and two 0.3 mV amplitude rf signals with frequency $f = 4$ MHz were superimposed in quadrature on the dc gate voltages. Only in some regions of the (U_1^{dc}, U_2^{dc})-plane did the current exceed the noise. In the top panel of Fig. 15 we show the result of a scan of U_1^{dc}, U_2^{dc} being kept constant. This panel represents a cut through the current surface which is represented in the bottom panel, the position of the cut indicated by a dotted line. For clarity, we have represented the "hills" and "basins" of the current using the following convention: A white dot means a positive current of between $+0.8\,ef$ and $1.05\,ef$ (upper pair of dashed lines in top panel), a black dot means a negative current of between $-0.8\,ef$ and $-1.05\,ef$ (lower pair of dashed lines) and the absence of dots means a current between $+0.8\,ef$ and $-0.8\,ef$. Apart from a slight global deformation and an overall translation, the pattern of hills and basins reproduces the honeycomb pattern of Fig. 14b, a hill (basin) corresponding to a P-type (N-type) triple point. We attribute the slight deformation of this experimental honeycomb pattern to the cross-capacitance C_x between islands and gates. We have calculated the ground state diagram for arbitrary $C_1, C_2, C_x, (C+C'')/C'$ and $(C-C'')/C'$. The best fit to the data is shown in the bottom panel of Fig. 15, and corresponds to $C_1 = 74 \pm 2\,aF$, $C_2 = 61 \pm 2\,aF$, $C_x = 7 \pm 1\,aF$, $(C+C'')/C' = 2.1 \pm 0.5$, and $(C-C'')/C' = -0.3 \pm 0.06$. In the fitting, we have of course allowed an overall translation of the diagram, corresponding to offset charges appearing on the gate capacitors. Although these offset charges were found to be constant on a time scale of one hour, abrupt shifts of the pattern on longer time intervals were observed. The capacitances that we obtain from the fit agree with the values estimated from the geometry of the nanolithographic mask. Fig. 16 shows the bias voltage dependence of the current at a "P type" triple point with $f = 4$ MHz and for two phase shifts separated by π; plateaus are observed near the center of the $I - V$ curve at currents close to the expected value $I = \pm ef$. The sign of the height of the plateau reverses abruptly as the phase shift between the rf voltages is varied continuously from $+\pi/2$ to $-\pi/2$. We also show for comparison the $I - V$ curve with no rf, and no Coulomb gap appears, as expected at a triple point. As in the turnstile, the plateaus are slightly rounded, but the deviations can be predicted almost analytically, as shown below.

5.4. Transfer accuracy

The transfer accuracy in the pump is limited by missed transitions, unwanted thermally activated tunneling and by co-tunneling.

5.4.1. Missed transitions

The relative error ϵ on the current resulting from missed transitions is also given by Eq. (19), but with $a = 4 \times 10^{-2}$. This error is negligible in the frequency range of the experiments reported here.

5.4.2. Thermally activated tunnel events

As opposed to the turnstile, no electron heating occurs in the pump at low frequencies because the pump is a reversible device. Using Eqs. (5) and (8), one can show that the heating power W is approximately

$$W \simeq \frac{e^2}{C} f^{3/2} (R_T C)^{1/2}. \tag{20}$$

The resulting electron temperature is

$$T_e \simeq \left(\frac{e^2 f^{3/2} (R_T C)^{1/2}}{\Sigma \Omega C} + T_{\text{ph}}^5 \right)^{1/5}. \tag{21}$$

One finds in our case that this heating almost explains the observed rounding of the plateaus at frequencies above 10 MHz. Below that frequency, co-tunneling is the principal source of the rounding.

5.4.3. Co-tunneling

Co-tunneling events can result in an error at each stage of the transfer. For example, the (0,0) to (1,0) transition can occur by a single tunnel event across the left junction or by co-tunneling across the two right junctions. At zero bias voltage and zero temperature, both transitions become possible together when the representative point in the (U_1, U_2) plane crosses the (0,0)-(1,0) border, because the initial and final configurations have the same energy for both cases. The rates are however different: the single tunneling transition rate is proportional to the distance to the border of the stable domain, while the co-tunneling rate is proportional to the third power of this distance. An important consequence is that the probability that an electron enters from the wrong side increases with the cycle speed, and is zero for slow cycles. The error due to co-tunneling at $V = 0$ can be calculated, and one obtains for an optimal cycle that

$$\epsilon \simeq -10 R_K C f. \tag{22}$$

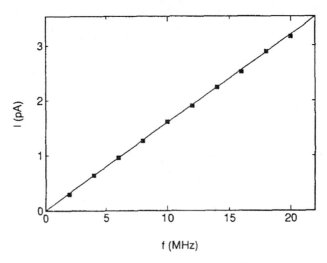

Figure 17. Current, measured at the inflexion point of the plateau, versus frequency in the single electron pump. Full line is $I = ef$.

The calculation can be extended to finite bias voltage, but numerical simulations are better suited for a detailed comparison with experimental results. The predicted $I - V$ characteristic is shown in Fig. 16 (full line). All parameters used in this calculation were taken from previous measurements and fall within the experimental error bars, except the value of the phase shift, which was fitted as it had only been determined by rf measurements at room temperature. The width of the plateau on either side of the inflexion point is seen to be well explained by the theory. At the precision level of the simulation, which was one part in 10^3, the tangent to the calculated curve at the inflexion point is the line $I = ef$. As shown in Fig. 17, we find that this current satisfies the expected relation $I = ef$ within the experimental uncertainty of $\Delta I = 0.05$ pA.

6. Is metrological accuracy achievable?

The turnstile and the pump we have discussed above are clearly not devices of metrological accuracy, and furthermore produce very small currents. Nevertheless, metrological applications have already been considered, with the hope that more elaborate devices will be developed.

6.1. Metrological applications

Two metrological applications of small junction devices have been considered up to now and will be discussed in this section.

6.1.1. Precise measurement of the fine structure constant α

In this experiment, proposed by E. Williams and J. Martinis from NIST, one charges a previously calibrated capacitance C with a given number N of electrons produced by a pump device, and one measures the resulting voltage $V = Ne/C$. This measurement is done using a double junction electrometer as a precision comparator with a voltage source calibrated using the ac Josephson effect. From the comparison of C with a calculable capacitor [12] along which light takes a time τ to propagate, one obtains $C = a\epsilon_0 c\tau$, where a is a known numerical factor and c the speed of light in vacuum. From the measurement of V with the ac Josephson effect, one obtains a frequency $\nu = V/\Phi_0$ where $\Phi_0 = h/2e$ is the flux quantum. One deduces from both measurements the relation $\alpha = e^2/2\epsilon_0 hc = a\nu\tau/4N$, which provides a determination of the fine structure constant α. This experiment requires a number $N \simeq 10^8$ electrons (for $C \simeq 1$ pF) transferred with an average error of less than one electron in a time shorter than one minute to avoid undesired drifts.

6.1.2. Realisation of a current standard producing a current $I = ef$ large enough to power a quantum Hall effect device

In the quantum Hall regime, the transverse voltage across a sample is related to the longitudinal current by $V = R_K I/n$ where n is the integer index of the quantum Hall plateau. In this combined device experiment, V would directly be related to f by the relation $V = (h/ne)f = (2/n)\Phi_0 f$, which is $(2/n)$ times the Josephson voltage $\Phi_0 f$ one would obtain in an ac Josephson effect experiment at the frequency f. The precise test of this relation would be an important consistency check of the representations of the volt and of the ohm based respectively on the ac Josephson effect and on the quantum Hall effect. This experiment requires a device producing a current $I = ef$ in the nA range with an accuracy of the order of 10^{-8}.

6.2. Improving the accuracy of charge transferring devices

Transfer errors in the turnstile and in the pump involve successive intermediate configurations of higher energy than the final configuration, whether the errors originate from thermal activation or from co-tunneling. Clearly, the reduction of the error rate due to both types of processes requires an increase in the energy and in the number of these intermediate configurations, since thermal activation depends mainly on the higher intermediate energy and co-tunneling is proportional to the inverse of the product of all the intermediate energies. Since the energy barrier for the passage of a single charge across p junctions is approximately $(p/4)(e^2/2C)$, the probability of an error due to thermally activated events only is proportional to $\exp(-pe^2/8k_B TC)$. In practice, this probability can be made negligible in a charge transferring device, even if there is significant heating of the electrons. One can also show that the errors due to thermally activated transitions combined with co-tunneling transitions can be made negligible. Co-tunneling thus constitutes the main limitation to the accuracy of charge transferring devices built with small junctions. The N-turnstile and the N-pump are devices based on the same working principles as the turnstile and the pump, but with a larger number N of junctions in order to decrease the error probability during the transfer.

6.2.1. N-turnstile

The N-turnstile, whose schematic representation is shown in Fig. 18a, is similar to the turnstile but with $N/2 > 2$ junctions in each arm and with a nominal gate capacitance of $2C/N$. Its operation is the same as for the turnstile, where a single

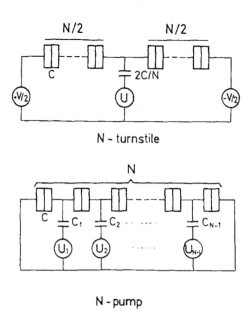

Figure 18. Circuit schematic of the N-turnstile and of the N-pump. The larger the value of N, the smaller the probability of transfer errors induced by co-tunneling.

electron is transferred during each cycle of the rf signal applied to the gate. When the transition threshold is reached in one arm, a cascade of single tunneling events takes place and a single electron enters (or leaves) the central island to which the gate is connected.

As in the turnstile, electric energy is transformed into electron kinetic energy. The average energy dissipated in each island of the device during each cycle is e^2/NC at low frequencies, and increases with frequency. Heating is expected in this device, but should not limit the accuracy if N is large enough. Co-tunneling is difficult to analyze because many different co-tunneling transitions are possible during the transfer cycle. At zero temperature, the relative error $\epsilon = (I - ef)/ef$ in the determination of the current at the plateau of the $I - V$ characteristic for an optimal cycle has an upper bound given by

$$|\epsilon| \leq a(N) \, \frac{1}{R_K C f} \left(\frac{R_K}{R_T} \right)^{1+N/2}, \qquad (23)$$

where $a(N)$ is a numerical coefficient depending on N. The error decreases with frequency because less time is spent in configurations unstable with respect to co-tunneling. The maximum frequency is as always limited by the finite time required to complete the tunneling transitions, and decreases with the number N. The N-turnstile with $N \simeq 10$ could deliver a current in the pA range with metrological accuracy, if the offset charges in the intermediate islands are compensated using auxiliary gates. This current is too low for practical applications, but if offset charges were not present, auxiliary gates would not be necessary and a large number of N-turnstiles could be operated in parallel in order to increase the current. Although such a project is not impossible using present day technology, it clearly requires an involved effort.

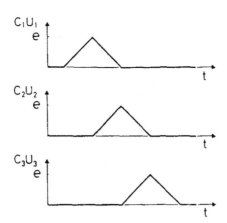

Figure 19. Evolution of the gate voltages in the 4-pump during a transfer cycle of a single electron. The positive triangular shaped pulses propagate a potential well containing a single electron along the array.

6.2.2. N-pump

The N-pump is a linear array of $N > 3$ junctions with $(N-1)$ gates as shown in Fig. 18b. A single charge is transferred through the array by applying shifted triangular shaped pulses to the gates so that a potential well containing a single electron is propagated from one end of the array to the other. A transfer sequence for $N = 4$ is indicated in Fig. 19. The transfer accuracy is limited mainly by co-tunneling and can be quantitatively estimated. At zero bias voltage, an electron can enter the first left island by a single tunneling event through the first left junction or by co-tunneling through the $(N-1)$ right junctions. The calculation of the co-tunneling rate needs to be modified because an island is charged in the final configuration. The product of the intermediate state energies which enters in the calculation of the rate cannot any longer be performed analytically, but one can derive the approximate expression

$$\Gamma'_c(\tilde{Q}_1) \simeq \left(\frac{N}{N-1}\right)^{2(N-2)} \Gamma_c\left(N-1, (N-1)e(\tilde{Q}_1 - e/2)/NC\right) \qquad (24)$$

where $\Gamma_c(N, E)$ is given by Eq. (11). The error due to co-tunneling is then easily calculated for this first transition, and can be calculated in a similar way for the later steps. The total relative error $\epsilon = (I - ef)/ef$ for an optimal cycle at zero temperature and at zero bias voltage is

$$\epsilon = -b(N)\left(\frac{R_K C f}{\pi^2}\right)^{N-2}, \qquad (25)$$

where $b(N)$ is a numerical coefficient. Although R_T does not appear in this formula, the calculation of the co-tunneling rate assumes that $R_T \gg R_K$ because it takes into account only the leading term in the perturbative treatment of the tunneling Hamiltonian. We have estimated that an accuracy of 10^{-8} is achievable with a 5-pump, for which $b(5) < 2.5 \times 10^6$, up to a frequency of 20 MHz for $C = 0.25$ fF. A 5-pump thus seems well suited to the first application previously discussed, namely the measurement of the fine structure constant α. The ability to reverse the transfer direction is an attractive feature because the number of electrons transferred to the calibrated capacitance can be controlled by a transfer of the same number of electrons in the opposite direction.

6.2.3. A reasonable bet

Although no experimental proof of the feasibility of metrological applications has yet been performed, the known effects which limit the accuracy of charge transferring devices do not preclude these applications.

7. Conclusion

With the single electron box, the turnstile and the pump, we have implemented the basic function of single electronics, namely the transfer of a single charge into an island or through a circuit. In all these devices, the control of the electron transfer is based on the quantization of the charge on an island. We have seen that the exactness of this quantization is limited by thermal fluctuations with a temperature possibly higher than the phonon temperature, by quantum fluctuations across one junction and by co-tunneling through two or more junctions. We have also seen that the present understanding of these phenomena is far from complete and certainly deserves further investigation. In particular, the following questions arise: How do the sawtooth oscillations observed in the single electron box vanish when R_T becomes smaller than or of the order of R_K? How do tunneling and co-tunneling coexist when both processes are possible? What is the maximum sensitivity a double junction electrometer can achieve? Although seemingly disconnected, all these questions involve a treatment of the tunneling Hamiltonian beyond the leading order in the perturbative expansion, and possibly involve non-perturbative calculations. Other possible experiments have also emerged from "playing with toys", such as the transfer of a single Cooper pair into an island despite the presence of quasiparticles. We have also discussed how to design charge transferring devices able to meet the accuracy requirements of metrological applications, and have proposed two types of devices, the N-turnstile and the N-pump.

Finally, let us note that the transfer of a single charge is certainly not the only function one can implement in a small junction circuit. Technologies other than metallic junctions, such as 2D electron gases discussed in Chap. 5, could further enlarge the possibilities. However, whether or not applications to electrometry, metrology and possibly other fields will become a reality is still an open question (see also Chap. 9).

ACKNOWLEDGEMENTS. The Quantronics Group at Saclay, which includes M. H. Devoret, P. Lafarge, P. Orfila, H. Pothier, C. Urbina and myself, has greatly benefited from collaborations with other physicists. The single electron box experiment was completed while E. Williams was visiting from NIST. The turnstile experiment resulted from fruitful interactions with L. J. Geerligs, V. F. Anderegg and J. E. Mooij from T.U. Delft. Our long-standing collaboration with H. Grabert proved once more to be important to our theoretical understanding. Many ideas developed in this chapter arose from confronting all the different points of view. I warmly thank A. Cleland, M. H. Devoret, H. Grabert, C. Urbina and all those who helped to improve this text.

References

[1] P. Lafarge, H. Pothier, E. R. Williams, D. Esteve, C. Urbina, and M. H. Devoret, Z. Phys. B **85**, 327 (1991).

[2] L. J. Geerligs, V. F. Anderegg, P. Holweg, J. E. Mooij, H. Pothier, D. Esteve, C. Urbina, and M. H. Devoret, Phys. Rev. Lett. **64**, 2691 (1990).

[3] H. Pothier, P. Lafarge, P. F. Orfila, C. Urbina, D. Esteve, and M. H. Devoret, Physica B **169**, 573 (1991); H. Pothier, P. Lafarge, C. Urbina, D. Esteve, and M. H. Devoret, Europhys. Lett. **17**, 249 (1992).

[4] T. A. Fulton and G. J. Dolan, Phys. Rev. Lett. **59**, 109 (1987).

[5] M. H. Devoret, D. Esteve, H. Grabert, G.-L. Ingold, H. Pothier, and C. Urbina, Phys. Rev. Lett. **64**, 1824 (1990).

[6] K. K. Likharev, N. S. Bakhvalov, G. S. Kazacha, and S. I. Serdyukova, IEEE Trans. Magn. **25**, 1436 (1989).

[7] D. V. Averin and A. A. Odintsov, Phys. Lett. A **140**, 251 (1989); L. J. Geerligs, D. V. Averin, and J. E. Mooij, Phys. Rev. Lett. **65**, 3037 (1990).

[8] D. V. Averin and K. K. Likharev, J. Low Temp. Phys. **62**, 345 (1986).

[9] K. A. Matveev, Zh. Eksp. Teor. Fiz. **99**, 1598 (1991) [Sov. Phys. JETP **72**, 892 (1991)].

[10] T. A. Fulton, P. L. Gammel, D. J. Bishop, L. N. Dunkleberger, and G. J. Dolan, Phys. Rev. Lett. **63**, 1307 (1989).

[11] M. L. Roukes, M. R. Freeman, R. S. Germain, R. C. Richardson, and M. B. Ketchen, Phys. Rev. Lett. **55**, 422 (1985); F. C. Wellstood, C. Urbina, and J. Clarke, Appl. Phys. Lett. **54**, 2599 (1989).

[12] J. Q. Shields, R. F. Dziula, and H. P. Layer, IEEE Trans. Instrum. Meas. **38**, 249 (1989).

Chapter 4

Josephson Effect in Low-Capacitance Tunnel Junctions

M. TINKHAM

Department of Physics, Harvard University, Cambridge
MA 02138, USA

1. Introduction

In a broad discussion of Josephson junctions, many regimes exist, distinguished by the relative magnitudes of various parameters. Characteristic energies are the Josephson energy E_J, the charging energy $E_c = e^2/2C$, and the thermal energy $k_B T$. Equally important are the characteristic resistances: the normal state resistance R_n, the quantum resistance[1] $R_Q = h/4e^2 = 6453$ Ω, and the impedance of free space $Z_o = 377$ Ω. The fact that $Z_o \ll R_Q$ implies that quantum effects are hard to observe, even in high resistance junctions, unless the junction is buffered from the low electromagnetic impedance of its leads. Recently, several groups have accomplished this to a considerable extent by inserting physically small isolating resistors of high resistance in series with the leads in the immediate vicinity of the junction. Because of stray capacitance effects, however, effective high impedance filters at microwave frequencies are not simple to achieve, and the extensive and systematic early experiments of the Harvard group were made with essentially direct connection to leads, which presumably acted as transmission lines characterized by characteristic impedances of order Z_o. These data could be, and were, explained in terms of quantum effects related to delocalization of phase by single electron charging energies. Now, with the general recognition of the importance of transmission line effects and the development of more comprehensive theories, the interpretation of these data needs to be reexamined.

In this chapter, we review the behavior of Josephson junctions in many regimes, with emphasis on small junctions of low capacitance and high resistance, using relatively simple phenomenological models and arguments to give as much physical insight as

[1] In this chapter only superconducting metals are discussed and the quantum resistance $R_Q = h/4e^2$ corresponding to a charge of $2e$ is used instead of $R_K = h/e^2 = 4R_Q$ employed in previous chapters.

Single Charge Tunneling, Edited by H. Grabert and
M.H. Devoret, Plenum Press, New York, 1992

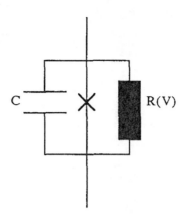

Figure 1. Schematic diagram of the RCSJ model.

possible. Our intention is to establish a basic foundation upon which more detailed and sophisticated treatments can be developed.

This chapter is divided into two major parts. The first deals with small classical Josephson junctions, including considerable discussion of the effects of thermal fluctuations and of the embedding impedance of the attached circuitry, in order to understand the range of phenomena which can be accounted for without resort to specifically quantum effects. The second part considers junctions small enough that the charging energies associated with individual electrons may cause additional effects due to the quantum incompatibility of exact specification of both phase and particle number.

2. Classical Josephson junctions

2.1. The RCSJ model of a Josephson junction

We describe the Josephson junction itself by a generalized RCSJ (Resistively and Capacitively Shunted Junction) model, in which current is carried by three parallel channels, as sketched in Fig. 1. The supercurrent channel is described by the Josephson relation

$$I_s = I_{co} \sin \varphi \tag{1}$$

where φ is the gauge-invariant phase difference between the Ginsburg-Landau order parameters on the two sides of the junction. I_{co} is often well-approximated by the result of Ambegaokar and Baratoff [1] for ideal tunnel junctions

$$I_{co}(T) = \frac{\pi \Delta(T)}{2eR_n} \tanh \left[\frac{\Delta(T)}{2k_B T} \right] \tag{2}$$

where $\Delta(T)$ is the BCS energy gap and R_n is the junction resistance in the normal state. At temperatures far below T_c, this reduces to $I_{co} \simeq \pi \Delta(0)/2eR_n$, where $\Delta(0) = 1.76\, k_B T_c$. (We assume throughout that the junctions are small compared to the Josephson penetration depth, so that it plays no role.) The resistive channel is characterized by a nonlinear temperature-dependent resistance R stemming from quasiparticle tunneling across the junction. Although no simple analytic expression covers all cases, a useful semiquantitative approximation is obtained by setting $R = R_n$ for $V > V_g = 2\Delta/e$, the gap voltage, and $R = R_n e^{\Delta/k_B T}$ for $V < V_g$. The latter expression takes account of the dominant exponential temperature dependence arising from the freeze-out of quasiparticles at low temperatures, but not the weaker effect of the singular density of states at the gap edge in BCS theory. Finally, the capacitive channel describes the displacement current associated primarily with the geometric shunting capacitance of the two electrodes of the junction.

Within the RCSJ model, the time dependence of the phase φ in the presence of an externally supplied bias current can be derived by equating the total junction current to the bias current I, as follows:

$$I = I_{co} \sin \varphi + \frac{V}{R} + C\frac{dV}{dt} \, . \tag{3}$$

Eliminating V in favor of φ using the Josephson relation $2eV = \hbar d\varphi/dt$, we obtain the second order differential equation

$$\frac{d^2\varphi}{d\tau^2} + \frac{1}{Q}\frac{d\varphi}{d\tau} + \sin\varphi = \frac{I}{I_{co}} \tag{4}$$

in which we have introduced a dimensionless time variable $\tau = \omega_p t$, $\omega_p = (2eI_c/\hbar C)^{1/2}$ being the so-called *plasma frequency* of the junction, and the quality factor $Q = \omega_p RC$. (This Q is identical with $\beta_c^{1/2}$, where β_c is a frequently used damping parameter introduced by Stewart and McCumber [2]).

So long as $I < I_{co}$, a static solution of (4) exists, with $\varphi = \sin^{-1}(I/I_{co})$. However, if $I > I_{co}$, only time-dependent solutions exist, and they determine even the dc $I - V$ curve of the junction. For example, if C is small so that $Q \ll 1$, the dc $I - V$ curve has the simple limiting form $V = R(I^2 - I_{co}^2)^{1/2}$, which smoothly interpolates between $V = 0$ for $I < I_{co}$, and Ohm's law $V = IR$ for $I \gg I_{co}$. When C is large enough that $Q > 1$, however, the $I - V$ curve becomes hysteretic. Upon increasing I from zero, $V = 0$ until I_{co}, at which point V jumps discontinuously up to a finite voltage V, corresponding to a "running state" in which the phase difference φ increases steadily at the rate $2eV/\hbar$. In an ideal tunnel junction at $T \ll T_c$, this voltage is near the energy gap voltage $V_g = 2\Delta/e$ [see Fig. 2]. If I is now reduced below I_{co}, V does not drop back to zero until a "retrapping current" $I_r \approx 4I_{co}/\pi Q$ is reached. (An approximate derivation of this formula is given later in this chapter.) The physical reason for the $1/Q$ dependence is that the energy stored in the capacitance during the high voltage (phase slip) part of the cycle must be dissipated in ~ 1 cycle of the oscillation of φ in order to prevent another phase slip from occuring, and thus to allow the system to retrap back into the zero-voltage state.

2.2. Effect of thermal fluctuations

When thermally-activated processes are taken into account, the $I - V$ curves of small Josephson junctions are strongly modified from the simple picture described above. This can be described mathematically by adding a Johnson noise current term to the bias current in (4). A more suggestive alternative approach for discussing these issues is provided

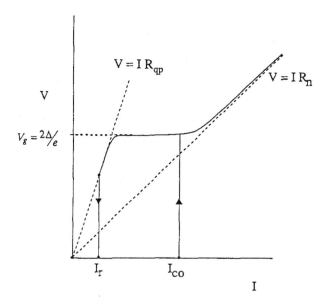

Figure 2. Schematic diagram of $I - V$ curve of hysteretic Josephson junction, showing the underlying quasiparticle tunneling resistance and the normal state $I - V$ curve.

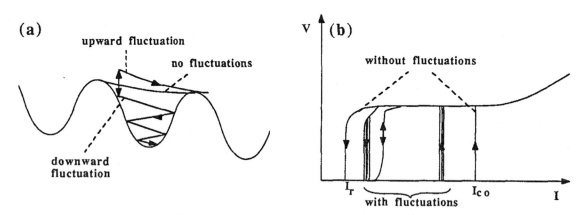

Figure 3. (a) Schematic diagram of tilted washboard potential, with trajectories showing effect of fluctuations on the retrapping process. In the trajectory, total energy is plotted vs. phase. (b) Sketch of the effect of thermal fluctuations in reducing and eventually eliminating the hysteresis in an underdamped Josephson junction.

by the so-called "tilted washboard" model, which provides a graphical representation of the energies involved in the problem. In this model, one recognizes that the equation of motion (4) is the same as that of a particle of mass $(\hbar/2e)^2 C$ moving along the φ axis in an effective potential

$$U(\varphi) = -E_J \, \cos\varphi - (\hbar I/2e)\varphi \tag{5}$$

and subjected to a viscous damping force $(\hbar/2e)^2(1/R)\mathrm{d}\varphi/\mathrm{d}t$. This potential is sketched in Fig. 3a. We use the vertical axis to indicate the total energy of a particle in its trajectory, so that the height of the trajectory above the cosine curve represents the instantaneous kinetic energy $(m/2)(\mathrm{d}\varphi/\mathrm{d}t)^2 = (C/2)V^2$. In the absence of damping, the trajectory is horizontal. Clearly, the characteristic energy scale in (5) is the Josephson coupling energy $E_J = (\hbar/2e)I_{co}$. The geometric significance of the critical current is that when $I = I_{co}$, the local minima of the cosine become horizontal inflection points in the otherwise continually downward slope, so that for $I \geq I_{co}$ no stable equilibrium point exists. Thermal noise randomly shifts the system energy up and down by an amount of order $k_B T$ on a time scale set by the relaxation time RC; the probability of a large excursion by an amount ΔE falls as $e^{-\Delta E / k_B T}$.

We can distinguish two regimes, depending on whether $k_B T$ is large or small compared to the barrier height $2E_J$. It follows from (2) that at $T \ll T_c$, the ratio $2E_J / k_B T = 1.76 \ (R_Q / R_n)(T_c / T)$, where the quantum resistance $R_Q = h/4e^2 = 6453 \ \Omega$. Thus, if $R_n \geq R_Q$ thermal fluctuations effects are prominent even well below T_c, although one can always minimize thermal fluctuations by going to sufficiently low temperatures.

When $k_B T \ll E_J$, thermally activated escapes from one potential minimum over the barrier to the next have a small probability $\sim e^{-\Delta U(I)/k_B T}$ at each attempt, and the attempt frequency is $\sim \omega_p / 2\pi$ as the representative phase point oscillates back and forth in the well at that characteristic frequency of (4). The current-dependent barrier height from (5) can be approximated by

$$\Delta U(I) \approx 2E_J(1 - I/I_{co})^{3/2} \ . \tag{6}$$

Thus, in an upward current sweep searching for the "critical current" I_c, the probability per unit time of escape from a local energy minimum rises exponentially from a very low value $\sim (\omega_p / 2\pi)e^{-2E_J/k_B T}$ at small currents to a large value $\sim \omega_p / 2\pi \sim 10^{10} \ \mathrm{s}^{-1}$ near I_{co}. We refer to this event as an "escape" because, with the presumed low damping associated with $Q > 1$, once the particle surmounts the barrier by a thermal fluctuation it will accelerate down the washboard until it reaches a terminal velocity, and never retrap in another energy minimum. This velocity is determined by the damping, which sets in strongly when $V = (\hbar/2e)\mathrm{d}\varphi/\mathrm{d}t$ reaches V_g, and R abruptly drops to R_n. Since this escape is a stochastic process, a distribution of values of I_c will be measured on successive upward current sweeps. This distribution is characterized by its width δI_c and by its mean depression below the "unfluctuated" critical current I_{co}. Although fully quantitative results require numerical computations (see Fulton and Dunkleberger [3]), one can show that the mean depression of I_c can be approximated by the formula

$$\langle I_c \rangle = I_{co} \left\{ 1 - \left[\frac{k_B T}{2E_J} \ln \left(\frac{\omega_p \Delta t}{2\pi} \right) \right]^{2/3} \right\} \tag{7}$$

where Δt is the time spent sweeping the current through the dense part of the distribution of observed I_c values. Since this Δt will typically be of the order of 1 s the logarithm will typically be of order $\ln(10^{10}) \approx 23 \gg 1$, and only logarithmically sensitive to the current sweep rate. Because this logarithm is so large, fluctuation effects cause a major reduction in I_c as soon as $k_B T$ is as large as 5% of E_J. The width δI_c of the switching distribution is approximately given by the mean depression of I_c divided by this same logarithmic factor.

Although more attention has traditionally been focused on I_c, in recent years it has been recognized that the retrapping current I_r, at which the junction drops back into the zero voltage state upon reducing the bias current, provides a uniquely direct probe of the damping experienced by the junction. As stated above, in underdamped ($Q > 1$) junctions in the absence of fluctuations, it can be shown that $I_{ro} \approx 4I_{co}/\pi Q$, so that I_r is directly proportional to the damping $\sim 1/Q$. If there were no damping for $V < V_g$, the phase point in Fig. 2 would not retrap as the tilt ($\sim I$) was reduced until I reached zero and incrementally reversed sign, corresponding to I_r being zero. With finite damping, I_r is fixed by the current (or tilt) at which the energy dissipated per cycle is exactly replaced by the work done by the drive current. For example, if the representative point starts at the top of one maximum of the tilted washboard, where it has zero velocity, it should just exactly reach the next maximum, again with zero velocity. This condition is the basis for deriving the formula $I_{ro} \approx 4I_{co}/\pi Q$.

When one takes account of thermal fluctuations, one finds the perhaps surprising result that they *increase* I_r above the unfluctuated I_{ro}, whereas they *decreased* I_c below I_{co}. The reason for this can be understood by the following physical argument. Assume $I = I_{ro}$, and consider a trajectory which just grazes successive maxima, as described above and shown in Fig. 3a. A single downward fluctuation in energy leads to a trajectory that spirals down to rest at the local minimum because there is dissipation, but no net energy input. On the other hand, a single upward fluctuation relaxes back to the initial trajectory, because the increased energy dissipation $\sim V^2$ outweighs the increased energy input $\sim V$. Thus $I = I_{ro}$ now leads to certain retrapping rather than being marginal, and the actual mean retrapping current I_r must be *greater* than I_{ro}. Of course, retrapping will also occur at a distribution of values on successive current downsweeps, because of the stochastic nature of the process. An analysis resembling that leading to (7) has been made by Ben-Jacob et al. [4] to estimate this effect, despite the fact that here one is dealing with a driven *nonequilibrium* regime. In this case the characteristic energy is that of the (downward) energy fluctuation needed to initiate trapping, rather than the upward fluctuation needed to initiate escape.

In Fig. 3b, we schematically summarize the effect of thermal fluctuations on the $I-V$ curve of an underdamped Josephson junction. In a low resistance junction ($R \ll R_Q$) or at low temperatures, fluctuation effects are unimportant, and I_c and I_r are observed to occur at sharply defined values near the unfluctuated values I_{co} and I_{ro}. In higher resistance junctions (or higher temperatures), I_c and I_r acquire a distribution of values, the means of which converge toward each other as the effect of fluctuations becomes more prominent. Finally, when $k_B T > E_J$ and fluctuations dominate, the system is rapidly jumping back and forth between the trapped and running states, leading, on a laboratory time scale, to a broadened *nonhysteretic* resistive transition at a current value intermediate between I_{co} and I_{ro}, at which the two states occur with equal frequency.

2.3. Effect of lead impedance

2.3.1. General considerations

For simplicity in the discussion above, we have treated the bias circuit as a pure current source, i.e., a circuit with infinite impedance at all frequencies so that it exerts no damping or capacitive loading effect on the junction. This can be an excellent approximation at dc, if the current source is a voltage source feeding through a sufficiently high fixed resistance such as 10^9 Ω. However, the characteristic frequencies involved in the dynamic processes leading to escape and retrapping are high, typically $\omega_p \sim 10^{11}$ s^{-1} and $\omega_g = 2eV_g/\hbar \sim 10^{12}$ s^{-1}. At such frequencies, the dc resistance of the remote current source plays no role. What is important is the embedding impedance of the bias circuit, as seen from the junction, at these high frequencies (at which the wavelength is $\sim 1\,\mathrm{mm}$ or less). Unless extremely small high impedance elements are inserted in the leads directly at the junction, the leads will typically show a characteristic impedance $Z_l \sim Z_o/2\pi = 60\,\Omega$, apart from logarithmic corrections for geometric factors. (Here $Z_o \approx 377\,\Omega$ is the characteristic impedance of free space.) For an ideal infinite transmission line, or one terminated in the same Z_l, this impedance is real and independent of frequency. If the line has impedance discontinuities at finite distances from the junction, its impedance shows resonances and becomes complex. Unless the microwave properties of the circuitry are well characterized, it is hard to predict the effects on the junction properties of these general complex frequency-dependent embedding impedances.

One case which is relatively manageable is that in which the line is sufficiently lossy at these frequencies that an outgoing electromagnetic wave is largely absorbed before it reaches any discontinuities which reflect it back to the junction. In this case, resonances are largely suppressed, and the line presents a real, nearly constant, high-frequency impedance Z_l, as if the line were infinite.[2] Since this Z_l is expected to be of order $100\,\Omega$, which is orders of magnitude less than the quasiparticle resistance in small, high-resistance ($R_n > R_Q$) junctions, the transmission line will greatly increase the damping, which should have a major impact on the retrapping current in such junctions. Just such an effect was reported recently by Johnson et al., as will be outlined below. It should be emphasized, however, that for typical lower resistance junctions with $R_n < 100\,\Omega$ and $I_c > 10\,\mu\mathrm{A}$, the effect of the leads is relatively minor, so that these effects were not given a great deal of attention in the past.

2.3.2. Modified model of Johnson et al.

The work of Johnson et al. [5] was triggered by the observation of a highly anomalous temperature dependence of I_r in the small high resistance $\mathrm{Sn} - \mathrm{SnO_x} - \mathrm{Sn}$ tunnel junctions studied down to dilution refrigerator temperatures by Iansiti et al. [6] and in additional ones studied by Johnson. The measured $I_r(T)$ for a typical junction with $R_n = 8.3\,\mathrm{k\Omega}$ is shown by the open circles in Fig. 4. At $T > 1\,\mathrm{K}$, $I_r(T)$ follows the expected exponential temperature dependence associated with the freezing out of the quasiparticle damping as $\mathrm{e}^{-\Delta/k_B T}$ with a concomitant increase in Q; below $\sim 1\,\mathrm{K}$, however, $I_r(T)$ *abruptly* stops decreasing and remains constant down to the lowest temperatures ($\sim 20\,\mathrm{mK}$). Because of the direct relation of I_r to damping, this observation implies that a new damping mechanism which does *not* freeze out at low temperatures suddenly comes into play at this crossover temperature. We find that this mechanism is the onset of pair-breaking tunneling processes when the voltage from which the junction retraps rises up to the gap voltage V_g. This interpretation is supported by the observation that in this regime the retrapping does occur with an abrupt drop from $V \approx V_g$ to 0. But why does the retrapping occur from such a high voltage in these junctions, and why has this phenomenon not been observed before?

To simplify the analysis, Johnson et al. used an energy-balance argument considering currents at only two frequencies: dc and the Josephson frequency $\omega_J = 2eV/\hbar$, where V is the dc average voltage across the junction. The energy input from the dc current

---

[2] Of course this approximation will break down for such high frequencies (e.g. optical) that the transmission line can no longer be treated as propagating only a single mode, and for dc (where the high resistance of the current bias source is dominant).

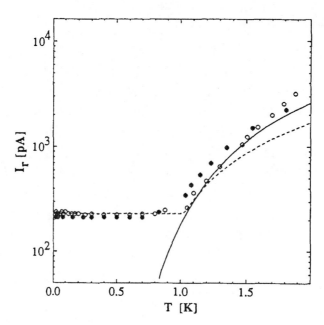

Figure 4. Retrapping current I_r vs. temperature for a junction with $R_n = 8.3$ kΩ. The open circles are data; the solid circles are results of analog simulations. The dashed line is an analytic approximation to the simulation, while the solid line shows the results of a standard model with frequency independent damping.

source is IV; the energy dissipation includes two terms: that dissipated by the ac super-current $I_{co}\sin(\omega_J t)$ generated by the junction flowing into a total impedance $Z(\omega_J)$, i.e., $(I_{co}^2/2)\mathrm{Re}[Z(\omega_J)]$, and that dissipated by the dc quasiparticle current[3] V^2/R_{qp}, so long as $V < V_g$. $Z(\omega_J)$ is obtained from the parallel contributions of the junction capacitance C, the junction quasiparticle resistance R_{qp}, and the real impedance Z_l of the line. That is,

$$1/Z(\omega_J) = i\omega_J C + 1/R_{qp} + 1/Z_l \tag{8}$$

so that

$$\mathrm{Re}[Z(\omega_J)] = \frac{R_{qp}Z_l/(R_{qp} + Z_l)}{1 + [\omega_J C R_{qp} Z_l/(R_{qp} + Z_l)]^2} . \tag{9}$$

Equating the input power to the dissipated power, we obtain

$$IV = \frac{I_{co}^2}{2}\,\mathrm{Re}[Z(\omega_J)] + \frac{V^2}{R_{qp}} . \tag{10}$$

Since $\omega_J = 2eV/\hbar$ and $Z(\omega)$ is known, (10) can then be solved to find

$$I = \frac{V}{R_{qp}} + \frac{I_{co}^2}{2V}\mathrm{Re}[Z] . \tag{11}$$

From the form of the right side of the equation, it is clear that there is a smallest value of I for which the equation can be satisfied. We identify this as the retrapping current

[3] This dc term was omitted in the original work of Johnson et al.

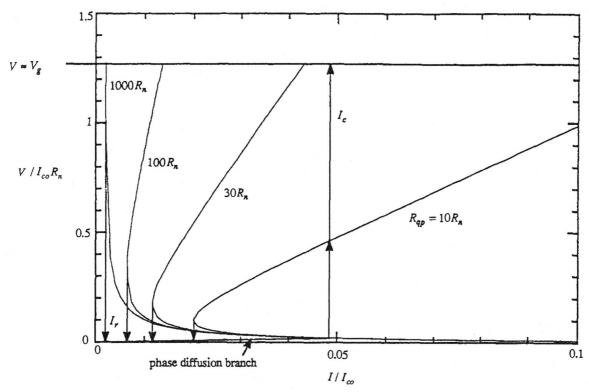

Figure 5. Graphs showing $I - V$ curves predicted by the modified model of Johnson et al., for a representative case in which $R_n = 500$ Re[Z] for ratios $R_{qp}/R_n = 10, 30, 100, 1000$, which correspond roughly to $T/T_c = 0.77, 0.52, 0.38,$ and 0.26. The phase diffusion branch is sketched in schematically at low voltage, but is not part of the basic model.

I_{ro} and the corresponding voltage as the retrapping voltage V_{ro}. Because Re[Z] depends on V through ω_J, we first solve for V_{ro}, and then I_{ro}, finding

$$V_{ro} = I_{co}(R_{qp} \, \mathrm{Re}[Z/2])^{1/2} = I_{ro}R_{qp}/2 \tag{12}$$

$$I_{ro} = I_{co}(2\mathrm{Re}[Z]/R_{qp})^{1/2} \tag{13}$$

in terms of which we can write the complete solution of (11) as

$$V(I) = \frac{IR_{qp}}{2}\left\{1 \pm \left[1 - (\frac{I_{ro}}{I})^2\right]^{1/2}\right\} . \tag{14}$$

Examples of such curves are displayed in Fig. 5. To test the accuracy of this estimation procedure, we use it to calculate the value of I_{ro} for the conventional RCSJ model, in which a single linear resistor of value R describes the dissipation at high frequencies as well as the slope of the quasiparticle $I - V$ curve below the gap. In that case, we can set $R_{qp} = R$ and $Z_l = \infty$ in (9), so that Re[Z] $= R/[1 + (\omega_J C R)^2]$, in which case (13) becomes

$$I_{ro} = I_{co}\left(\frac{2}{1 + \omega_J^2 R^2 C^2}\right)^{1/2} . \tag{15}$$

In the limit when $Q = \omega_J RC \gg 1$, this is $I_{ro} = 2^{1/2} I_{co}/Q$, which agrees with the exact result $4I_{co}/\pi Q$ within 10%. The corresponding retrapping voltage is $V_{ro} = (\hbar\omega_p/2e)/2^{1/4}$, independent of the value of R, and typically well below the gap voltage, so that our assumption about quasiparticle dissipation is self-consistent.

We now repeat this analysis, but with the shunting impedance Z_l of the line included. Since $Z_l \approx 100\ \Omega \ll R_{qp}$, (9) shows that

$$\text{Re}[Z] = \frac{Z_l}{1 + (\omega_J C Z_l)^2} \tag{16}$$

so that (13) becomes

$$I_{ro} = I_{co} \left[\frac{2Z_l}{R_{qp}\left(1 + \omega_J^2 Z_l^2 C^2\right)} \right]^{1/2}. \tag{17}$$

The second term in the denominator is typically somewhat smaller than unity if $C \leq 3$ fF and $Z_l \leq 100\ \Omega$. If we can neglect it, (17) simplifies to

$$I_{ro} \approx I_{co}(2Z_l/R_{qp})^{1/2} \tag{18}$$

which in any case provides an upper bound. According to (12) the retrapping voltage V_{ro} rises without limit as $R_{qp}^{1/2}$, as R_{qp} increases with decreasing T, rather than having a fixed value set by the plasma frequency as in the classic RCSJ model. This result is self-consistent so long as this $V_{ro} \leq V_g$; but when (12) gives a value of V_{ro} which exceeds V_g, it is not self-consistent, since above the gap voltage the dissipation jumps to approximately V^2/R_n which is *much* greater than V^2/R_{qp}. This keeps the voltage pinned at V_g over a wide range of current. A detailed treatment then requires dealing with the sharply nonlinear response at the onset of pair-breaking tunneling. But since R_{qp} is rising rapidly with decreasing temperature, at a slightly lower temperature one reaches the situation where V_g^2/R_{qp} in (10) is negligible, and I_{ro} is pinned at the value

$$I_{ro} \approx \frac{I_{co}^2 Z_l}{2V_g \left[1 + (4\Delta C Z_l/\hbar)^2\right]} \tag{19}$$

which is independent of T at low temperatures, as observed experimentally. In this limit, it is not safe to assume the second term in the denominator can be neglected since the capacitive reactance $\hbar/4\Delta C$ has a numerical value $\sim 300\ \Omega$ for a 1 fF tin junction, which is already in the range expected for Z_l. On the other hand, Z_l itself may be modified by the strongly nonlinear quasiparticle response right at V_g. Thus, (19) may be sensitive to such effects.

For junctions of a given material, $V_g = 2\Delta/e$ is a fixed value for $T \ll T_c$, and Z_l is expected to have a value $\sim 100\ \Omega$ regardless of details of the leads. Thus, this model predicts that I_{ro} should scale with $I_{co}^2 \sim 1/R_n^2$, insofar as the capacitive term in the denominator of (19) can be neglected. In fact, Iansiti et al. [6], had reported an unexplained trend of proportionality to R_n^2 of the apparent limiting value of R at low temperatures, noted after the fact in analyzing measurements of $I - V$ curves of a series of ~ 10 junctions with R_n ranging from 0.5 kΩ to 140 kΩ (see Fig. 6). If we identify this apparent maximum value of R with the limiting low temperature ratio of V_g/I_{ro} given

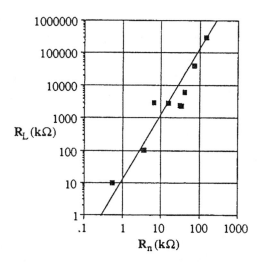

Figure 6. Nominal low-temperature limit R_L (called R_{max} in the text) of quasiparticle resistance in the presence of retrapping for all junctions measured by Iansiti et al. The solid line has a slope of 2, as predicted by (20), and absolute value corresponding to Re[Z] = 500 Ω.

by (19), and insert the low temperature limit BCS result $I_{co}R_n = \pi\Delta/2e = (\pi/4)V_g$, we obtain

$$R_{max} = \frac{32}{\pi^2} \frac{R_n^2}{Z_l} \left[1 + \left(\frac{4\Delta C Z_l}{\hbar} \right)^2 \right] . \tag{20}$$

If one drops the final factor which should be of order unity for small junctions, this result accounts for the observed trend of R_{max} over almost 5 orders of magnitude, and a fit to the data implies a value of Z_l of $\sim$ 500 Ω, within a factor of 3 uncertainty on either side. This spread in values may well stem from the neglect of the capacitive term and other approximations, and from the difficulty in measuring values of I_r in the picoampere range from $I - V$ curves which were taken for other purposes. We also note that if the low temperature limit had not yet been reached at the lowest temperature at which data were taken, (20) would yield too large a value of Z_l. For comparison, Johnson et al. [5], using data (such as those shown in Fig. 4) which were taken more recently as part of a more systematic study of I_r in a similar experimental configuration, obtained a value of $Z_l \approx 90$ Ω using an analog simulation which should be quite reliable; in fact it was these simulator experiments which provided the original motivation for setting up the analytic model. Use of the model presented here yields a similar value. Given the crudeness of the model and the limited accuracy of the data, this sort of order of magnitude consistency may be as much as can be expected. Not only does this model account for the observed sudden bottoming out of I_r at low temperatures, it also accounts for the temperature at which it occurs and the limiting value of I_r, all with a reasonable value for the line impedance Z_l as the only parameter. Further confirmation of the essential correctness of the model is provided by the experimental observation that the retrapping in these junctions *does* occur from a high voltage near V_g.

It should be noted that in this model, for simplicity, no account is taken of thermal noise fluctuations, which in the classic RCSJ model substantially increase I_r. Presumably they would have less effect when the low value of Z_l is taken into account, because the lower resistance has the effect of raising the cutoff frequency of the noise spectrum from

$1/RC$ to $1/Z_lC \sim 10^{13}$ s$^{-1} \gg \omega_J$. This causes much of the noise power density to occur at frequencies so high that the current averages out over the time the phase point spends in one minimum, thus making it less effective in inducing retrapping.

2.4. The phase diffusion branch

Another consequence of the strong high-frequency damping of a high resistance junction by the low resistive impedance of the leads is the existence of a regime of classical phase diffusion which gives rise to a resistive voltage in the nominally zero-resistance state below I_c. This was pointed out by Martinis and Kautz [7] and elaborated in extensive computer simulations by Kautz and Martinis [8]. Such a regime with finite thermally-activated resistance below I_c is a familiar property of nonhysteretic *overdamped* junctions. For example, in the classic work of Ambegaokar and Halperin [9], it is shown that the resistance at $I \approx 0$ is given by $R_o/R_n = [I_0(u/2)]^{-2} \propto u\,e^{-u}$, where $u = \hbar I_{co}/ek_BT$ is the normalized activation energy, I_0 is the modified Bessel function, and the exponential dependence holds when $u \gg 1$. The essential point of the new work was the recognition that a similar steady-state phase diffusion without runaway could occur in a small high resistance junction even if it is *underdamped* at *low* frequencies and shows hysteresis, provided that it is *overdamped* at the *high* frequencies relevant for repeated retrapping into a minimum of the cosine potential.

The simplest estimate of the phase diffusion resistance is that obtained by using the Josephson relation $V = (\hbar/2e)\mathrm{d}\varphi/\mathrm{d}t$, and finding $\mathrm{d}\varphi/\mathrm{d}t$ by taking the difference in the rate of jumping the barrier in downward and upward directions along the tilted washboard. Assuming an attempt frequency ω_A and barrier heights $(E_b \pm hI/4e)$ in the two directions, and assuming that damping is sufficient that each jump stops at the nearest adjacent well where $\Delta\varphi = \pm 2\pi$, standard transition rate theory yields

$$R_o = \frac{V}{I} = \frac{h}{4e^2}\frac{\hbar\omega_A}{k_BT}e^{-E_b/k_BT} \ . \tag{21}$$

The work of Kautz and Martinis shows that this formula qualitatively fits their data if one takes the usual values $\omega_A = \omega_p$ and $E_b = 2E_J$ in (21), but gives a resistance value that is about an order of magnitude too low. Their detailed simulations show that this discrepancy stems largely from the neglect of multiple-well phase-slip events, which become more important at higher temperatures. In fact, (21) is expected to be completely inappropriate when T is high enough that the exponential is no longer small compared to unity, and it clearly fails to reduce to R_n at T_c, where $E_b = 0$. In the extreme low temperature limit, (21) also fails because phase slips would become so infrequent that none occurs during the measurement period, and the statistical averaging implied in (21) becomes non-physical. However, quantum tunneling through the barrier between wells appears to provide a non-zero resistance at $T = 0$, as will be discussed in the second half of this chapter.

As the bias current is increased, the voltage rises above the linear approximation to the resistance of the phase diffusion regime described above, and eventually abruptly switches up to the gap voltage on the quasiparticle branch of the $I - V$ curve at a switching current I_c. This switching current is determined by the frequency-dependent damping at $\omega = \omega_p$, which will be dominated by the transmission line Z_l, and essentially independent of the quasiparticle resistance R of the junction itself. This switching current condition is analyzed in considerable depth by Kautz and Martinis [8] by numerical simulations and by analytic approximations for the case of a junction with isolating resistors, and found to be quite complex.

A less ambitious attack can be made using the simple energy balance arguments introduced in the discussion of the model of Johnson et al. [5]. Here one argues that the dc input power on the phase-diffusion branch can not exceed the maximum power that can be disposed of by the high-frequency current produced by the Josephson junction; in this case the quasiparticle dissipation V^2/R_{qp} is negligible since $R_o \ll R_{qp}$. If more power than this maximum is fed in, it can only be dissipated by having the voltage jump up to the quasiparticle tunneling branch. That is, if the junction is to remain on the phase diffusion branch, IV must be less than $I_{co}^2 \mathrm{Re}[Z/2]$. With the further approximations that R_o holds all the way to I_c, so that $I_c V_c \approx I_c^2 R_o$, and that the capacitive shunting of the leads is negligible at these low voltages (and hence frequencies), this constraint leads to the relation

$$I_c \approx I_{co}(\mathrm{Re}[Z]/2R_o)^{1/2} \tag{22}$$

with $\mathrm{Re}[Z]$ given by (9). Insofar as this crude approximation holds, it predicts that

$$(I_c/I_{co})^2 R_o \approx \mathrm{Re}[Z/2] \tag{23}$$

independent of the values of R_n and T. In fact, when the data of Iansiti et al. [6] and additional data reported in the thesis of Johnson [10] (including data from Ono et al. [11] and from Martinis and Kautz [7]) are analyzed, a trend in semiquantitative agreement with this relation is found, as shown in Fig. 7. This plot indicates that data from 7 samples of 2 different superconductors from 2 laboratories covering a $50:1$ temperature range and ranging over 4 orders of magnitude in R_o are fitted with a value of $\mathrm{Re}[Z] \approx 12\,\Omega$ within a factor of 3 in either direction. In view of the extreme simplicity of the model, this degree of "data collapse" is quite impressive. On the other hand, the recent data of Kuzmin et al. [12], yield a larger value $\sim 700\,\Omega$, presumably reflecting their higher degree of resistive isolation. Because this energy balance relation should hold regardless of whether the phase-slip process proceeds by thermal activation or by quantum tunneling, this experimental result is unable to distinguish between the two mechanisms. Thus, it should continue to hold even at temperatures so low that quantum tunneling effects might be expected to dominate over thermal activation.

Another point should be made about I_c : Experimentally it is found that the measured values of I_c, although greatly depressed below I_{co}, show very little dispersion in the values measured on successive current sweeps. This is completely *inconsistent* with the behavior associated with the depression of I_c by "premature switching" from the zero-voltage state described by (7), since if this were the mechanism, a large depression of $\langle I_c \rangle$ would always be accompanied by a large dispersion of the measured values of I_c. By contrast, in the case of phase diffusion, the phase is constantly slipping and retrapping, and I_c is simply the well-defined point at which the driving force is sufficiently great that this dynamic stability is lost.

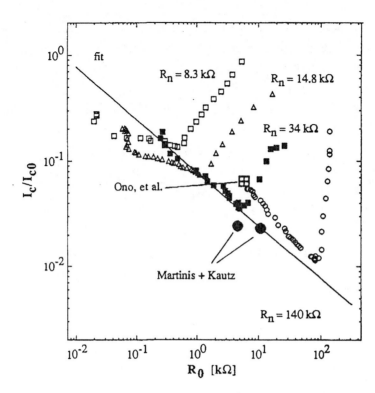

Figure 7. I_c/I_{co} vs. R_o for 5 junctions from the Harvard work, and 3 from NIST. The sudden breakaway of the curves after the minimum of I_c appears to stem from the disappearance of hysteresis in the $I - V$ curve at about that temperature because of the onset of substantial quasiparticle dissipation in addition to the transmission line loss. The solid line is a plot of $(I_c/I_{co})^2 R_o = 6\ \Omega$.

2.5. Recapitulation of classical regime

In the classic RCSJ model, in which R is frequency-independent from dc to microwave frequencies, it is well established that a phase-slip event from the zero-voltage branch launches an irreversible transition to the quasiparticle branch of the $I - V$ curve. At zero temperature, this first occurs when the current is increased to I_{co}, but at finite temperatures, premature switching to the voltage state can occur at a distribution of lower currents centered on $I_c \leq I_{co}$. On sweeping the current down again, the junction retraps into the zero voltage state at a current $I_{ro} \approx 4I_{co}/\pi Q$ (where $Q = \omega_p RC$) at $T = 0$, and at a distribution of current values centered at $I_r \geq I_{ro}$ at non-zero temperatures. In the presence of intense thermal fluctuations, the distributions of I_c and I_r merge together and the transition becomes rounded and non-hysteric, but under no circumstances are two distinct finite voltage regimes possible at the same current value.

Experimentally, however, in small high-resistance junctions, *hysteretic* $I - V$ curves are commonly observed in which the low-voltage branch shows a measurable resistive voltage. As shown by the NIST group, this phenomenon can be explained in classical terms if one introduces frequency-dependent damping, which arises naturally from the fact that at *high* frequencies the junction is loaded by the characteristic impedance of the leads (typically $\sim 100\ \Omega$) even if the current bias circuit has a dc impedance of $10^9\ \Omega$. For the low-voltage phase-diffusion branch to be stable, one requires sufficient damping at $\omega \approx \omega_p$ to give $Q(\omega_p) \approx \omega_p R(\omega_p) C \approx \omega_p Z_l C \leq 1$. Since typically $\omega_p \sim 10^{11}\ s^{-1}$ and $Z_l \approx 100\ \Omega$, this requires $C \leq 10^{-13}\ F = 100\ fF$, i.e., a junction with area $\leq 4\ \mu m^2$. In addition, for the resistance to be appreciable in magnitude, thermal activation must occur constantly, so one needs $k_B T \geq E_J$; for this to hold over a reasonably large temperature range below T_c, one requires $R_n \geq R_Q \approx 6$ kΩ. From these numerical estimates, it is clear why the phase-diffusion regime was not observed until studies were made of small area, high resistance (and hence low critical current) junctions.

3. Quantum effects in Josephson junctions

3.1. Introduction

In the first half of this chapter, we have treated the Josephson junction as a purely classical system, in which the phase difference φ across the junction and the charge $Q = CV$ on the junction were treated as classical variables which could be specified to arbitrary precision. In fact there is a quantum uncertainty relation $\Delta\varphi \, \Delta N \geq 1$, where N is the number of Cooper pairs transferred across the junction, which limits this classical precision and thereby modifies the results of the classical theory of the Josephson effect. To get some feeling for these effects in an elementary way, we start by restricting the theoretical discussion to $T = 0$ and zero damping. The restriction to $T = 0$ is not unreasonable, because the classical theory works quite well at higher temperatures; the key questions to be answered are: does R_o remains finite at $T = 0$, or go to zero as predicted by the classical theory? And, does I_c remain depressed at $T = 0$, or recover its full unfluctuated value I_{co}? The restriction to zero damping is much more limiting but it is needed to permit an elementary treatment because to introduce damping into the quantum mechanics, it is necessary to include the heat bath in the Hamiltonian of the system being treated, as shown by Caldeira and Leggett [13], for example. In fact, damping from the low impedance leads may largely quench the "static" quantum effects, but introduce quantum noise which induces a "dynamic" phase uncertainty $\langle (\Delta\varphi)^2 \rangle$.

Before going into any theory, let us summarize the experimental situation as found in the work of Iansiti et al. [6], on a series of very small $Sn - SnO_x - Sn$ junctions, with normal resistance up to 140 kΩ and capacitances of order 1 fF. The behavior from T_c (3.7 K) down to below ~ 1 K was found to be well described by the classical theory discussed above. Below about 2 K, hysteresis is found, with a low-voltage phase diffusion branch and a quasiparticle branch at the gap voltage, as shown in Fig. 8. If quantum effects are observable, they are probably most evident in the phase-diffusion branch, which is described by an initial slope R_o, and a maximum current I_c. In classical theory, one would expect R_o to go to zero at $T = 0$, since all thermally activated processes should freeze out. The experiments summarized in Fig. 9 show that R_o falls far below R_n at low temperatures, but (at least for $R_n > 6$ kΩ) appears to extrapolate at $T = 0$ to a finite value which increases systematically with R_n. This behavior suggests that the phase diffusion process continues by a macroscopic tunneling type of process after the thermally-activated channel freezes out, but this is hard to reconcile with the expected supression of MQT by strong damping. Similarly, at very low temperatures I_c is found to rise, but it appears to reach a limiting value at $T \approx 0$ which is still an order of magnitude below I_{co}. The temperature dependence of both I_c and I_r are shown for a 70 kΩ junction in Fig. 10, and $I_c(T)$ for seven junctions are systematically compared in Fig. 11. As noted earlier, the relation $(I_c/I_{co})^2 R_o \approx$ constant $\sim \mathrm{Re}[Z_l/2]$ suggested by (23) appears

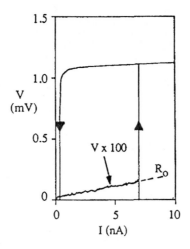

Figure 8. $I - V$ characteristic for a junction with $R_n = 14.8$ kΩ at $T = 0.6$ K. The phase-diffusion branch is replotted on a 100 times expanded scale.

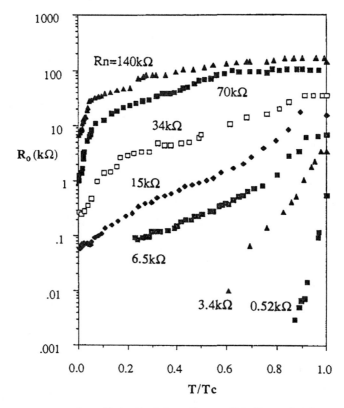

Figure 9. $R_o(T)$ for seven samples. Note that junctions with $R_n > 6$ kΩ appear to have $R_o(0) > 0$.

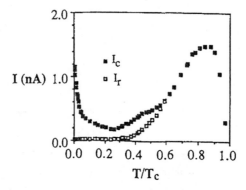

Figure 10. $I_c(T)$ and $I_r(T)$ for a 70 kΩ junction, showing the region of hysteresis at lower temperatures. These values may be compared with the Ambegaokar-Baratoff value $I_{co}(0) \approx 13$ nA.

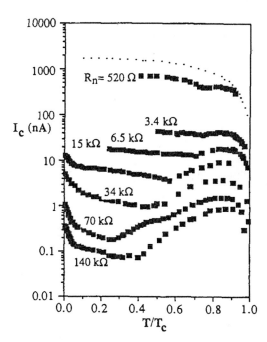

Figure 11. $I_c(T)$ for seven samples. For comparison, the dotted line shows the unfluctuated, or Ambegaokar-Baratoff values, of $I_{co}(T)$ for the 550 Ω sample. The temperature dependence of I_{co} is the same for all.

to hold, at least qualitatively (see Fig. 7), suggesting that indeed the stability of the phase diffusion branch may be maintained by damping due to the transmission line loading. If this is the case, I_c and R_o are not independent quantities, and an explanation of either one would suffice to clarify the situation considerably.

3.2. The isolated junction

For the simplest possible starting point, we consider an isolated junction with no leads attached, so that $I = 0$. After making the operator replacement for the charge in the φ-representation, namely $Q/2e \longrightarrow i\partial/\partial\varphi$, the form of the Hamiltonian based on (5) but including the charging energy is

$$H = -E_J \cos\varphi - 4E_c \frac{\partial^2}{\partial\varphi^2} \tag{24}$$

where $E_c = e^2/2C$ is the charging energy of the junction for a single electron charge. This H describes only the Cooper pairs, neglecting the quasiparticle degrees of freedom, which should be unimportant at $T = 0$ and $I = 0$.

We see that the parameter $y \equiv 4E_c/E_J$ provides a measure of the relative importance of the charging energy of pairs in forcing a delocalization of the phase, away from the minimum potential energy point at $\varphi = 0$. To get a feel for the numbers, it is convenient to recall that $E_c/k_B \approx 1$ K for $C = 1$ fF, and varies inversely with C, while $E_J/k_B \approx 0.88$ $(R_Q/R_n)T_c$, with $R_Q = h/4e^2$. The ratio $y \equiv 4E_c/E_J = 4.53(R_n/R_Q)(E_c/k_BT_c)$, so that for small junctions with $C \approx 1$ fF and $T_c \approx 4$ K, $y \approx (R_n/R_Q)$, and $R_n \approx R_Q$ is a reasonable indicator of the crossover point from one regime to the other. For $y \ll 1$, the ground state is a narrowly peaked wavefunction $\Psi(\varphi)$ with width of order $y^{1/4}$, and there are a number of higher states in the potential minimum, resembling the excited states of a harmonic oscillator. By contrast, when $y \gg 1$, the term in E_c is dominant, and Ψ approaches a constant to minimize it. For this case, one can no longer ignore the periodicity of the potential term $-E_J \cos\varphi$ and the question of whether φ should be viewed as an extended variable, or a cyclic one such that φ and $\varphi + 2\pi$ are physically indistinguishable. From the former point of view, $\Psi(\varphi)$ had the form of a Bloch function $u(\varphi)e^{iq\varphi}$, where q is a "charge" or "pair number" variable, and $u(\varphi)$ is periodic with period 2π. From the latter point of view, $\Psi(\varphi)$ is only defined between $-\pi$ and $+\pi$, and it must satisfy appropriate boundary conditions at those points. So long as we restrict our attention to the absolute ground state, which we expect to correspond to $q = 0$ in the Bloch picture, and to the boundary conditions $\Psi(\pi) = \Psi(-\pi)$ and $\Psi'(\pi) = \Psi'(-\pi) = 0$ in the single cell picture, both pictures yield the same eigenvalue problem, and the same energy eigenvalues E.

3.2.1. Variational approach to the ground state

Since this problem is one-dimensional, it is easy to solve exactly by numerical means. However, one gets a bit more insight by a variational approach to find an approximation to the ground state, using trial functions appropriate to the limiting cases of $y \ll 1$ and $y \gg 1$, respectively. For $y \ll 1$, one asssumes a Gaussian trial function,

$$\Psi(\varphi) \sim \exp(-\varphi^2/4\sigma^2) \qquad (25)$$

where σ, the rms spread in φ, is chosen to minimize the expectation value of (24). The resulting minimum energy is

$$E = E_J(1 - \sigma^2)\exp(\sigma^2) \qquad (26)$$

where σ has the valued determined by the solution of the transcendental equation

$$\sigma^4 \exp(-\sigma^2/2) = y/2 \qquad (27)$$

For $y \ll 1$, (26) and (27) lead to the analytic approximation

$$E = -E_J \left[1 - (y/2)^{1/2}\right] . \qquad (28)$$

In the other limiting case of $y \gg 1$, an appropriate trial function which satisfies the boundary conditions at the edge of the cell is

$$\Psi(\varphi) \sim (1 + a\cos\varphi) . \qquad (29)$$

Minimization of the expectation value of the energy with respect to the parameter a leads to the condition that

$$a = y \left[\left(1 + 2/y^2\right)^{1/2} - 1\right] . \qquad (30)$$

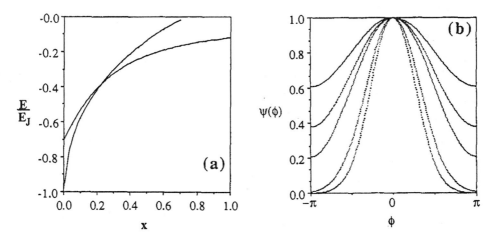

Figure 12. (a) Binding energy vs. $x = E_c/E_J$ for Gaussian and cosinusoidal trial functions, as explained in text. (b) Ground state wave functions $\Psi(\varphi)$ for $x = y/4 = 1, 0.5, 0.3, 0.1,$ and 0.05.

For this value of a, the energy is

$$E = -E_J(\frac{y}{2}) \left[(1 + \frac{2}{y^2})^{1/2} - 1 \right] .$$

$$(31)$$

For $y \gg 1$, this has the limiting form

$$E \approx -E_J/2y = -E_J^2/8E_c$$

$$(32)$$

where the second form shows explicitly that in this limit the binding energy is *second order* in E_J, in contrast to the *first order* binding energy in the classical limit.

These variational approximations to the ground state energy are plotted in Fig. 12a. The tight-binding approximation (26) and (27) gives a lower (more accurate) energy for $y < 1$, and the loose-binding approximation (29) gives a better energy for $y > 1$. Numerical solutions in the cross-over region near $y = 1$ show that the exact binding energy exceeds the better of the two approximations by less than 5%, even in the worst case. The wave functions $\Psi(\varphi)$ for values of $y(\equiv 4x)$ ranging from 0.2 to 4 are shown in Fig. 12b. Qualitatively, it is clear that for $y > 1$, (e.g., for $C \sim 1\,\text{fF}$ and $R_n > R_Q$) the probability density for the phase variable φ is sufficiently delocalized that it is no longer a good approximation to treat φ as a semiclassical variable as we did in the first half of this chapter. We must still investigate to what extent these quantum effects lead to consequences which might be observable in the low temperature limit of the experimental data, which were, of course, taken on junctions which were not isolated but had leads attached. But first, let us examine these same quantum states of the isolated junction using the more appropriate, if less familiar, number (charge) representation instead of the φ-representation we have been using.

3.2.2. Particle number eigenstates

Going back to BCS, a superconducting ground state with definite phase φ but ill-defined particle number is given by [14]

$$|\varphi\rangle = \prod_k \left(|u_k| + |v_k|\, e^{i\varphi} c_{k\uparrow}^* c_{-k\downarrow}^* \right) |0\rangle \tag{33}$$

where the u_k and v_k are given by the BCS theory, and $|0\rangle$ represents the electron vacuum. Such states have the same energy for any value of φ, and can be combined in any linear combination and still represent a possible BCS ground state with the same energy. As pointed out by Anderson many years ago, the linear combination which has a precise number N of Cooper pairs is

$$|N\rangle = (1/2\pi) \int_0^{2\pi} \mathrm{d}\varphi\, e^{-iN\varphi} |\varphi\rangle \;. \tag{34}$$

This number eigenfunction is the natural choice for discussing problems dominated by charging energy, just as the φ-representation is the natural choice for discussing situations dominated by the phase-dependent Josephson coupling energy.

For example, in a single isolated superconductor, the minimum energy state would be of the form (34) with N chosen so that the charge of the electrons exactly cancels the charge of the positive ion cores, in order that the system would be electrically neutral and have no capacitive self-energy. If a second isolated superconductor is present but far away from the first, it will also be in such a state with M pairs, with M chosen so it too will be electrically neutral. If we now bring these two superconductors close enough together that Cooper pairs can tunnel back and forth, the system could lower its energy by the Josephson effect by having the phases of the two superconductors lock together, but this is impossible in a number eigenstate like (34) in which all phases are equally represented.

To gain the benefit of the Josephson coupling energy, a superposition of $|N\rangle$ states must be set up which has a non-zero expectation value of $\cos\varphi$. To avoid any net charge on the combined system, the system state must be made up of a superposition of product states of the form

$$|a\rangle = \sum a_j |N + j\rangle |M - j\rangle \tag{35}$$

where in the jth term the charge on the junction capacitance is $2je$. Clearly the Coulomb energy in the jth term will be $4j^2 E_c$. The energy trade-off is clear: to gain a well-defined phase to optimize the Josephson effect requires a large number of terms in (35), which increases the Coulomb energy. Finding the state of lowest total energy by choosing the set of a_j is an alternative way of minimizing the expectation value of the Hamiltonian (24) or solving the equivalent Schrödinger equation.

In the weakly coupled case, only a few terms in (35) are important, since large charge transfers are energetically prohibitive. For example, the trial function (29) is equivalent to

$$|a\rangle = |N\rangle|M\rangle + (a/2)[|N + 1\rangle|M - 1\rangle + |N - 1\rangle|M + 1\rangle] \tag{36}$$

for which, by inspection, one sees that

$$\langle (\delta N)^2 \rangle = \frac{a^2}{2 + a^2} \approx \frac{a^2}{2} . \tag{37}$$

For $y = 4E_c/E_J \gg 1$, (30) yields the approximate result $a = 1/y$, so that $\langle (\delta N)^2 \rangle \approx 1/2y^2 \ll 1$, and on average only a small fraction of a single Cooper pair is (virtually) exchanged in the entire macroscopic junction area. In the other, low y limit ($E_c \ll E_J$), (27) shows that $\langle (\delta N)^2 \rangle \approx (8y)^{-1/2}$, so that to have $\langle (\delta N)^2 \rangle$ be even as large as unity, y must be reduced to a value corresponding to $E_J = 32E_c$.

To find the Josephson coupling contribution to the energy, one must evaluate the expectation value $\langle \cos \varphi \rangle$, where φ is the phase *difference* between the two superconductors in the state $|a\rangle$ in (36). Using (34) one finds

$$\langle \cos \varphi \rangle = \frac{2a}{2 + a^2} \approx a \tag{38}$$

in agreement with what is found from (29). Using (38) to evaluate the Josephson energy and (37) to find the charging energy, and varying a to minimize the sum of the energies, one obtains $a = 1/y$, confirming by an independent procedure the values found earlier for a and the minimum energy, for $y \gg 1$.

3.3. Estimation of the critical current

While the calculation of the binding energy in the absolute ground state of the isolated junction can be done with some confidence by the procedure outlined above, this is not a directly measureable quantity. We now seek to estimate the critical current if this junction could be connected to leads of infinite impedance at all frequencies, so that damping would remain zero. Three distinct approaches were discussed by Iansiti et al. [6] as a guide to interpretation of actual data taken with less ideal leads.

In the first, one simply argues that I_c will be reduced by the quantum phase fluctuations by the same factor as is the binding energy. This is motivated by the argument that the maximum slope $dE/d\langle \varphi \rangle$, which determines the maximum tilt in the washboard potential consistent with a stable minimum point, will scale with the effective well depth, since the period in $\langle \varphi \rangle$ of E will remain 2π. In the weak coupled limit ($y \gg 1$), (32) then leads to

$$I_c = I_{co}/2y = I_{co}(E_J/8E_c) \tag{39}$$

a result which scales with E_J^2 rather than with E_J as in the classical limit. The physical significance of this result is admittedly problematic because time-dependent eigenfunctions can not be found for the tilted washboard problem, because of quantum tunneling to the next lower well in the washboard. A value exactly twice as large as (39) was estimated by Mirhashem and Ferrell [15], using a linear response theory approach which assumes that I_c scales with the inverse kinetic inductance, as in classical Josephson junctions.

Since a current can only flow by transfer of Cooper pairs from one electrode to the other, and since in the loose coupled case, charging energies force a careful accounting

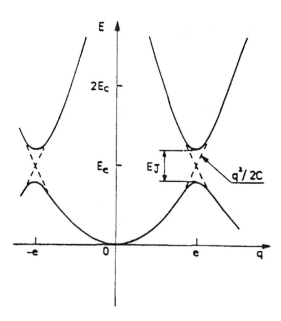

Figure 13. Schematic diagram of energy bands as described by Likharev and Zorin (Ref. [16]).

of Cooper pair number to minimize Coulomb energy in the junction capacitance, a more appropriate theory should focus on number instead of the complementary variable, phase. Such states are of the form (35). For example, (36) represents a state with zero *expectation value* for the number of Cooper pairs or charges transferred, but nonetheless with non-zero charging energy because of the virtual exchange of pairs in *both* directions with equal amplitudes a. By adjusting the coefficients a_j, one can adjust the charge on the junction and work out the energy as a function of charge (treated as a continuous variable), obtaining an energy band picture, with Bloch functions in which the role of crystal momentum k in ordinary band theory is played by the dimensionless charge variable q. Such a picture was elaborated in detail by Likharev and Zorin [16], in a path-breaking paper, with results sketched in Fig. 13. In this picture, a dc current source forces a steady evolution of the charge variable q, just as a steady electric field forces a steady evolution of k in ordinary band theory. Similarly, dE/dq fixes the voltage on the capacitor which determines $d\varphi/dt$, just as dE/dk determines the particle velocity in ordinary band theory. In the equivalent of the free electron parabolic energy band, the band energy is $q^2(2e)^2/2C$, with $-1/2 \leq q \leq 1/2$. The two zero-order zone boundary states, which differ by transfer of one Cooper pair, have the same energy. When the Josephson term of the Hamiltonian is taken into account, they combine to form two eigenstates which have zero expectation value for charge and voltage, and differ in energy by a band gap of width E_J.

Given this picture, if a small constant current is applied, q evolves steadily in time, and the expectation value of charge and voltage oscillate periodically in time (Bloch oscillation) with dE/dq. Each time as the system passes through the zone boundary, it has two alternatives: transfer a Cooper pair and repeat the same cycle, or jump across the gap to the next higher band in a variant of the Zener tunneling which is familiar from ordinary band theory [17]. By transcribing the usual result to the new variables, one finds that the probability of Zener tunneling on each cycle is

$$P_{\text{Zener}} = e^{-I_Z/I} \tag{40}$$

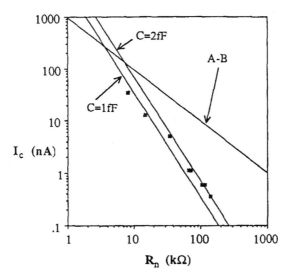

Figure 14. Measured criti... l current I_c at $H = 0$ and $T = 30$ mK for six samples shown by black squares. The A-B line shows the Ambegaokar-Baratoff prediction for I_{co}, while the other two lines correspond to (39) with two representative values of C. Although there is reasonable agreement, the trend is significantly incorrect when the systematic increase in C with decreasing R_n is taken into account.

where the characteristic current level at which the probability is $1/e$ is $I_Z = \pi e E_J^2/8\hbar E_c$ $= (\pi/4y)I_{co}$, which is just $\pi/2$ times the estimate for I_c found in (39).

Thus, there are three approaches, all of which give values of a characteristic critical current $\sim I_{co}(E_J/E_c)$ with similar prefactors. This form predicts $I_c \propto C/R_n^2$ for a given superconductor, provided $E_c \gg E_J$. The data of Iansiti et al. [6], on I_c extrapolated to $T = 0$ for 7 tin-tin junctions taken down to 30 mK, shown in Fig. 14, appear to show a trend of $I_c \propto R_n^{-2}$ over two orders of magnitude, with a coefficient in agreement with (39) if C is taken in the expected range $1 - 2$ fF. On the face of it, this agreement made it plausible that the quantum effects on which (39) and (40) are based are responsible for the fact that I_c remains more than an order of magnitude below I_{co} even at temperatures far below both E_J and E_c. On closer inspection, however, one notes that the data points for the *smaller*, higher resistance junctions tend to fall on the theoretical line for the *larger* capacitance, so that the agreement with (39) is not so convincing as it first appeared. Thus, the apparent good agreement may well be fortuitous, particularly in view of the fact that the experiments were carried out with leads having low high-frequency impedance, not the infinite impedance assumed in the analysis above.

The need for this cautious appraisal is supported by our recent tantalizing observation that the same low-temperature I_c data over a wide range of samples is fitted even better, as shown in Fig. 15, by the distinctly different functional dependence

$$I_c/I_{co} = (1/\pi^2)(2E_J/E_c)^{1/2} \tag{41}$$

for which no convincing argument is available. Two very different conclusions could be drawn from this remarkable fit: (a) This formula must have some physical basis because it fits so well, and theoretical justification should be sought, or (b) one must be skeptical of even impressive numerical fits to data if there is no feature to give a clean "signature" of the underlying model, such as the sharp energy-gap-induced break in the temperature

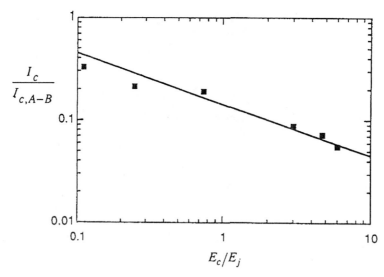

Figure 15. Comparison of the same data shown in Fig. 14 (plotted with different axes) with (41), which gives remarkably good fit to the entire trend.

dependence of I_r in a large variety of samples which is explained by the model of Johnson et al., as described above.

Recently, formal theoretical analyses of the single junction case with a more realistic range of assumptions about the circuit impedance have been made by Averin et al. and by Falci et al. [18]. It appears that these calculations elaborated in Chap. 2 may be providing numerical results which can account for the order of magnitude and general temperature dependence of our experimental results. Unfortunately these results are not yet sufficiently detailed to allow quantitative comparison to be made with our data.

3.4. Estimation of R_o

The value given by (21) for R_o arising from phase diffusion in a classical junction, namely

$$R_o = \frac{V}{I} = \frac{h}{4e^2} \frac{\hbar \omega_A}{k_B T} e^{-E_b/k_B T} \tag{42}$$

goes to zero exponentially at $T = 0$ for any finite barrier $E_b \approx 2E_J$, whereas our experiments indicate a nonzero limiting value as T approaches zero. To evaluate this situation more quantitatively, it is convenient to note that the dominant exponential factor can be written

$$\exp(-2E_J/k_B T) = \exp(-1.76\, R_Q/R_n t) = \exp(-11.4\,\text{k}\Omega/R_n t) \tag{43}$$

where $t = T/T_c$, and BCS relations have been assumed. Thus, the exponential drop in R_o will not really start until $t \leq 11.4\,\text{k}\Omega/R_n$, which corresponds to $T \approx 300$ mK for a 140 kΩ junction, the highest resistance studied by Iansiti et al. For such high resistance junctions the strong freeze out of R_o is predicted to take place at still lower temperatures. If one inserts the numerical prefactors in (21), one predicts a value of R_o of less than an Ohm at 20 mK (the lowest accessible temperature in the experiments), whereas the experiments on this junction appear to yield a low temperature limit close

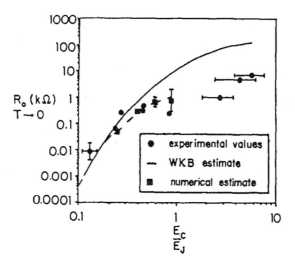

Figure 16. Comparison of estimates and measured values for the low-temperature low-voltage resistance R_o as a function of E_c/E_J. Estimates are described in the text.

to the quantum resistance 6 kΩ. This discrepancy suggests that a quantum tunneling process may be at work at this low temperature.

Iansiti et al. [6] followed up this idea by carrying out calculations of the probability of quantum tunneling from one minimum to the next in the limit of very small current (or small tilt of the potential). In the WKB approximation, the tunneling probability takes the place of the thermal activation probability in (21), and one obtains within numerical factors

$$R_o \sim \frac{h}{4e^2} \exp\left[-(\frac{2\pi E_J}{E_c})^{1/2}\right] .$$ (44)

Note that the exponential factors in (44) and (21) are equal at a crossover temperature given by

$$k_B T_{\text{cr}} = \hbar\omega_p/2\pi$$ (45)

if we recall that $E_b = 2E_J$ and $\hbar\omega_p = (8E_J E_c)^{1/2}$; below this T_{cr}, tunneling should dominate. This WKB procedure was fairly accurate so long as $E_c/E_J < 1/2$, but for larger values of this ratio, numerical methods had to be used to get a better approximation to the tunneling rate. Finally, for $E_c > E_J$, the lifetime-determined level widths become comparable with the well depth, and the method fails since localization of phase is lost. Despite these limitations, the calculations yielded values of R_o which increased with E_c/E_J, and appeared to level off at $R_o \approx 6$ k$\Omega \approx h/4e^2$, the quantum resistance which sets the scale for the expression. Moreover, the experimental data for R_o at the lowest temperatures showed the same trend, with good agreement also in numerical values, as shown in Fig. 16. Unfortunately, this analysis fails to take account of the strong damping expected from the electromagnetic environment and of the effect of multiple potential wells. Whether or not the theoretical analysis is trustworthy, however, the experimental data of Iansiti et al. seem to be well described by it, and suggest some sort of quantum limit for the low temperature resistance.

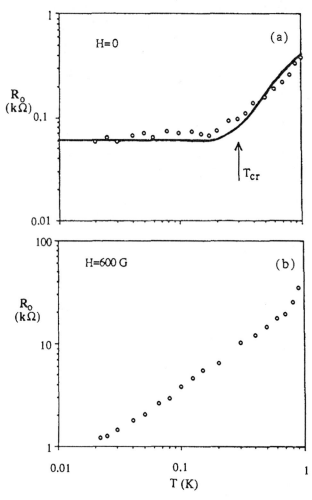

Figure 17. R_o vs. T for a sample with $R_n = 14.8$ kΩ. (a) for $H = 0$, and (b) for $H = 600$ G. The experimental values are the hollow circles, while the solid curve is a theoretical fit. While the data taken at $H = 0$ show a clear plateau attributed to quantum tunneling, the data taken at 600 G appear to be thermally activated for the complete temperature range. The comparison of these two curves seems to exclude thermometry or heating problems as a possible explanation for the low-temperature plateau in (a).

In subsequent work, Iansiti et al. [19] attempted a more detailed comparison of theory and experimental data on $R_o(T)$ for a particular junction, with $R_n = 14.8$ kΩ and $C \approx 2.5$ fF. This relatively low value of R_n was chosen to obtain a sample where $(E_c/E_J) \approx 1/4$, so that the phase is sufficiently localized that the semi-classical language still makes sense. From the lowest accessible temperatures (20 mK) up to 0.35 K, Iansiti et al. measured a value of $R_o(= 60\,\Omega)$ which was temperature independent and in good agreement with the value computed by the MQT approach, using the light-damping approximation with damping resistance ~ 35 kΩ. At higher temperatures, R_o rose rapidly with increasing temperature, indicating thermally activated processes were dominating. This part of the behavior was also calculated theoretically, taking account of quantum corrections and damping from the same 35 kΩ resistance in the thermal activation theory, and again found to agree well with the data, as shown in Fig. 17a. The crossover temperature of 0.035 K was also found in good agreement with the value estimated from (45).

To test whether this crossover was a false indicator, stemming from a failure of the sample temperature to follow the thermometer temperature down to lower temperatures, a magnetic field of 600 G was applied to reduce the crossover temperature. When this was done, R_o increased to a higher value, and continued to decrease smoothly all the way down to 20 mK, as shown in Fig. 17b, indicating that the sample then remained in the classical thermally activated regime, and that the sample temperature did indeed follow the thermometer. On the face of it, the agreement of data and calculations is so good over a wide range that this fit is hard to ignore. Still, no convincing explanation has yet been provided for such a large value for the damping resistance as 35 kΩ at ω_p, when a value of 100 Ω is more plausible for any sort of leads. In addition, the model of Johnson et al. discussed earlier seems to require a loading impedance of the expected order ~ 100 Ω to fit the retrapping current data, and also to account qualitatively for the empirical relation $(I_c/I_{co})^2 R_o \approx 6$ Ω for a large number of samples. The resolution of this puzzle remains unclear but may follow naturally from the sort of calculations presented in Sec. 5 of Chap. 2.

ACKNOWLEDGEMENTS. The assistance of T. S. Tighe in the preparation of figures for this manuscript is gratefully acknowledged. This research was supported in part by NSF grant DMR-89-12927, by ONR grant N00014-J-1565, and by JSEP grant N00014-89-J-1023.

References

[1]   V. Ambegaokar and A. Baratoff, Phys. Rev. Lett. **10**, 486 (1963).

[2]   W. C. Stewart, Appl. Phys. Lett. **12**, 277 (1968); D.E. McCumber, J. Appl. Phys. **39**, 3113 (1968).

[3]   T. Fulton and L. N. Dunkleberger, Phys. Rev. B **9**, 4760 (1974).

[4]   E. Ben-Jacob, D. J. Bergman, B. J. Matkowsky, and Z. Schuss, Phys. Rev. A **26**, 2805 (1982).

[5]   A. T. Johnson, C. J. Lobb, and M. Tinkham, Phys. Rev. Lett. **65** , 1263 (1990).

[6]   M. Iansiti, M. Tinkham, A. T. Johnson, W. F. Smith, and C. J. Lobb, Phys. Rev. B **39**, 6465 (1989).

[7]   J. M. Martinis and R. L. Kautz, Phys. Rev. Lett. **63**, 1507 (1989).

[8]   R. L. Kautz and J. M. Martinis, Phys. Rev. B **42**, 9903 (1990).

[9]   V. Ambegaokar and B. I. Halperin, Phys. Rev. Lett. **22**, 1364 (1969).

[10] A. T. Johnson, Jr. Dissertation, Harvard University, 1990, (unpublished).

[11] R. H. Ono, M. J. Cromar, R. L. Kautz, R. J. Soulen, J. H. Colwell, and W. E. Fogle, IEEE Trans. Magn. **MAG-23**, 1670 (1987).

[12] L. S. Kuzmin, Yu. V. Nazarov, D. B. Haviland, P. Delsing, and T. Claeson, Phys. Rev. Lett. **67**, 1161 (1991).

[13] A. O. Caldeira and A. J. Leggett, Phys. Rev. Lett. **46**, 211 (1981).

[14] M. Tinkham, *Introduction to Superconductivity*, McGraw-Hill 1975; reprinted by Krieger 1980, 1985.

[15] B. Mirhashem and R. A. Ferrell, Phys. Rev. Lett. **61**, 483 (1989).

[16] K. K. Likharev and A. B. Zorin, J. Low Temp. Phys. **59**, 347 (1985).

[17] See, for example, J. M. Ziman, *Principles of the Theory of Solids*, Cambridge, 1964, pp. 163-168.

[18] D. V. Averin, Yu. V. Nazarov, and A. A. Odintsov, Physica B **165&166**, 945 (1990); G. Falci, V. Bubanja, and G. Schön, Z. Phys. B **85**, 451 (1991); for background, see G. Schön and A. D. Zaikin, Phys. Rep. **198**, 237 (1990).

[19] M. Iansiti, A. T. Johnson, C. J. Lobb, and M. Tinkham, Phys. Rev. B **40**, 11370 (1989).

Chapter 5

Coulomb-Blockade Oscillations
in Semiconductor Nanostructures

H. VAN HOUTEN, C. W. J. BEENAKKER,

and A. A. M. STARING

Philips Research Laboratories, 5600 JA Eindhoven, The Netherlands

1. Introduction

1.1. Preface

Coulomb-blockade oscillations of the conductance are a manifestation of single-electron tunneling through a system of two tunnel junctions in series (see Fig. 1) [1]–[5]. The conductance oscillations occur as the voltage on a nearby gate electrode is varied. This setup is the SET transistor described in Sec. 6 of Chap. 2. The number N of conduction electrons on an island (or dot) between two tunnel barriers is an integer, so that the charge $Q = -Ne$ on the island can only change by discrete amounts e. In contrast, the electrostatic potential difference of island and leads changes continuously as the electrostatic potential ϕ_{ext} due to the gate is varied. This gives rise to a net charge imbalance $C\phi_{\mathrm{ext}} - Ne$ between the island and the leads, which oscillates in a saw-tooth pattern with gate voltage (C is the mutual capacitance of island and leads). Tunneling is blocked at low temperatures, except near the degeneracy points of the saw-tooth, where the charge imbalance jumps from $+e/2$ to $-e/2$. At these points the Coulomb blockade of tunneling is lifted and the conductance exhibits a peak. In metals treated in the previous chapters, these "Coulomb-blockade oscillations" are essentially a classical phenomenon [6, 7]. Because the energy level separation ΔE in the island is much smaller than the thermal energy $k_{\mathrm{B}}T$, the energy spectrum may be treated as a continuum. Furthermore, provided that the tunnel resistance is large compared to the resistance quantum h/e^2, the number N of electrons on the island may be treated as a sharply defined classical variable.

Coulomb-blockade oscillations can now also be studied in semiconductor nanostructures, which have a discrete energy spectrum. Semiconductor nanostructures are fabricated by lateral confinement of the two-dimensional electron gas (2DEG) in Si-inversion layers, or in GaAs-AlGaAs heterostructures. At low temperatures, the conduction elec-

Single Charge Tunneling, Edited by H. Grabert and
M.H. Devoret, Plenum Press, New York, 1992

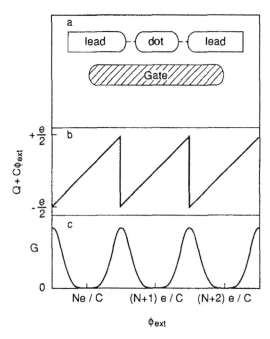

Figure 1. (a) Schematic illustration of a confined region (dot) which is weakly coupled by tunnel barriers to two leads. (b) Because the charge $Q = -Ne$ on the dot can only change by multiples of the elementary charge e, a charge imbalance $Q + C\phi_{ext}$ arises between the dot and the leads. This charge imbalance oscillates in a saw-tooth pattern as the electrostatic potential ϕ_{ext} is varied (ϕ_{ext} is proportional to the gate voltage). (c) Tunneling is possible only near the charge-degeneracy points of the saw-tooth, so that the conductance G exhibits oscillations. These are the "Coulomb-blockade oscillations".

trons in these systems move over large distances (many μm) without being scattered inelastically, so that phase coherence is maintained. Residual elastic scattering by impurities or off the electrostatically defined sample boundaries does not destroy this phase coherence. The Fermi wavelength $\lambda_F \sim 50$ nm in these systems is comparable to the size of the smallest structures that can now be made using electron-beam lithography. This has led to the discovery of a variety of quantum size effects in the ballistic transport regime. These effects may be adequately understood without considering electron-electron interactions [8].

The first type of semiconductor nanostructure found to exhibit Coulomb-blockade oscillations is a narrow disordered wire, defined by a split-gate technique [9]–[14]. As shown in Fig. 2a, such a quantum wire may break up into disconnected segments if it is close to pinch-off. Conduction at low temperatures proceeds by tunneling through the barriers delimiting a segment, which plays the role of the central island in Fig. 1. The dominant oscillations in a wire typically have a well-defined periodicity, indicating that a single segment limits the conductance. Nevertheless, the presence of additional segments may give rise to multiple periodicities and to beating effects.

The second type of nanostructure exhibiting Coulomb-blockade oscillations is a small artificially confined region in a 2DEG (a quantum dot), connected by tunnel barriers either to narrow leads (Fig. 2b) [15, 16], or to wide electron reservoirs (Fig. 2c) [17]. The distinction between these two types of nanostructures is not fundamental, since a segment of a quantum wire delimited by two particularly strong scattering centers can be seen as a naturally formed quantum dot. Both types of structure are of interest:

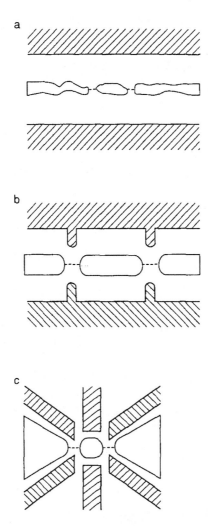

Figure 2. Schematic top-view of three semiconductor nanostructures exhibiting Coulomb-blockade oscillations. Hatched regions denote gates, electron gas regions are shaded. Dashed lines indicate tunneling paths. (a) Disordered quantum wire with a single conductance limiting segment. (b) Quantum dot in a narrow channel. (c) Quantum dot between wide regions with separate sets of gates to modulate the tunnel barriers, and to vary the external potential of the dot.

Whereas artificially defined quantum dots are more suited to a study of the effect under relatively well-controlled conditions, the significance of the phenomenon of *periodic* conductance oscillations in *disordered* quantum wires lies in its bearing on the general problem of transport in disordered systems. It contradicts the presumed ubiquity of random conductance fluctuations in mesoscopic systems, and directly demonstrates the predominant role of electrostatic interactions in a disordered conductor [18].

In a typical experiment, the segment of the wire, or the quantum dot, contains $N \sim 100$ electrons, with an average energy level separation $\Delta E \sim 0.2$ meV. At temperatures below a few Kelvin, the level spacing ΔE exceeds the thermal energy $k_{\mathrm{B}}T$, so that transport through the quantum dot proceeds by resonant tunneling. Resonant tunneling can by itself also lead to conductance oscillations as a function of gate voltage or Fermi energy. The interplay of resonant tunneling and the Coulomb blockade occurs when ΔE and the charging energy e^2/C are of comparable magnitude (which is the case experimentally, where $e^2/C \sim 1$ meV). This chapter reviews our current understanding of this interplay in semiconductor nanostructures. After a brief introduction to the properties of a 2DEG (based on Ref. [8]) we present in Sec. 2 a discussion of the key results of a linear response theory for Coulomb-blockade oscillations in a quantum dot [19, 20]. In Sec. 3 we review experimental results on quantum dots [15]–[17] and disordered quantum wires [9]–[14] in the absence of a magnetic field, and discuss to what extent they are now understood.

Kastner and collaborators [9, 10, 15, 21] originally suggested that the conductance oscillations which they observed were due to the formation of a charge density wave or "Wigner crystal". They inferred from a model due to Larkin and Lee [22], and Lee and Rice [23], that the conductance would be thermally activated because of the pinning of the charge density wave by impurities in the narrow channel. The activation energy would be determined by the most strongly pinned segment in the channel, and periodic oscillations in the conductance as a function of gate voltage or electron density would reflect the condition that an integer number of electrons is contained between the two impurities delimiting that specific segment. A Wigner crystal is a manifestation of long-range order neglected in the theory of Coulomb-blockade oscillations. In a quantum wire with weak disorder (no tunnel barriers), a Wigner crystal may well be an appropriate description of the ground state [24]. The point of view adopted in this chapter, following Ref. [25], is that the Coulomb blockade model is adequate for the present experiments in systems with artificial or natural tunnel barriers. We limit ourselves to a discussion of that model, and refer the reader to Ref. [11] for an exposition of the alternative point of view of Kastner and collaborators.

The Coulomb blockade and Wigner crystal models have in common that electron-electron interactions play a central role. In contrast, some authors have argued that resonant tunneling of non-interacting electrons can by itself explain the observed conductance oscillations [26, 27]. We stress that one cannot discriminate between these two models on the basis of the periodicity of the oscillations. Conductance oscillations due to resonant tunneling through non-degenerate levels as well as Coulomb-blockade oscillations both have a periodicity corresponding to the addition of a single electron to the confined region. Other considerations (notably the absence of spin-splitting of the peaks in a magnetic field, and the large activation energy — by far exceeding ΔE) are necessary to demonstrate the inadequacy of a model based on resonant tunneling of non-interacting electrons.

Semiconductor nanostructures offer the additional intriguing possibility to study single-electron tunneling in the quantum Hall effect regime. This is the subject of Sec. 4. In this regime of a strong magnetic field, the one-electron states are extended along equipotential contours [8]. The contours of subsequent states within the same Landau level enclose one extra flux quantum h/e. States at the Fermi level are edge states circulating along the circumference of the quantum dot. If charging effects are negligible, oscillations in the conductance of the dot are observed as a function of gate voltage or magnetic field, due to resonant tunneling through circulating edge states [28]. This is a manifestation of the Aharonov-Bohm effect, normally associated with magnetoconductance oscillations in a ring, rather than a dot. Circulating edge states, however, make the dot behave effectively as a ring [29] — at least for non-interacting electrons. As we will discuss, the single-electron charging energy can cause a "Coulomb blockade" of the Aharonov-Bohm effect in a quantum dot [30, 31]. The magnetoconductance oscillations are suppressed when e^2/C becomes comparable to the Landau level separation $\hbar\omega_c$ (with

$\omega_{\mathrm{c}} = eB/m$). However, the periodic oscillations as a function of gate voltage remain. This difference illustrates how in the presence of charging effects magnetic and electrostatic fields play fundamentally different roles [12], in contrast to the equivalent roles played in the diffusive or ballistic transport regimes.[1] An additional topic covered in Sec. 4 is the effect of a magnetic field on the amplitude and position of the oscillations, from which detailed information can be obtained on the one-electron energy spectrum of the quantum dot [32].

In this chapter we consider the Coulomb-blockade oscillations in zero magnetic field and in the *integer* quantum Hall effect regime. The generalization to the *fractional* quantum Hall effect is still an open problem, at least experimentally. Some theoretical considerations have been given [33], but will not be considered here. We limit ourselves to the linear response regime, and do not discuss the non-linear current-voltage characteristics [34, 35]. In metallic tunnel junctions with very different tunnel rates through the two barriers one finds steps in the current as a function of source-drain voltage [1, 2]. This "Coulomb staircase" discussed in Sec. 6 of Chap. 2 has recently also been observed in a quantum dot [36]. A third limitation is to stationary transport phenomena, so that we do not consider the effects of radio-frequency modulation of the source-drain or gate voltages. A new development in metals is the realization of a "turnstile clocking" of the current through an array of junctions at a value ef, with f the frequency of the modulation of the voltage on a gate [37, 38]. These effects described in Sec. 4 of Chap. 3 have very recently also been observed in a quantum dot [36]. Concerning the types of sample, we limit ourselves to quantum dots and wires defined by a split-gate in a two-dimensional electron gas. Quantum dots may also be defined by etching a pillar out of a quantum well [39, 40]. Such "vertical" structures have the advantage over the planar structures considered here that the thickness and height of the potential barriers separating the quantum dot from the leads can be tailored to a great precision during the epitaxial growth. A disadvantage is that it is more difficult to change the carrier density in the dot by means of a gate electrode [41]. In the planar structures based on a 2DEG not only the electron density, but also the geometry can be varied continuously using gates.

1.2. Basic properties of semiconductor nanostructures

Electrons in a two-dimensional electron gas (2DEG) are constrained to move in a plane, due to a strong electrostatic confinement at the interface between two semiconductor layers (in the case of a GaAs-AlGaAs heterostructure), or at the interface between a semiconductor and an insulator (in the case of a Si-inversion layer, where the insulator is SiO_2). The areal density n_{s} may be varied continuously by changing the voltage on a gate electrode deposited on the top semiconductor layer (in which case isolation is provided automatically by a Schottky barrier) or on the insulator. The gate voltage is defined with respect to an ohmic contact to the 2DEG. The density under a gate electrode of large area changes linearly with the electrostatic potential of the gate $\phi_{\mathbf{gate}}$, according to

---

[1] Examples of this equivalence are the fluctuations in the conductance as a function of gate voltage or magnetic field due to quantum interference, and the sequence of quantized conductance plateaux (at integer multiples of e^2/h) as a result of magnetic or electrostatic depopulation of one-dimensional subbands [8].

the plate capacitor formula

$$\delta n_{\rm s} = \frac{\epsilon}{ed}\delta\phi_{\rm gate} \, , \tag{1}$$

where ϵ is the dielectric constant of the material of thickness d between gate and 2DEG. For GaAs $\epsilon = 13\epsilon_0$, whereas SiO_2 has $\epsilon = 3.9\epsilon_0$.

A unique feature of a 2DEG is that it can be given any desired shape using lithographic techniques. The shape is defined by etching a pattern (resulting in a permanent removal of the electron gas), or by electrostatic depletion using a patterned gate electrode (which is reversible). A local (partial) depletion of the 2DEG under a gate is associated with a local increase of the electrostatic potential, relative to the undepleted region. At the boundaries of the gate a potential step is thus induced in the 2DEG. The potential step is smooth, because of the large depletion length (of the order of 100 nm for a step height of 10 meV). This large depletion length is at the basis of the split-gate technique, used to define narrow channels of variable width with smooth boundaries.

The energy of non-interacting conduction electrons in an unbounded 2DEG is given by

$$E(k) = \frac{\hbar^2 k^2}{2m} \, , \tag{2}$$

as a function of momentum $\hbar k$. The effective mass m is considerably smaller than the free electron mass $m_{\rm e}$ as a result of interactions with the lattice potential (for GaAs $m = 0.067 m_{\rm e}$, for Si $m = 0.19 m_{\rm e}$, both for the (100) crystal plane). The density of states $\rho_{\rm 2D}(E) \equiv {\rm d}n(E)/{\rm d}E$ is the derivative of the number of electronic states $n(E)$ (per unit surface area) with energy smaller than E. In k-space, these states fill a circle of area $A = 2\pi m E/\hbar^2$ [according to Eq. (2)], containing a number $g_{\rm s} g_{\rm v} A/(2\pi)^2$ of states. The factors $g_{\rm s}$ and $g_{\rm v}$ account for the spin and valley-degeneracy, respectively (in GaAs $g_{\rm v} = 1$, in Si $g_{\rm v} = 2$; $g_{\rm s} = 2$ in zero magnetic field). One thus finds $n(E) = g_{\rm s} g_{\rm v} m E/2\pi\hbar^2$, so that the density of states per unit area,

$$\rho_{\rm 2D} = g_{\rm s} g_{\rm v} \frac{m}{2\pi\hbar^2} \, , \tag{3}$$

is *independent* of the energy. In equilibrium, the states are occupied according to the Fermi-Dirac distribution function

$$f(E - E_{\rm F}) = \left[1 + \exp\left(\frac{E - E_{\rm F}}{k_{\rm B} T} \right) \right]^{-1} . \tag{4}$$

At low temperatures $k_{\rm B} T \ll E_{\rm F}$, the Fermi energy (or chemical potential) $E_{\rm F}$ of a 2DEG is thus directly proportional to its sheet density $n_{\rm s}$, according to

$$E_{\rm F} = n_{\rm s}/\rho_{\rm 2D} \, . \tag{5}$$

The Fermi wave number $k_{\rm F} \equiv (2m E_{\rm F}/\hbar^2)^{1/2}$ is related to the density by $k_{\rm F} = (4\pi n_{\rm s}/g_{\rm s} g_{\rm v})^{1/2}$. Typically, $E_{\rm F} \sim 10$ meV, so that the Fermi wavelength $\lambda_{\rm F} \equiv 2\pi/k_{\rm F} \sim 50$ nm.

If the 2DEG is confined laterally to a narrow channel, then Eq. (2) only represents the kinetic energy from the free motion (with momentum $\hbar k$) *parallel* to the channel axis. Because of the lateral confinement, the conduction band is split itself into a series of one-dimensional (1D) subbands, with band bottoms at E_n, $n = 1, 2, \ldots$. The total energy $E_n(k)$ of an electron in the n-th 1D subband is given by

$$E_n(k) = E_n + \frac{\hbar^2 k^2}{2m} \, , \tag{6}$$

in zero magnetic field. Two frequently used potentials to model analytically the lateral confinement are the square well potential (of width W), and the parabolic potential well (described by $V(x) = \frac{1}{2} m \omega_0{}^2 x^2$). The confinement levels are given by $E_n = (n\pi\hbar)^2/2mW^2$, and $E_n = (n - \frac{1}{2})\hbar\omega_0$, respectively.

Transport through a very short quantum wire (of length $L \sim 100$ nm, much shorter than the mean free path) is perfectly ballistic. When such a short and narrow wire forms a constriction between two wide electron gas reservoirs, one speaks of a quantum point contact [42]. The conductance G of a quantum point contact is quantized in units of $2e^2/h$ [43, 44]. This effect requires a unit transmission probability for all of the occupied 1D subbands in the point contact, each of which then contributes $2e^2/h$ to the conductance (for $g_s g_v = 2$). Potential fluctuations due to the random distribution of ionized donors have so far precluded any observation of the conductance quantization in longer quantum wires (even if they are considerably shorter than the mean free path in wide 2DEG regions). Quantum wires are extremely sensitive to disorder, since the effective scattering cross-section, being of the order of the Fermi wavelength, is comparable to the width of the wire. Indeed, calculations demonstrate [45] that a quantum wire close to pinch-off breaks up into a number of isolated segments. The Coulomb-blockade oscillations in a quantum wire discussed in Sec. 3 are associated with tunneling through the barriers separating these segments (see Fig. 2a).

A quantum dot is formed in a 2DEG if the electrons are confined in all three directions. The energy spectrum of a quantum dot is fully discrete. Transport through the discrete states in a quantum dot can be studied if tunnel barriers are defined at its perimeter. The quantum dots discussed in Sec. 3 are connected by quantum point contacts to their surroundings (see Figs. 2b and 2c). The quantum point contacts are operated close to pinch-off ($G < 2e^2/h$), where they behave as tunnel barriers of adjustable height and width. The shape of such barriers differs greatly from that encountered in metallic tunnel junctions: the barrier height typically exceeds the Fermi energy by only a few meV, and the thickness of the barrier at E_F is large, on the order of 50 nm. This may lead to a strong energy dependence of the tunnel rates, not encountered in metals.

2. Theory of Coulomb-blockade oscillations

Part of the interest in quantum dots derives from the fact that their electronic structure mimics that of an isolated atom — with the fascinating possibility to attach wires to this "atom" and study transport through its discrete electronic states. In this section we address this problem from a theoretical point of view, following Ref. [19].

2.1. Periodicity of the oscillations

We consider a quantum dot, which is weakly coupled by tunnel barriers to two electron reservoirs. A current I can be passed through the dot by applying a voltage difference V between the reservoirs. The linear response conductance G of the quantum dot is defined as $G \equiv I/V$, in the limit $V \to 0$. Since transport through a quantum dot proceeds by tunneling through its discrete electronic states, it will be clear that for small V a net current can flow only for certain values of the gate voltage (if $\Delta E \gg k_{\mathrm{B}}T$). In the absence of charging effects, a conductance peak due to resonant tunneling occurs when the Fermi energy E_{F} in the reservoirs lines up with one of the energy levels in the dot. This condition is modified by the charging energy. To determine the location of the conductance peaks as a function of gate voltage requires only consideration of the equilibrium properties of the system [19, 30], as we now discuss.

The probability $P(N)$ to find N electrons in the quantum dot in equilibrium with the reservoirs is given by the grand canonical distribution function

$$P(N) = \text{constant} \times \exp\left(-\frac{1}{k_{\mathrm{B}}T}\left[F(N) - NE_{\mathrm{F}}\right]\right), \qquad (7)$$

where $F(N)$ is the free energy of the dot and T the temperature. The reservoir Fermi energy E_{F} is measured relative to the conduction band bottom in the reservoirs. In general, $P(N)$ at $T = 0$ is non-zero for a *single* value of N only (namely the integer which minimizes the thermodynamic potential $\Omega(N) \equiv F(N) - NE_{\mathrm{F}}$). In that case, $G \to 0$ in the limit $T \to 0$. As pointed out by Glazman and Shekhter [5], a non-zero G is possible only if $P(N)$ and $P(N+1)$ are both non-zero for some N. Then a small applied voltage is sufficient to induce a current through the dot, via intermediate states $N \to N+1 \to N \to N+1 \to \cdots$. To have $P(N)$ and $P(N+1)$ both non-zero at $T = 0$ requires that both N and $N+1$ minimize Ω. A necessary condition is $\Omega(N+1) = \Omega(N)$, or

$$F(N+1) - F(N) = E_{\mathrm{F}}. \qquad (8)$$

This condition is also sufficient, unless Ω has more than one minimum (which is usually not the case).

Equation (8) expresses the equality of the electrochemical potential of dot and leads. The usefulness of this result is that it maps the problem of determining the location of the conductance peaks onto the more familiar problem of calculating the electrochemical potential $F(N+1) - F(N)$ of the quantum dot, i.e. the energy cost associated with the addition of a single electron to the dot. This opens the way, in principle, to a study of exchange and correlation effects on conductance oscillations in a quantum dot (e.g. along the lines of work by Bryant [46] and by Maksym and Chakraborty [47]).

At $T = 0$ the free energy $F(N)$ equals the ground state energy of the dot, for which we take the simplified form $U(N) + \sum_{p=1}^{N} E_p$. Here $U(N)$ is the charging energy, and E_p ($p = 1, 2, \ldots$) are single-electron energy levels in ascending order, measured relative to the bottom of the potential well in the quantum dot. The term $U(N)$ accounts for the charge imbalance between dot and reservoirs. The sum over energy levels accounts for the internal degrees of freedom of the quantum dot, evaluated in a mean-field approximation (cf. Ref. [48]). Each level contains either one or zero electrons. Spin degeneracy, if present, can be included by counting each level twice, and other degeneracies can be included similarly. The energy levels E_p depend on gate voltage and magnetic field, but are assumed to be independent of N, at least for the relevant range of values of N. We conclude from Eq. (8) that a peak in the low-temperature conductance occurs whenever

$$E_N + U(N) - U(N-1) = E_F, \tag{9}$$

for some integer N (we have relabeled N by $N-1$).

We adopt the simple approximation of the orthodox model [4] of taking the charging energy into account macroscopically. We write $U(N) = \int_0^{-Ne} \phi(Q')\,dQ'$, where

$$\phi(Q) = Q/C + \phi_{\text{ext}} \tag{10}$$

is the potential difference between dot and reservoir, including also a contribution ϕ_{ext} from external charges (in particular those on a nearby gate electrode). The capacitance C is assumed to be independent of N (at least over some interval). The charging energy then takes the form

$$U(N) = (Ne)^2/2C - Ne\phi_{\text{ext}}. \tag{11}$$

To make connection with some of the literature [3, 49] we mention that $Q_{\text{ext}} \equiv C\phi_{\text{ext}}$ plays the role of an "externally induced charge" on the dot, which can be varied continuously by means of an external gate voltage (in contrast to Q which is restricted to integer multiples of e). In terms of Q_{ext} one can write

$$U(N) = (Ne - Q_{\text{ext}})^2/2C + \text{constant},$$

which is equivalent to Eq. (11). We emphasize that Q_{ext} is an externally controlled variable, via the gate voltage, regardless of the relative magnitude of the various capacitances in the system.

Substitution of Eq. (11) into Eq. (9) gives

$$E_N^* \equiv E_N + \left(N - \frac{1}{2}\right)\frac{e^2}{C} = E_F + e\phi_{\text{ext}} \tag{12}$$

as the condition for a conductance peak. The left-hand-side of Eq. (12) defines a renormalized energy level E_N^*. The renormalized level spacing $\Delta E^* = \Delta E + e^2/C$ is enhanced above the bare level spacing by the charging energy. In the limit $e^2/C\Delta E \to 0$, Eq. (12) is the usual condition for resonant tunneling. In the limit $e^2/C\Delta E \to \infty$, Eq. (12) describes the periodicity of the classical Coulomb-blockade oscillations in the conductance versus electron density [3]–[7].

In Fig. 3 we have illustrated the tunneling of an electron through the dot under the conditions of Eq. (12). In panel (a) one has $E_N + e^2/2C = E_F + e\phi(N-1)$, with N referring to the lowest unoccupied level in the dot. In panel (b) an electron has tunneled into the dot. One now has $E_N - e^2/2C = E_F + e\phi(N)$, with N referring to the highest occupied level. The potential difference ϕ between dot and reservoir has decreased by

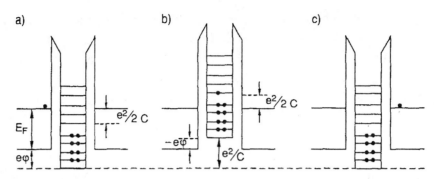

Figure 3. Single-electron tunneling through a quantum dot, under the conditions of Eq. (12), for the case that the charging energy is comparable to the level spacing. An infinitesimally small voltage difference is assumed between the left and right reservoirs. (From Beenakker et al. [31].)

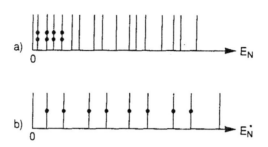

Figure 4. Diagram of the bare energy levels (a) and the renormalized energy levels (b) in a quantum dot for the case $e^2/C \approx 2\langle\Delta E\rangle$. The renormalized level spacing is much more regular than the average bare level spacing $\langle\Delta E\rangle$. Note that the spin degeneracy of the bare levels is lifted by the charging energy. (From Staring et al. [12].)

e/C (becoming negative), because of the added electron. Finally, in panel (c) the added electron tunnels out of the dot, resetting the potentials to the initial state of panel (a).

Let us now determine the periodicity of the oscillations. Theoretically, it is convenient to consider the case of a variation of the Fermi energy of the reservoirs at constant ϕ_{ext}. The periodicity ΔE_F follows from Eq. (12),

$$\Delta E_F = \Delta E^* \equiv \Delta E + \frac{e^2}{C} \, . \tag{13}$$

In the absence of charging effects, ΔE_F is determined by the irregular spacing ΔE of the single-electron levels in the quantum dot. The charging energy e^2/C *regulates* the spacing, once $e^2/C \gtrsim \Delta E$. This is illustrated in Fig. 4, for the case that there is no valley degeneracy. The spin degeneracy of the levels is lifted by the charging energy. In a plot of G versus E_F this leads to a doublet structure of the oscillations, with a spacing alternating between e^2/C and $\Delta E + e^2/C$.

Experimentally, one studies the Coulomb-blockade oscillations as a function of gate voltage. To determine the periodicity in that case, we first need to know how E_F and the set of energy levels E_p depend on ϕ_{ext}. In a 2DEG, the external charges are supplied by ionized donors and by a gate electrode (with an electrostatic potential difference ϕ_{gate} between gate and 2DEG reservoir). One has

$$\phi_{\text{ext}} = \phi_{\text{donors}} + \alpha\phi_{\text{gate}} \, , \tag{14}$$

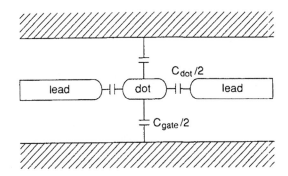

Figure 5. Equivalent circuit of quantum dot and split gate. The mutual capacitance of leads and gate is much larger than that of the dot and the split gate (C_{gate}), or the dot and the leads (C_{dot}), and can be neglected.

where α (as well as C) is a rational function of the capacitance matrix elements of the system. The value of α depends on the geometry. Here we consider only the geometry of Figs. 2a, b in detail, for which it is reasonable to assume that the electron gas densities in the dot and in the leads increase, on average, equally fast with ϕ_{gate}. For equidistant energy levels in the dot we may then assume that $E_F - E_N$ has the same value at each conductance peak. The period of the oscillations now follows from Eqs. (12) and (14),

$$\Delta\phi_{\text{gate}} = \frac{e}{\alpha C} . \tag{15}$$

To clarify the meaning of the parameters C and α, we represent the system of dot, gates and leads in Figs. 2a, b by the equivalent circuit of Fig. 5. The mutual capacitance of gates and leads does not enter our problem explicitly, since it is much larger than the mutual capacitances of gate and dot (C_{gate}) and dot and leads (C_{dot}). The capacitance C determining the charging energy e^2/C is formed by C_{gate} and C_{dot} in parallel,

$$C = C_{\text{gate}} + C_{\text{dot}} . \tag{16}$$

The period of the oscillations corresponds in our approximation of equidistant energy levels ($E_F - E_N = \text{constant}$) to the increment by e of the charge on the dot with no change in the voltage across C_{dot}. This implies $\Delta\phi_{\text{gate}} = e/C_{\text{gate}}$, or

$$\alpha = C_{\text{gate}}/(C_{\text{gate}} + C_{\text{dot}}) . \tag{17}$$

Thus, in terms of the electrostatic potential difference between gate and 2DEG reservoirs, the period of the conductance oscillations is $\Delta\phi_{\text{gate}} = e/C_{\text{gate}}$. Note that this result applies regardless of the relative magnitudes of the bare level spacing ΔE and the charging energy e^2/C.

In an experiment the gate voltage is the *electrochemical* potential difference V_{gate} between gate and leads, i.e. the difference in Fermi level, whereas so far we have discussed the period of the oscillations in terms of the *electrostatic* potential difference ϕ_{gate}, i.e. the difference in conduction band bottoms. In one period, the change in Fermi energy in the dot and leads (measured with respect to their local conduction band bottom) is approximately equal to ΔE. The change in Fermi energy in the (metal) gate is negligible, because the density of states in a metal is much larger than in a 2DEG. We thus find that the oscillation period ΔV_{gate} in the geometry of Figs. 2a, b is

$$\Delta V_{\text{gate}} = \frac{\Delta E}{e} + \Delta \phi_{\text{gate}} = \frac{\Delta E}{e} + \frac{e}{C_{\text{gate}}} \,. \tag{18}$$

Note that C_{dot} does not affect the periodicity. In many of the present experiments ΔE is a factor of 10 below e^2/C_{gate}, so that the differences between $\Delta \phi_{\text{gate}}$ and ΔV_{gate} are less than 10 %. Even in such a case, these differences are quite important, since their study yields direct information on the energy spectrum of the quantum dot.

In the case of a two-fold spin-degeneracy, the level separation $E_{p+1} - E_p$ in a dot of area A alternates between 0 and $\Delta E \sim 2\pi\hbar^2/mA$ [cf. Eq. (3)]. As mentioned above, this leads to a doublet structure of the oscillations as a function of E_{F}. To determine the peak spacing as a function of gate voltage we approximate the change in E_{F} with ϕ_{gate} by $\partial E_{\text{F}}/\partial \phi_{\text{gate}} \sim \Delta E C_{\text{gate}}/2e$. We then obtain from Eqs. (12), (14), (16), and (17) that the spacing alternates between two values:

$$\Delta \phi_{\text{gate}}^{(1)} = \left(\frac{e}{C_{\text{gate}}} \right) \frac{e^2/C}{\Delta E/2 + e^2/C} \,, \tag{19}$$

$$\Delta \phi_{\text{gate}}^{(2)} = \left(\frac{e}{C_{\text{gate}}} \right) \frac{\Delta E + e^2/C}{\Delta E/2 + e^2/C} \,. \tag{20}$$

The average spacing equals e/C_{gate}, in agreement with Eq. (15) [derived for non-degenerate equidistant levels]. To obtain ΔV_{gate} one has to add $\Delta E/2e$ to the factor e/C_{gate} between brackets in Eqs. (19) and (20). If the charging energy dominates ($e^2/C \gg \Delta E$), one has equal spacing $\Delta \phi_{\text{gate}}^{(1)} = \Delta \phi_{\text{gate}}^{(2)} = e/C_{\text{gate}}$, as for non-degenerate levels. In the opposite limit $\Delta E \gg e^2/C$, one finds instead $\Delta \phi_{\text{gate}}^{(1)} = 0$, and $\Delta \phi_{\text{gate}}^{(2)} = 2e/C_{\text{gate}}$. Thus, the period is effectively doubled, corresponding to the addition of *two* electrons to the dot, instead of one. This is characteristic for resonant tunneling of non-interacting electrons through two-fold spin-degenerate energy levels. An external magnetic field will resolve the spin-degeneracy, leading to a splitting of the conductance peaks which increases with the field.

2.2. Amplitude and lineshape

Equation (12) is sufficient to determine the periodicity of the conductance oscillations, but gives no information on their amplitude and width, which requires the solution of a kinetic equation. For the linear response conductance in the resonant tunneling regime an analytical solution has been derived by Beenakker [19], which generalizes earlier results by Kulik and Shekhter [7] in the classical regime. Equivalent results have been obtained independently by Meir, Wingreen, and Lee [20]. Related work on the non-linear current-voltage characteristics has been performed by Averin, Korotkov, and Likharev [34], and by Groshev [35]. In this sub-section we summarize the main results of Ref. [19], along with the underlying assumptions.

A continuum of states is assumed in the reservoirs, which are occupied according to the Fermi-Dirac distribution (4). The tunnel rate from level p to the left and right reservoirs is denoted by Γ_p^{l} and Γ_p^{r}, respectively. We assume that $k_{\text{B}}T \gg h(\Gamma^{\text{l}} + \Gamma^{\text{r}})$ (for all levels participating in the conduction), so that the finite width $h\Gamma = h(\Gamma^{\text{l}} + \Gamma^{\text{r}})$ of the transmission resonance through the quantum dot can be disregarded. This assumption allows us to characterize the state of the quantum dot by a set of occupation numbers, one for each energy level. (As we will discuss, in the classical regime $k_B T \gg \Delta E$ the condition $\Delta E \gg h\Gamma$ takes over from the condition $k_B T \gg h\Gamma$ appropriate for the resonant tunneling regime.) We assume here that inelastic scattering takes place exclusively in the reservoirs — not in the quantum dot. (The effects of inelastic scattering in the dot for $k_B T \gg h\Gamma$ are discussed in Ref. [19].)

The equilibrium distribution function of electrons among the energy levels is given by the Gibbs distribution in the grand canonical ensemble:

$$P_{eq}(\{n_i\}) = Z^{-1} \exp\left[-\frac{1}{k_B T}\left(\sum_{i=1}^{\infty} E_i n_i + U(N) - N E_F\right)\right],\tag{21}$$

where $\{n_i\} \equiv \{n_1, n_2, \ldots\}$ denotes a specific set of occupation numbers of the energy levels in the quantum dot. (The numbers n_i can take on only the values 0 and 1.) The number of electrons in the dot is $N \equiv \sum_i n_i$, and Z is the partition function,

$$Z = \sum_{\{n_i\}} \exp\left[-\frac{1}{k_B T}\left(\sum_{i=1}^{\infty} E_i n_i + U(N) - N E_F\right)\right].\tag{22}$$

The joint probability $P_{eq}(N, n_p = 1)$ that the quantum dot contains N electrons *and* that level p is occupied is

$$P_{eq}(N, n_p = 1) = \sum_{\{n_i\}} P_{eq}(\{n_i\})\delta_{N,\sum_i n_i}\delta_{n_p,1}.\tag{23}$$

In terms of this probability distribution, the conductance is given by

$$G = \frac{e^2}{k_B T}\sum_{p=1}^{\infty}\sum_{N=1}^{\infty}\frac{\Gamma_p^l \Gamma_p^r}{\Gamma_p^l + \Gamma_p^r}P_{eq}(N, n_p = 1)$$

$$\times\left[1 - f(E_p + U(N) - U(N-1) - E_F)\right].\tag{24}$$

This particular product of distribution functions expresses the fact that tunneling of an electron from an initial state p in the dot to a final state in the reservoir requires an occupied initial state and empty final state. Equation (24) was derived in Ref. [19] by solving the kinetic equation in linear response. This derivation is presented in the appendix. The same formula has been obtained independently by Meir, Wingreen, and Lee [20], by solving an Anderson model in the limit $k_B T \gg h\Gamma$.

We will now discuss some limiting cases of the general result (24). We first consider the conductance of the individual barriers and the quantum dot in the high temperature limit $k_B T \gg e^2/C, \Delta E$ where neither the discreteness of the energy levels nor the charging energy are important. The conductance then does not exhibit oscillations as a function of gate voltage. The high temperature limit is of interest for comparison with the low temperature results, and because its measurement allows a straightforward estimate of the tunnel rates through the barriers. The conductance of the quantum dot in the high temperature limit is simply that of the two tunnel barriers in series

$$G = \frac{G^l G^r}{G^l + G^r} \ , \ \text{if } \Delta E, \ e^2/C \ll k_B T \ll E_F. \tag{25}$$

The conductances G^l, G^r of the left and right tunnel barriers are given by the thermally averaged Landauer formula

$$G^{l,r} = -\frac{e^2}{h} \int_0^\infty dE \ T^{l,r}(E) \frac{df}{dE} \ . \tag{26}$$

The transmission probability of a barrier $T(E)$ equals the tunnel rate $\Gamma(E)$ divided by the attempt frequency $\nu(E) = 1/h\rho(E)$,

$$T^{l,r}(E) = h\Gamma^{l,r}(E)\rho(E) \ . \tag{27}$$

If the height of the tunnel barriers is large, the energy dependence of the tunnel rates and of the density of states ρ in the dot can be ignored (as long as $k_B T \ll E_F$). The conductance of each barrier from Eq. (26) then becomes

$$G^{l,r} = (e^2/h)T^{l,r} = e^2\Gamma^{l,r}\rho \tag{28}$$

(where T, Γ, and ρ are evaluated at E_F), and the conductance of the quantum dot from Eq. (25) is

$$G = e^2\rho\frac{\Gamma^l\Gamma^r}{\Gamma^l + \Gamma^r} = \frac{e^2}{h}\frac{T^l T^r}{T^l + T^r} \equiv G_\infty \ , \ \text{if } \Delta E, \ e^2/C \ll k_B T \ll E_F. \tag{29}$$

The conductance G_∞ in the high temperature limit depends only on the barrier height and width (which determine T), not on the area of the quantum dot (which determines ρ and Γ, but cancels in the expression for G_∞).

The validity of the present theory is restricted to the case of negligible quantum fluctuations in the charge on the dot [4]. Since charge leaks out of the dot at a rate $\Gamma^l + \Gamma^r$, the energy levels are sharply defined only if the resulting uncertainty in energy $h(\Gamma^l + \Gamma^r) \ll \Delta E$. In view of Eq. (27), with $\rho \sim 1/\Delta E$, this requires $T^{l,r} \ll 1$, or $G^{l,r} \ll e^2/h$. In the resonant tunneling regime of comparable ΔE and $k_B T$, this criterion is equivalent to the criterion $h\Gamma \ll k_B T$ mentioned earlier. In the classical regime $\Delta E \ll k_B T$, the criterion $h\Gamma \ll \Delta E$ dominates. The general criterion $h\Gamma \ll \Delta E, k_B T$ implies that the conductance of the quantum dot $G \ll e^2/h$.

As we lower the temperature, such that $k_B T < e^2/C$, the Coulomb-blockade oscillations become observable. This is shown in Fig. 6. The classical regime $\Delta E \ll k_B T$ was first studied by Kulik and Shekhter [6, 7]. In this regime a continuum of energy levels in the confined central region participates in the conduction. If $\Delta E \ll k_B T \ll e^2/C$, only the terms with $N = N_{min}$ contribute to the sum in Eq. (24), where N_{min} minimizes the absolute value of $\Delta(N) = U(N) - U(N-1) + \bar{\mu} - E_F$. [Here $\bar{\mu}$ is the equilibrium chemical potential of the dot, measured relative to the bottom of the potential well.]

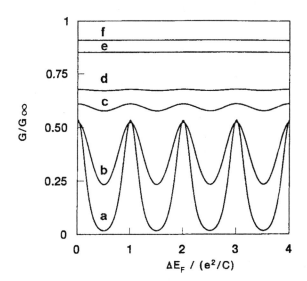

Figure 6. Temperature dependence of the Coulomb-blockade oscillations as a function of Fermi energy in the classical regime $k_B T \gg \Delta E$. Curves are calculated from Eq. (24) with $\Delta E = 0.01 e^2/C$, for $k_B T/(e^2/C) = 0.075$ (a), 0.15 (b), 0.3 (c), 0.4 (d), 1 (e), and 2 (f). Level-independent tunnel rates are assumed, as well as equidistant non-degenerate energy levels.

We define $\Delta_{\min} \equiv \Delta(N_{\min})$. For energy-independent tunnel rates and density of states $\rho \equiv 1/\Delta E$, one obtains a line shape of individual conductance peaks given by

$$G/G_{\max} = \frac{\Delta_{\min}/k_B T}{\sinh(\Delta_{\min}/k_B T)} \approx \cosh^{-2}\left(\frac{\Delta_{\min}}{2.5\,k_B T}\right) , \qquad (30)$$

$$G_{\max} = \frac{e^2}{2\Delta E} \frac{\Gamma^l \Gamma^r}{\Gamma^l + \Gamma^r} . \qquad (31)$$

The second equality in Eq. (31) is approximate, but holds to better than 1%. A plot of $G/G_{\max}$ versus $\Delta_{\min}$ is shown for an isolated peak in Fig. 7 (dashed curve).

Whereas the width of the peaks increases with T in the classical regime, the peak height (reached at $\Delta_{\min} = 0$) is temperature independent (compare traces (a) and (b) in Fig. 6). The reason is that the $1/T$ temperature dependence associated with resonant tunneling through a particular energy level is canceled by the T dependence of the number $k_B T/\Delta E$ of levels participating in the conduction. This cancellation holds only if the tunnel rates are energy independent within the interval $k_B T$. A temperature dependence of the conductance may result from a strong energy dependence of the tunnel rates. In such a case one has to use the general result (24). This is also required if peaks start to overlap for $k_B T \sim e^2/C$, or if the dot is nearly depleted ($E_F \lesssim k_B T$). The latter regime does not play a role in metals, but is of importance in semiconductor nanostructures because of the much smaller E_F. The presence of only a small number $E_F/\Delta E$ of electrons in a quantum dot leads also to a gate voltage dependence of the oscillations in the classical regime $k_B T \gg \Delta E$.

Despite the fact that the Coulomb blockade of tunneling is lifted at a maximum of a conductance peak, the peak height $G_{\max}$ in the classical Coulomb-blockade regime $\Delta E \ll k_B T \ll e^2/C$ is a factor of two smaller than the conductance G_∞ in the high temperature regime $k_B T \gg e^2/C$ of negligible charging energy (in the case of energy-

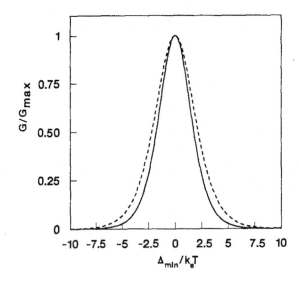

Figure 7. Comparison of the lineshape of a thermally broadened conductance peak in the resonant tunneling regime $h\Gamma \ll k_B T \ll \Delta E$ (solid curve) and in the classical regime $\Delta E \ll k_B T \ll e^2/C$ (dashed curve). The conductance is normalized by the peak height G_{max}, given by Eqs. (31) and (34) in the two regimes. The energy Δ_{min} is proportional to the Fermi energy in the reservoirs, cf. Eq. (32). (From Beenakker [19].)

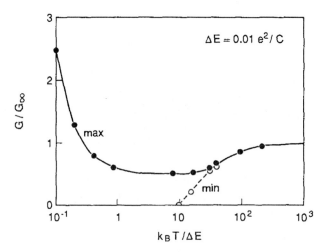

Figure 8. Temperature dependence of the maxima (max) and the minima (min) of the Coulomb-blockade oscillations, in the regime $h\Gamma \ll k_B T$. The calculation, based on Eq. (24), was performed for the case of equidistant non-degenerate energy levels (at separation $\Delta E = 0.01e^2/C$), all with the same tunnel rates Γ^l and Γ^r.

independent tunnel rates). The reason is a correlation between subsequent tunnel events, imposed by the charging energy. This correlation, expressed by the series of charge states $Q = -N_{min}e \to Q = -(N_{min} - 1)e \to Q = -N_{min}e \to \ldots$, implies that an electron can tunnel from a reservoir into the dot only half of the time (when $Q = -(N_{min} - 1)e$). The tunnel probability is therefore reduced by a factor of two compared to the high temperature limit, where no such correlation exists.

The temperature dependence of the maxima of the Coulomb-blockade oscillations as obtained from Eq. (24) is plotted in Fig. 8. Also shown in Fig. 8 are the minima, which are seen to merge with the maxima as $k_B T$ approaches e^2/C. In the resonant tunneling regime $k_B T \lesssim \Delta E$ the peak height increases as the temperature is reduced, due to the

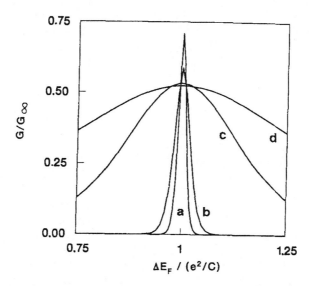

Figure 9. Lineshape for various temperatures, showing the crossover from the resonant tunneling regime (a and b) where both the width and the peak height depend on T, to the classical regime (c and d) where only the width of the peak depends on T. Curves are calculated from Eq. (24) with $\Delta E = 0.01 e^2/C$, and for $k_B T/\Delta E = 0.5$ (a), 1 (b), 7.5 (c), and 15 (d).

diminished thermal broadening of the resonance. The crossover from the classical to the quantum regime is shown in Fig. 9 [calculated directly from Eq. (24)].

In the case of well-separated energy scales in the resonant tunneling regime ($h\Gamma \ll k_B T \ll \Delta E$), Eq. (24) can again be written in a simplified form. Now the single term with $p = N = N_{\min}$ gives the dominant contribution to the sum over p and N. The integer $N_{\min}$ minimizes the absolute value of

$$\Delta(N) = E_N + U(N) - U(N - 1) - E_F. \tag{32}$$

We again denote $\Delta_{\min} \equiv \Delta(N_{\min})$. Equation (24) reduces to

$$G/G_{\max} = -4k_B T f'(\Delta_{\min}) = \cosh^{-2}\left(\frac{\Delta_{\min}}{2k_B T}\right), \tag{33}$$

$$G_{\max} = \frac{e^2}{4k_B T} \frac{\Gamma^l_{N_{\min}} \Gamma^r_{N_{\min}}}{\Gamma^l_{N_{\min}} + \Gamma^r_{N_{\min}}}. \tag{34}$$

As shown in Fig. 7, the lineshape in the resonant tunneling regime (full curve) is different from that in the classical regime (dashed curve), if they are compared at equal temperature. Equation (33) can be seen as the usual resonant tunneling formula for a thermally broadened resonance, generalized to include the effect of the charging energy on the resonance condition. Eqs. (33) and (34) hold regardless of the relative magnitude of ΔE and e^2/C. As illustrated in Fig. 8, the peak height in the resonant tunneling regime increases monotonically as $k_B T/\Delta E \to 0$, as long as $k_B T$ is larger than the resonance width $h\Gamma$.

No theory has been worked out for Coulomb-blockade oscillations in the regime $k_B T \lesssim h\Gamma$ (although the theory of Meir et al. [20] is sufficiently general to be applicable in principle). For *non-interacting* electrons, the transmission probability has the Breit-

Wigner form [49]–[51]

$$G_{\mathrm{BW}} = \mathcal{G}\frac{e^2}{h}\frac{\Gamma^{\mathrm{l}}\Gamma^{\mathrm{r}}}{\Gamma^{\mathrm{l}}+\Gamma^{\mathrm{r}}}\frac{\Gamma}{(\epsilon/\hbar)^2+(\Gamma/2)^2}. \tag{35}$$

Here $\mathcal{G}$ is the degeneracy of the resonant level, and ϵ is the energy separation of that level from the Fermi level in the reservoirs. In the presence of inelastic scattering with rate Γ_{in} one has to replace Γ by $\Gamma+\Gamma_{\mathrm{in}}$ [49]–[51]. This has the effect of reducing the conductance on resonance by a factor $\Gamma/(\Gamma+\Gamma_{\mathrm{in}})$, and to increase the width of the peak by a factor $(\Gamma+\Gamma_{\mathrm{in}})/\Gamma$. This is to be contrasted with the regime $h\Gamma \ll k_{\mathrm{B}}T \ll \Delta E$, where inelastic scattering has no effect on the conductance. [This follows from the fact that the thermal average $-\int G_{\mathrm{BW}}f'(\epsilon)\,\mathrm{d}\epsilon \approx \int G_{\mathrm{BW}}\,\mathrm{d}\epsilon/4kT$ is independent of Γ_{in}.] If inelastic scattering is negligible, and if the two tunnel barriers are equal, then the maximum conductance following from the Breit-Wigner formula is $\mathcal{G}e^2/h$ — a result that may be interpreted as the fundamental contact conductance of a $\mathcal{G}$–fold degenerate state [50, 52]. We surmise that the charging energy will lift the level degeneracy, so that the maximum peak height of Coulomb-blockade oscillations is $G_{\mathrm{max}} = e^2/h$ for the case of equal tunnel barriers.

A few words on terminology, to make contact with the resonant tunneling literature [49, 50]. The results discussed above pertain to the regime $\Gamma \gg \Gamma_{\mathrm{in}}$, referred to as the "coherent resonant tunneling" regime. In the regime $\Gamma \ll \Gamma_{\mathrm{in}}$ it is known as "coherent sequential tunneling" (results for this regime are given in Ref. [19]). Phase coherence plays a role in both these regimes, by establishing the discrete energy spectrum in the quantum dot. The classical, or incoherent, regime is entered when $k_{\mathrm{B}}T$ or $h\Gamma_{\mathrm{in}}$ become greater than ΔE. The discreteness of the energy spectrum can then be ignored.

We close this overview of theoretical results by a discussion of the activation energy of the minima of the conductance oscillations. It is shown in Ref. [19] that G_{min} depends exponentially on the temperature, $G_{\mathrm{min}} \propto \exp(-E_{\mathrm{act}}/k_{\mathrm{B}}T)$, with activation energy

$$E_{\mathrm{act}} = \tfrac{1}{2}(\Delta E + e^2/C) = \tfrac{1}{2}\Delta E^*. \tag{36}$$

This result holds for equal tunnel rates at two subsequent energy levels. The renormalized level spacing $\Delta E^* \equiv \Delta E + e^2/C$, which according to Eq. (13) determines the periodicity of the Coulomb-blockade oscillations as a function of Fermi energy, thus equals twice the activation energy of the conductance minima. The exponential decay of the conductance at the minima of the Coulomb blockade oscillations results from the suppression of tunneling processes which conserve energy in the intermediate state in the quantum dot. Tunneling via a *virtual* intermediate state is not suppressed at low temperatures, and may modify the temperature dependence of the minima if $h\Gamma$ is not much smaller than $k_{\mathrm{B}}T$ and ΔE [53, 54]. For $h\Gamma \ll k_{\mathrm{B}}T$, ΔE this co-tunneling or "macroscopic quantum tunneling of the charge" discussed in Chap. 6 can be neglected.

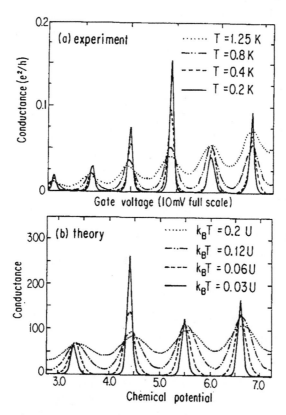

Figure 10. (a) Measured conductance as a function of gate voltage in a quantum dot in the 2DEG of a GaAs-AlGaAs heterostructure, with a geometry as shown in Fig. 2b. (Experimental results obtained by U. Meirav, M. Kastner, and S. Wind, unpublished; U. Meirav, PhD Thesis (M.I.T., 1990).) (b) Calculated conductance based on Eq. (24). The conductance is given in units $\Gamma_1 C$, and the chemical potential of the reservoirs in units of e^2/C. The level spacing was taken to be $\Delta E = 0.1 e^2/C$. The tunnel rates of the levels increase in a geometric progression $\Gamma_{p+1} = 1.5^p \Gamma_1$, with Γ_4 increased by an additional factor of 4 to simulate disorder. The temperature is quoted in units of e^2/C. (From Meir et al. [20].)

3. Experiments on Coulomb-blockade oscillations

3.1. Quantum dots

Coulomb-blockade oscillations in the conductance of a quantum dot were first studied by Meirav, Kastner, and Wind [15]. The geometry of their device is shown in Fig. 2b. A split-gate electrode with a 300 nm wide slit is used to define a narrow channel. Small protrusions on each part of the split gate are used to define quantum point contacts in the narrow channel, 1 μm apart. For sufficiently strong negative gate voltages the electron gas in the point contacts is depleted so that the channel is partitioned into a quantum dot, two tunnel barriers, and two leads. The width of the quantum dot is estimated to be 50 nm, whereas its length is about 1 μm. The conductance of this device exhibits conductance peaks periodic in the gate voltage, at temperatures between 50 mK and 1 K (see Fig. 10a). Based on estimates of the gate capacitance, it was concluded that one electron was added to the quantum dot in each oscillation period. This conclusion was supported by experiments on devices with different values for the tunnel barrier separation [15]. Meirav et al. have also shown that the lineshape of an isolated peak could be fitted very well by a function of the form $\cosh^{-2}(\gamma(V_{\text{gate}} - V_0)/2 k_B T)$. We note that, since the fit was done with γ and T as adaptable parameters, equally good agreement would have been obtained with the theoretical line shapes for the Coulomb-blockade oscillations in the classical or quantum regimes [Eqs. (31) and (34)].

Meirav et al. found that the temperature dependence of the peak width yielded an estimate for $e^2/2C$ that was a factor of 3.5 lower than the value inferred from the periodicity. One way to possibly resolve this discrepancy is to note that the width of the peaks, as well as the activation energy, is determined by the charging energy $e^2/2C$ with $C = C_{\text{dot}} + C_{\text{gate}}$ [Eq. (16)]. This energy is smaller than the energy $e^2/2C_{\text{gate}}$ obtained from a measurement of the periodicity $\Delta V_{\text{gate}} \simeq e/C_{\text{gate}}$ [Eq. (18)]. Alternatively, a strong energy dependence of the tunnel rates may play a role [20].

Meir, Wingreen, and Lee [20] modeled the experimental data shown in Fig. 10a by means of Eq. (24) (derived independently by these authors), using parameters consistent with experimental estimates ($\Delta E = 0.1$ meV, $e^2/C = 1$ meV). The results of their calculation are reproduced in Fig. 10b. The increasing height of successive peaks is due to an assumed increase in tunnel rates for successive levels ($\Gamma_{p+1} = 1.5^p \Gamma_1$). Disorder is simulated by multiplying Γ_4 by an additional factor of 4. No attempt was made to model the gate-voltage dependence of the experiment, and instead the chemical potential of the reservoirs was chosen as a variable in the calculations. Figs. 10a and 10b show a considerable similarity between experiment and theory. The second peak in the theoretical trace is the anomalously large Γ_4 peak, which mimics the fourth peak in the experimental trace. In both theory and experiment a peak adjacent to the anomalously large peak shows a non-monotonic temperature dependence. This qualitative agreement, obtained with a consistent set of parameter values, supports the interpretation of the effect as Coulomb-blockade oscillations in the regime of a discrete energy spectrum.

It is possible that at the lowest experimental temperatures in the original experiment of Meirav et al. [15] the regime $k_B T \lesssim h\Gamma$ of intrinsically broadened resonances is entered. An estimate of the average tunnel rates is most reliably obtained from the high-temperature limit, where the peaks begin to overlap. From Fig. 10a we estimate $G_\infty \sim 0.1 e^2/h$. For a symmetric quantum dot ($\Gamma^l = \Gamma^r$) Eq. (29) with $\rho \sim 1/\Delta E$ then implies $h\Gamma \equiv h(\Gamma^l + \Gamma^r) \sim 0.4\Delta E \sim 0.04$ meV. The condition $k_B T \lesssim h\Gamma$ thus yields a crossover temperature of 500 mK. Meirav et al. [15] reported a saturation of the linear temperature dependence of the width of the peaks to a much weaker dependence for $T \lesssim 500$ mK. It is thus possible that the approach of the intrinsically broadened regime $k_B T \lesssim h\Gamma$ is at the origin of the saturated width at low temperatures (current heating of the electron gas [15] may also play a role). Unfortunately, as noted in Sec. 2, a theory for the lineshape in this regime is not available.

We close the discussion of the experiments of Meirav et al. by noting that some of their samples showed additional periodicities in the conductance, presumably due to residual disorder. Thermal cycling of the sample (to room temperature) strongly affected the additional structure, without changing the dominant oscillations due to the quantum dot between the point contact barriers.

Williamson et al. [17] have studied the Coulomb-blockade oscillations using a quantum dot of the design shown in Fig. 2c. The device has three sets of gates to adjust the transmission probability of each tunnel barrier and the potential ϕ_{ext} of the dot. (Because of the proximity of the gates the adjustments are not independent.) The tunnel barriers are formed by quantum point contacts close to pinch-off. A device with multiple gates in a lay-out similar to that of Fig. 2b was studied by Kouwenhoven et al. [16]. From a measurement of the Coulomb-blockade oscillations for a series of values of the conductance of the individual quantum point contacts it has been found in both experiments that the oscillations disappear when the conductance of each point contact approaches the first quantized plateau, where $G^{l,r} = 2e^2/h$. It is not yet clear whether this is due to virtual tunneling processes, or to a crossover from tunneling to ballistic transport through the quantum point contacts. We note that this ambiguity does not arise in tunnel junctions between metals, where the area of the tunnel barrier is usually much larger than the Fermi wavelength squared, so that a barrier conductance larger than e^2/h can easily be realized *within* the tunneling regime. In semiconductors, tunnel barriers of large area can also be made — but it is likely that then e^2/C will become too small. A dynamical treatment is required in the case of low tunnel barriers, since the field across the barrier changes during the tunnel process [55]. Similar dynamic polarization effects are known to play a role in large-area semiconductor tunnel junctions, where they are related to image-force lowering of the barrier height.

3.2. Disordered quantum wires

Scott-Thomas et al. [9] found strikingly regular conductance oscillations as a function of gate voltage (or electron gas density) in a narrow disordered channel in a Si inversion layer. The period of these oscillations differed from device to device, and did not correlate with the channel length. Based on estimates of the sample parameters, it was concluded that one period corresponds to the addition of a single electron to a conductance-limiting segment of the disordered quantum wire.

Two of us have proposed that the effect is the first manifestation of *Coulomb-blockade oscillations* in a semiconductor nanostructure [25]. To investigate this phenomenon further, Staring et al. have studied the periodic conductance oscillations in disordered quantum wires defined by a split gate in the 2DEG of a GaAs-AlGaAs heterostructure [12, 13]. Other studies of the effect have been made by Field et al. [11] in a narrow channel in a 2D hole gas in Si, by Meirav et al. [10] in a narrow electron gas channel in an inverted GaAs-AlGaAs heterostructure, and by De Graaf et al. [14] in a very short split gate channel (or point contact) in a Si inversion layer. Here we will only discuss the results of Staring et al. in detail.

In a first set of samples [12], a delta-doping layer of Be impurities was incorporated during growth, in order to create strongly repulsive scattering centers in the narrow channel. (Be is an acceptor in GaAs; some compensation was also present in the narrow Si inversion layers studied by Scott-Thomas et al. [9].) A second set of samples [13] did not contain Be impurities. The mean free path in the Be-doped samples in wide regions adjacent to the channel is 0.7 μm. In the other samples it is 4μm. Close to pinch-off the channel will break up into a few segments separated by potential barriers formed by scattering centers. Model calculations have shown that statistical variations in the random positions of ionized donors in the AlGaAs are sufficient to create such a situation [45]. Indeed, both the samples with and without Be exhibited the Coulomb-blockade oscillations.

In Fig. 11a we reproduce representative traces of conductance versus gate voltage at various temperatures for a sample without Be [13]. Note the similarity to the results

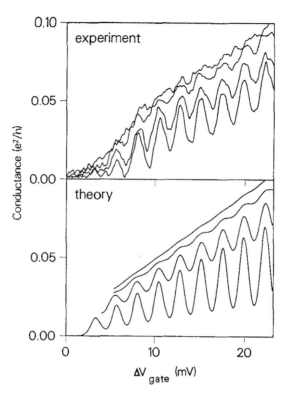

Figure 11. (a) Measured conductance of an unintentionally disordered quantum wire in a GaAs-AlGaAs heterostructure, of a geometry as shown in Fig. 2a; $T = 1.0$, 1.6, 2.5, and 3.2 K (from bottom to top). (b) Model calculations based on Eq. (24), for $\Delta E = 0.1$ meV, $e^2/C = 0.6$ meV, $\alpha = 0.27$, and $h\Gamma_p^{l,r} = 2.7 \times 10^{-2} p \Delta E$ (p labels spin-degenerate levels). (From Staring et al. [13].)

obtained for a single quantum dot shown in Fig. 10a. The oscillations generally disappear as the channel is widened away from pinch-off. No correlation was found between the periodicity of the oscillations and the channel length. At channel definition its width equals the lithographic width $W_{\text{lith}} = 0.5$ μm, and the sheet electron density $n_s = 2.9 \times 10^{11}$ cm^{-2}. As the width is reduced to 0.1 μm, the density becomes smaller by about a factor of 2. (The estimate for W is based on typical lateral depletion widths of 200 nm/V [8, 45, 56], and that for n_s on an extrapolation of the periodicity of the Shubnikov-De Haas oscillations.) A 3 μm long channel then contains some 450 electrons. Calculations for a split-gate channel [56] indicate that the number of electrons per unit length increases approximately linearly with gate voltage. The periodicity of the conductance oscillations as a function of gate voltage thus implies a periodicity as a function of density per unit length.

Our model for the Coulomb-blockade oscillations in a disordered quantum wire is essentially the same as that for a quantum dot, to the extent that a single segment limits the conductance. To calculate C_{dot} and C_{gate} is a rather complicated three-dimensional electrostatic problem, hampered further by the uncertain dimensions of the conductance limiting segment. Experimentally, the conductance peaks are spaced by $\Delta V_{\text{gate}} \sim 2.4$ mV, so that from Eq. (18) we estimate $C_{\text{gate}} \sim 0.7 \times 10^{-16}$ F. The length L of the quantum dot may be estimated from the gate voltage range $\delta V_{\text{gate}} \sim 1$ V between channel definition and pinch-off: $\delta V_{\text{gate}} \sim en_s W_{\text{lith}} L / C_{\text{gate}}$, where n_s is the sheet density in the channel at definition. From the above estimate of C_{gate} and using

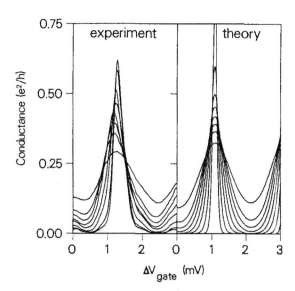

Figure 12. Experimental and theoretical lineshapes of an isolated conductance peak in a Be-doped disordered quantum wire in a GaAs-AlGaAs heterostructure, at $B = 6.7$ T, and $T = 110, 190, 290, 380, 490, 590, 710$, and 950 mK (from top to bottom). The theoretical curves have been calculated from Eq. (24), with $\Delta E = 0.044$ meV (non-degenerate), $e^2/C = 0.53$ meV, $h\Gamma = 0.13$ meV, and $\alpha = 0.27$. (From Staring et al. [13].)

$\delta V_{\text{gate}} \sim 1$ V, we estimate $L \sim 0.3$ μm.[2] The width of the dot is estimated to be about $W \sim 0.1$ μm in the gate voltage range of interest. The level splitting in the segment is $\Delta E \sim 2\pi\hbar^2/mLW \sim 0.2$ meV (for a 2-fold spin-degeneracy). Since each oscillation corresponds to the removal of a single electron from the dot, the maximum number of oscillations following from ΔE and the Fermi energy $E_{\text{F}} \sim 5$ meV at channel definition is given by $2E_{\text{F}}/\Delta E \sim 50$, consistent with the observations. From the fact that the oscillations are still observable at $T = 1.5$ K, albeit with considerable thermal smearing, we deduce that in our experiments $e^2/C + \Delta E \sim 1$ meV. Thus, $C \sim 2.0 \times 10^{-16}$ F, $C_{\text{dot}} = C - C_{\text{gate}} \sim 1.3 \times 10^{-16}$ F,[3] and the parameter $\alpha \equiv C_{\text{gate}}/C \sim 0.35$. In Fig. 11 we compare a calculation based on Eq. (24) with the experiment, taking the two-fold spin-degeneracy of the energy levels into account [13]. The tunnel rates were taken to increase by an equal amount $0.027\Delta E/h$ for each subsequent spin-degenerate level, at equal separation $\Delta E = 0.1$ meV. The capacitances were fixed at $e^2/C = 0.6$ meV and $\alpha = 0.25$. These values are consistent with the crude estimates given above. The Fermi energy was assumed to increase equally fast as the energy of the highest occupied level in the dot (cf. Sec. 2.1.). The temperature range shown in Fig. 11 is in the classical regime ($k_{\text{B}}T > \Delta E$).

The resonant tunneling regime $k_{\text{B}}T < \Delta E$ can be described qualitatively by Eq. (24), as shown in Fig. 12 for an isolated peak. The data was obtained for a different sample (with Be doping) in the presence of a magnetic field of 6.7 T. The parameter values

---

[2] The estimated values for C_{gate} and L are consistent with what one would expect for the mutual capacitance of a length L of a wire of diameter W running in the middle of a gap of width W_{lith} in a metallic plane (the thickness of the AlGaAs layer between the gate and the 2DEG is small compared to W_{lith}): $C_{\text{gate}} \sim 4\pi\epsilon L/2\text{arccosh}(W_{\text{lith}}/W) \sim 0.9 \times 10^{-16}$F (see Ref. [57]).

[3] The mutual capacitance of dot and leads may be approximated by the self-capacitance of the dot, which should be comparable to that of a two-dimensional circular disc of diameter L [57] $C_{\text{dot}} \sim 4\epsilon L \sim 1.4 \times 10^{-16}$F, consistent with the estimate given in the text.

used are $\Delta E = 0.045$ meV, $e^2/C = 0.53$ meV, $h\Gamma = 0.13$ meV, and $\alpha = 0.27$. A fully quantitative theoretical description of the experimental lineshapes in Fig. 12 is not yet possible, because the experiment is in the regime of intrinsically broadened resonances, $k_B T < h\Gamma$, for which the theory has not been worked out.

The semi-quantitative agreement between theory and experiment in Figs. 11 and 12, for a consistent set of parameter values, and over a wide range of temperatures, supports our interpretation of the conductance oscillations as Coulomb-blockade oscillations in the regime of comparable level spacing and charging energies. Note that $e^2/C_{\text{gate}} \sim 10\Delta E$, so that irregularly spaced energy levels would not easily be discernable in the gate voltage scans [cf. Eq. (18)]. Such irregularities might nevertheless play a role in causing peak height variations. Some of the data (not shown) exhibits beating patterns [12, 13], similar to those reported in Refs. [9] and [11]. These are probably due to the presence of multiple segments in the quantum wires [13]. Coulomb-blockade oscillations in arrays of tunnel junctions in the classical regime have been studied by several authors [58, 59].

As an alternative explanation of the conductance oscillations resonant tunneling of non-interacting electrons has been proposed [26, 27]. There are several compelling arguments for rejecting this explanation (which apply to the experiments on a quantum dot as well as to those on disordered quantum wires). Firstly, for resonant tunneling the oscillations would be irregularly spaced, due to the non-uniform distribution of the bare energy levels [cf. Eq. (20)]. This is in contradiction with the experimental observations [11]. Secondly [12], in the absence of charging effects the measured activation energy of the conductance minima would imply a level spacing $\Delta E \sim 1$ meV. Since the Fermi energy E_F in a typical narrow channel is about 5 meV, such a large level spacing would restrict the possible total number of oscillations in a gate voltage scan to $E_F/\Delta E \sim 5$, considerably less than the number seen experimentally [9, 12]. Thirdly, one would expect a spin-splitting of the oscillations by a strong magnetic field, which is not observed [11]. Finally, the facts that no oscillations are found as a function of *magnetic* field [11, 12] and that the spin-splitting does not occur all but rule out resonant tunneling of non-interacting electrons as an explanation of the oscillations as a function of gate voltage.

3.3. Relation to earlier work on disordered quantum wires

The disordered quantum wires discussed in this chapter exhibit *periodic* conductance oscillations as a function of gate voltage. The effect has been seen in electron and hole gases in Si [9, 11, 14] and in the electron gas in GaAs [10, 12, 13]. In contrast, previous work by Fowler et al. [60] and by Kwasnick et al. [61] on narrow inversion and accumulation layers in Si has produced sharp but *aperiodic* conductance peaks. How are these observations to be reconciled? We surmise that the explanation is to be found in the different strength and spatial scale of the potential fluctuations in the wire, as illustrated in Fig. 13.

Coulomb-blockade oscillations require a small number of large potential spikes, so that a single segment limits the conductance (Fig. 13a). The random conductance fluctuations seen previously [60, 61] are thought instead to be due to variable range hopping between a large number of localized states, distributed randomly along the length of the channel (Fig. 13b) [62]–[64]. No segment containing a large number of states (localized within the same region) is present in the potential of Fig. 13b, in contrast to the situation

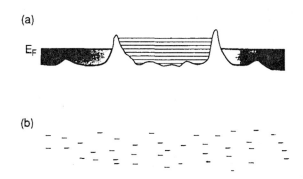

Figure 13. (a) Coulomb-blockade oscillations occur in a disordered quantum wire as a result of the formation of a conductance limiting segment which contains many localized states. (b) Random conductance fluctuations due to variable range hopping between localized states (indicated by dashes) are found in the absence of such a segment.

shown in Fig. 13a. At large Fermi energy a transition eventually occurs to the diffusive transport regime in either type of wire. Both the regular Coulomb-blockade oscillations, and the random conductance peaks due to variable range hopping are then replaced by "universal" conductance fluctuations caused by quantum interference [65, 66].

Fowler et al. [67] have also studied the conductance of much shorter channels than in Ref. [60] (0.5 μm long, and 1 μm wide). In such channels they found well-isolated conductance peaks, which were temperature independent below 100 mK, and which were attributed to resonant tunneling. At very low temperatures a fine structure (some of it time-dependent) was observed. A numerical simulation [68] of the temporal fluctuations in the distribution of electrons among the available sites also showed fine structure if the time scale of the fluctuations is short compared to the measurement time, but large compared to the tunnel time. It is possible that a similar mechanism causes the fine structure on the Coulomb-blockade oscillations in a disordered quantum wire (cf. Fig. 11).

There have also been experimental studies of the effect of a strong magnetic field on variable range hopping [69] and on resonant tunneling through single impurity states [70]. We briefly discuss the work on resonant tunneling by Kopley et al. [70], which is more closely related to the subject of this chapter. They observed large conductance peaks in a Si inversion layer under a split gate. Below the 200 nm wide slot in the gate the inversion layer is interrupted by a potential barrier. Pronounced conductance peaks were seen at 0.5 K as the gate voltage was varied in the region close to threshold. The peaks were attributed to resonant tunneling through single impurity states in the Si bandgap in the barrier region. The lineshape of an isolated peak could be fitted with the Breit-Wigner formula [Eq. (35)]. The amplitude of most peaks was substantially suppressed on applying a strong magnetic field. This was interpreted as a reduction of the tunnel rates because of a reduced overlap between the wavefunctions on the (asymmetrically placed) impurity and the reservoirs. The amplitude of one particular peak was found to be unaffected by the field, indicative of an impurity which is placed symmetrically in the barrier ($\Gamma^r = \Gamma^l$). The width of that peak was reduced, consistent with a reduction of Γ. This study therefore exhibits many characteristic features of resonant tunneling through a single localized site. Yet, one would expect Coulomb interactions of two electrons on the site to be important, and indeed they might explain the absence of spin-splitting of the peaks in a strong magnetic field [70]. Theoretical work indicates that Coulomb interactions also modify the lineshape of a conductance peak [68, 71]. The experimental evidence [63, 67, 69, 70] is not conclusive, however.

4. Quantum Hall effect regime

4.1. The Aharonov-Bohm effect in a quantum dot

The Aharonov-Bohm effect is a quantum interference effect which results from the influence of the vector potential on the phase of the electron wavefunction. Aharonov and Bohm [72] originally considered the influence of the vector potential on electrons confined to a multiply-connected region (such as a ring), within which the magnetic field is zero. The ground state energy of the system is periodic in the enclosed flux with period h/e, as a consequence of gauge invariance. Coulomb repulsion does not affect this periodicity.

In the solid state, the Aharonov-Bohm effect manifests itself as a periodic oscillation in the conductance of a sample as a function of an applied magnetic field B. A well-defined periodicity requires that the conducting paths through the sample enclose a constant area A, perpendicular to B. The periodicity of the oscillations is then $\Delta B = h/eA$, plus possibly harmonics (e.g. at $h/2eA$). The constant area may be imposed by confining the electrons electrostatically to a ring or to a cylindrical film [73, 74].

Entirely new mechanisms for the Aharonov-Bohm effect become operative in strong magnetic fields in the quantum Hall effect regime. These mechanisms do not require a ring geometry, but apply to singly-connected geometries such as a point contact [75] or a quantum dot [28, 29]. As discussed below, these geometries behave as if they were multiply connected, because of circulating edge states. Resonant tunneling through these states leads to magnetoconductance oscillations with a fundamental periodicity $\Delta B = h/eA$, governed by the addition to the dot of a single quantum of magnetic flux h/e.

An essential difference with the original Aharonov-Bohm effect is that in these experiments the magnetic field extends into the conducting region of the sample. Since the periodicity is now no longer constrained by gauge invariance, this opens up the possibility, in principle, of an influence of Coulomb repulsion. We will discuss in the next subsection that the Aharonov-Bohm effect may indeed be *suppressed* by charging effects [30]. In this subsection we will first introduce the case of negligible charging effects in some detail.

If one applies a magnetic field B to a metal, then the electrons move with constant velocity $v_\parallel$ in a direction parallel to B, and in a circular cyclotron orbit with tangential velocity $v_\perp$ in a plane perpendicular to B. The cyclotron frequency is $\omega_c = eB/m$, and the cyclotron radius is $l_{cycl} = v_\perp/\omega_c$. Quantization of the periodic cyclotron motion in a strong magnetic field leads to the formation of Landau levels

$$E_n(k_\parallel) = E_n + \frac{\hbar^2 k_\parallel^2}{2m} \,, \tag{37}$$

$$E_n = (n - \tfrac{1}{2})\hbar\omega_c \,, \tag{38}$$

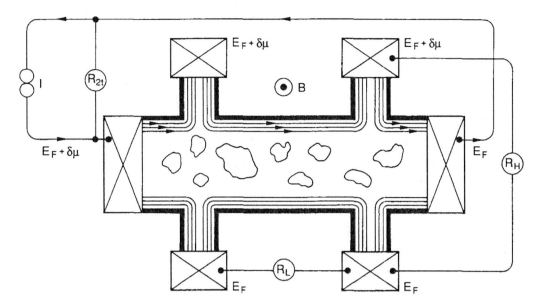

Figure 14. Measurement configuration for the two-terminal resistance R_{2t}, the four-terminal Hall resistance R_{H}, and the longitudinal resistance R_{L}. The N_{L} edge channels at the Fermi level are indicated, arrows point in the direction of motion of edge channels filled by the source contact at chemical potential $E_{\mathrm{F}} + \delta\mu$. The current $N_{\mathrm{L}} e \delta\mu / h$ is equipartitioned among the edge channels at the upper edge, corresponding to the case of local equilibrium. Localized states in the bulk do not contribute to the conductance. The resulting resistances are $R_{2t} = R_{\mathrm{H}} = h/N_{\mathrm{L}}e^2$, $R_{\mathrm{L}} = 0$. (From Beenakker and Van Houten [8].)

labeled by the Landau level index $n = 1, 2, \ldots$. In a field of 10 T (which is the strongest field that is routinely available), the Landau level separation $\hbar\omega_c$ is about 1 meV (for $m = m_e$). Consequently, in a metal the number of occupied Landau levels $N_{\mathrm{L}} \sim E_{\mathrm{F}}/\hbar\omega_c$ is a large number, of order 1000. Even so, magnetic quantization effects are important at low temperatures, since $\hbar\omega_c > k_{\mathrm{B}}T$ for $T < 10$ K. A familiar example is formed by the Shubnikov-De Haas oscillations in the magnetoresistance, which are caused by peaks in the density of states at the energies E_n which coincide with E_{F} for successive values of n as B is varied.

Because of the free motion along B, the density of states in a metal does not vanish at energies between two Landau levels. Consequently, in metals magnetic quantum effects are relatively small. The situation is different in a 2DEG. Here the energy spectrum of the electrons becomes fully discrete in a strong perpendicular magnetic field, since no free translational motion parallel to B is possible. The vanishing of the density of states between Landau levels is at the origin of the pronounced magnetic quantum effects in a 2DEG. Well known is the integer quantum Hall effect, characterized by a vanishing longitudinal resistance R_{L} and a quantized Hall resistance R_{H} at values of $h/N_{\mathrm{L}}e^2$. The distinction between a longitudinal and Hall resistance is topological (see Fig. 14): A four-terminal resistance measurement gives R_{H} if current and voltage contacts alternate along the boundary of the conductor, and R_{L} if that is not the case. There is no need to further characterize the contacts in the case of local equilibrium at the edge (in the opposite case the Hall resistance may take on anomalous values [8]). Frequently, the resistance of a sample is measured using only two contacts (which then act both as current and as voltage probes). In the quantum Hall effect regime, the two-terminal resistance $R_{2t} = R_{\mathrm{H}} + R_{\mathrm{L}} = R_{\mathrm{H}}$ is quantized at the same value as the Hall resistance.

The Fermi energy in a 2DEG is quite small (10 meV in conventional samples, 1 meV for samples with a very low density $n_s \sim 10^{10}$ cm^{-2}). Since, in addition, the effective mass is small, the extreme magnetic quantum limit $N_L = 1$ is accessible. This is the realm of the fractional quantum Hall effect, studied in high-mobility samples at milli-Kelvin temperatures, and of the Wigner crystallization of the 2DEG. Both phenomena are due to electron-electron interactions in a strong magnetic field. This chapter is limited to the integer quantum Hall effect.

To the extent that broadening of the Landau levels by disorder can be neglected, the density of states (per unit area) in an unbounded 2DEG can be approximated by a series of delta functions,

$$\rho(E) = g_s g_v \frac{eB}{h} \sum_{n=1}^{\infty} \delta(E - E_n) \,. \tag{39}$$

The spin-degeneracy g_s is removed in strong magnetic fields as a result of the Zeeman splitting $g\mu_B B$ of the Landau levels ($\mu_B \equiv e\hbar/2m_e$ denotes the Bohr magneton; the Lande g-factor is a complicated function of the magnetic field in these systems [76]).

In the modern theory of the quantum Hall effect [77], the longitudinal and Hall conductance (measured using two pairs of current contacts and voltage contacts) are expressed in terms of the transmission probabilities between the contacts for electronic states at the Fermi level. When E_F lies between two Landau levels, these states are *edge states* extended along the boundaries (Fig. 14). Edge states are the quantum mechanical analogue of *skipping orbits* of electrons undergoing repeated specular reflections at the boundary [8]. For a smooth confining potential $V(\mathbf{r})$, the edge states are extended along equipotentials of V at the guiding center energy E_G, defined by

$$E_G = E - (n - \tfrac{1}{2})\hbar\omega_c \,, \tag{40}$$

for an electron with energy E in the n-th Landau level ($n = 1, 2, \ldots$). The confining potential should be sufficiently smooth that it does not induce transitions between different values of n. This requires that $l_m V' \lesssim \hbar\omega_c$, with $l_m \equiv (\hbar/eB)^{1/2}$ the magnetic length (which plays the role of the wave length in the quantum Hall effect regime). Since the lowest Landau level has the largest guiding center energy, the corresponding edge state is located closest to the boundary of the sample, whereas the higher Landau levels are situated further towards its center.

In an open system, the single-electron levels with quantum number n form a 1D subband with subband bottom at $E_n = (n - \tfrac{1}{2})\hbar\omega_c$. These 1D subbands are referred to as *edge channels*. Each of the $N_L \sim E_F/\hbar\omega_c$ edge channels at the Fermi level contributes $2e^2/h$ to the Hall conductance if backscattering is suppressed. This happens whenever the Fermi level is located between two bulk Landau levels, so that the only states at E_F are those extended along the boundaries. Backscattering then requires transitions between edge states on *opposite* boundaries, which are usually far apart. In a very narrow channel, the Hall conductance may deviate from its quantized value $N_L e^2/h$ (and the longitudinal resistance may become non-zero) due to tunneling between opposite edges — a process that is strongly enhanced by disorder in the channel. The reason is that localized states at the Fermi energy may act as intermediate sites in a tunneling process from one edge to the other. We will come back to this point at the end of the section.

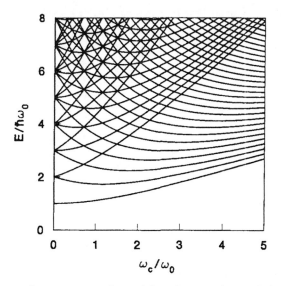

Figure 15. Energy spectrum of a quantum dot with a harmonic confining potential as a function of magnetic field, according to Eq. (41). Spin-splitting is neglected.

In a closed system, such as a quantum dot, the energy spectrum is fully discrete (for E_G less than the height E_B of the tunnel barriers which connect the dot to the leads). An example which can be solved exactly is a quantum dot defined by a 2D harmonic oscillator potential $V(r) = \frac{1}{2}m\omega_0^2 r^2$. The energy spectrum is given by [78, 79]

$$E_{nm} = \tfrac{1}{2}(n-m)\hbar\omega_c + \tfrac{1}{2}\hbar\left(\omega_c^2 + 4\omega_0^2\right)^{1/2}(n+m-1), \quad n,m = 1,2,\ldots \tag{41}$$

Each level has a two-fold spin-degeneracy, which is gradually lifted as B is increased. For simplicity, we do not take the spin degree of freedom into account. The energy spectrum (41) is plotted in Fig. 15. The asymptotes corresponding to the first few Landau levels are clearly visible.

In the limit $\omega_0/\omega_c \to 0$ of a smooth potential and a fairly strong magnetic field, Eq. (41) reduces to

$$E_{nm} = \hbar\omega_c\left(n - \tfrac{1}{2} + (n+m-1)(\omega_0/\omega_c)^2\right), \tag{42}$$

which may also be written as

$$E_{nm} = (n - \tfrac{1}{2})\hbar\omega_c + V(R_{nm}), \quad B\pi R_{nm}^2 = (m + \gamma_n)\frac{h}{e}, \tag{43}$$

with $\gamma_n = n - 1$. Equation (43) is equivalent to the requirement that the equipotential of the edge state, of radius R_{nm}, encloses $m + \gamma_n$ flux quanta. This geometrical requirement holds generally for smooth confining potentials, in view of the Bohr-Sommerfeld quantization rule

$$\frac{1}{h}\oint P dQ = m + \gamma_n. \tag{44}$$

The canonically conjugate variables P and Q, in the present case, are proportional to the guiding center coordinates $\mathbf{R} = (X, Y)$, defined by

$$X = x - v_y/\omega_c ,$$
(45)

$$Y = y + v_x/\omega_c ,$$
(46)

in terms of the position $\mathbf{r} = (x, y)$ and velocity $\mathbf{v} = (v_x, v_y)$ of the electron. If one identifies $Q \equiv X$, $P \equiv eBY$, one can verify the canonical commutation relation $[Q, P] = i\hbar$ (using $m\mathbf{v} = \mathbf{p} + e\mathbf{A}$, $[x, p_x] = [y, p_y] = i\hbar$, $[p_y, A_x] - [p_x, A_y] = i\hbar B$). The Bohr-Sommerfeld quantization rule thus becomes

$$\Phi = B \oint Y \, dX = \frac{h}{e}(m + \gamma_n) ,$$
(47)

which is the requirement that the flux Φ enclosed by the guiding center drift is quantized in units of the flux quantum. To close the argument, we compute the guiding center drift $\dot{\mathbf{R}} = B^{-2}\mathbf{E}(\mathbf{r}) \times \mathbf{B} \sim B^{-2}\mathbf{E}(\mathbf{R}) \times \mathbf{B}$, in the approximation that the electric field $\mathbf{E}$ does not vary strongly over the cyclotron radius $|\mathbf{r} - \mathbf{R}|$. In this case of a smoothly varying V, the motion of $\mathbf{R}$ is along equipotentials at the guiding center energy $E_G = E - (n - \frac{1}{2})\hbar\omega_c$. The Bohr-Sommerfeld quantization rule can thus be written in the general form

$$E_{nm} = (n - \tfrac{1}{2})\hbar\omega_c + E_G(n, m) ,$$
(48)

where $E_G(n, m)$ is the energy of the equipotential which encloses $m + \gamma_n$ flux quanta. For the harmonic oscillator potential, $\gamma_n = n - 1$. For other smooth confining potentials γ_n may be different. (Knowledge of γ_n is not important if one only considers states within a single Landau level.)

Equation (48) does not hold for a hard-wall confining potential. An exact solution exists in this case for a circular disc [80] of radius R, defined by $V(r) = 0$ for $r < R$, and $V(r) = \infty$ for $r > R$. The case of a square disc was studied numerically by Sivan et al. [29]. In Fig. 16a we show the energy spectrum as a function of B for the circular disc. (Fig. 16b is discussed in the following subsection.) The asymptotes correspond to the bulk Landau levels $E_n = (n - \frac{1}{2})\hbar\omega_c$. The first two Landau levels ($n = 1, 2$) are visible in Fig. 16a. The states between the Landau levels are edge states, which extend along the perimeter of the disc. These circulating edge states make the geometry effectively doubly connected — in the sense that they enclose a well-defined amount of flux. Resonant tunneling through these states is the mechanism leading to the Aharonov-Bohm magnetoconductance oscillations in a quantum dot.

Three cases of interest are illustrated in Fig. 17. In a strong magnetic field, only edge states with $n = 1$ corresponding to the first Landau level are occupied (Fig. 17a). As the field is reduced, also the second Landau level, $n = 2$, is occupied, as indicated in Fig. 17b. Tunneling through the quantum dot still occurs predominantly through the $n = 1$ edge states, which have the largest tunnel probability through the barriers. If the height E_B of the potential barriers is reduced, the $n = 1$ edge states near the Fermi level may have $E_G > E_B$, so that they form an extended edge channel. The edge states with $n > 1$ may still have $E_G < E_B$, and remain bound in the dot as before. As

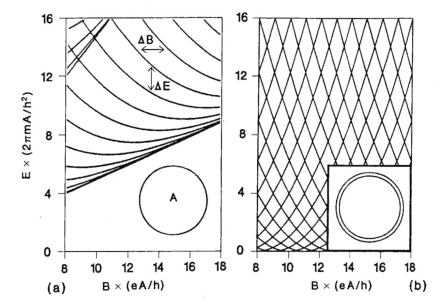

Figure 16. Comparison of the energy levels in a disc and a ring. (a) Circular hard-wall disc (after Geerinckx et al. [80]). (b) Circular channel or ring of width $W \ll l_m$ (after Büttiker et al. [81]). The levels in (b) are plotted relative to the energy of the bottom of the one-dimensional subband in the channel. The case $W \gtrsim l_m$ is qualitatively the same as long as the area S of the annulus is much smaller than the area A. Spin-splitting is disregarded. (From Beenakker et al. [30].)

illustrated in Fig. 17c, resonant tunneling now occurs predominantly through the edge states belonging to the second Landau level.

In the quantum Hall effect regime scattering between edge channels can be neglected on length scales comparable to the diameter of the dot [82] (this is known as adiabatic transport [8]). The edge channels may then be treated as independent parallel conduction paths. The edge channels with $E_G > E_B$ contribute e^2/h to the conductance. Resonant tunneling through the edge states with $E_G < E_B$ gives an oscillating contribution to the conductance of the quantum dot as a function of magnetic field. The periodicity of the conductance oscillations can be deduced from the result (48) for the edge state energy spectrum. Resonant tunneling from the reservoir with Fermi energy E_F into an edge state in the quantum dot is possible when $E_F = E_{nm}$ for certain quantum numbers n and m. For the edge states in the n-th Landau level the condition for resonant tunneling is that the equipotential at the guiding center energy $E_G \equiv E_F - (n - \frac{1}{2})\hbar\omega_c$ should enclose $m + \gamma_n$ flux quanta, for some integer m. Let $A(B)$ denote the (magnetic field dependent) area of the equipotential at energy E_G. The m-th conductance peak occurs at a magnetic field B_m determined by $B_m A(B_m) = (h/e)(m + \gamma_n)$. The periodicity $\Delta B \equiv B_{m+1} - B_m$ of the conductance oscillations from the n-th Landau level is obtained by expanding $A(B)$ around B_m,

$$\Delta B = \frac{h}{e}[A(B_m) + B_m A'(B_m)]^{-1}$$

$$\equiv \frac{h}{e}\frac{1}{A_{\text{eff}}(B_m)}. \tag{49}$$

The effective area $A_{\text{eff}}(B)$ can differ substantially from the geometrical area $A(B)$ in the case of a smooth confining potential [28]. The magnetoconductance oscillations are

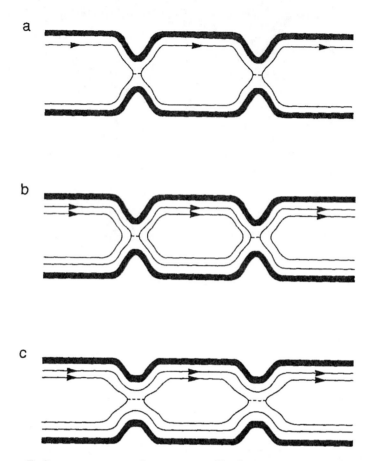

Figure 17. Aharonov-Bohm magnetoconductance oscillations may occur due to resonant tunneling through circulating edge states. Tunneling paths are indicated by dashed lines. (a) Only the first Landau level is occupied. If the capacitance of the dot is sufficiently small, the Coulomb blockade suppresses the Aharonov-Bohm oscillations. (b) Two Landau levels are occupied. Resonant tunneling through the dot occurs predominantly through the first (outer) Landau level. The Aharonov-Bohm effect is not suppressed by the charging energy. (c) Two Landau levels are occupied, one of which is fully transmitted. Since the number of electrons in the dot is not discretized, no Coulomb blockade of the Aharonov-Bohm effect is expected.

approximately periodic in B if the change in $A_{\mathrm{eff}}(B)$ in one period ΔB is much smaller than the effective area itself. Since the change in A_{eff} is of order h/eB per period, while $A_{\mathrm{eff}} \sim mh/eB$, approximately periodic oscillations occur for $m \gg 1$. This is the Aharonov-Bohm effect in the quantum Hall regime, first observed by Van Wees et al. [28] Their experimental results (reproduced in Fig. 18) correspond to the situation of Fig. 17c with one (or more) fully transmitted edge channels.

We close this subsection by mentioning that resonant backscattering (or resonant reflection) can cause similar Aharonov-Bohm oscillations as those caused by resonant transmission. Resonant backscattering may occur via a localized state bound on a potential maximum, created artificially (for example in a ring) or created by the presence of disorder [83]. The mechanism is illustrated in Fig. 19. Resonant backscattering leads to a periodic suppression of the conductance, in contrast to the periodic enhancement considered above.

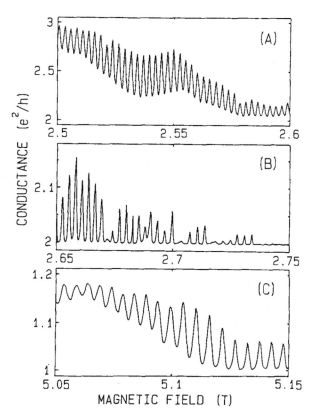

Figure 18. Magnetoconductance of a quantum dot in the 2DEG of a GaAs-AlGaAs heterostructure of 1.5 μm diameter, with point contacts at entrance and exit serving as tunnel barriers. The temperature is 30 mK. (a) and (b) Aharonov-Bohm magnetoconductance oscillations due to resonant tunneling through bound states belonging to the third (spin-split) edge channel. The first two (spin-split) Landau levels are fully transmitted (cf. Fig. 17c). (c) Resonant tunneling through bound states belonging to the second (spin-split) edge channel. The first (spin-split) edge channel is fully transmitted. (From Van Wees et al. [28].)

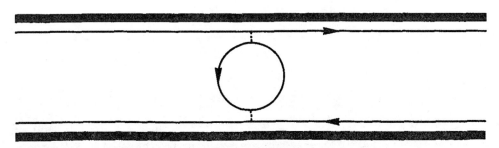

Figure 19. A circulating edge state bound on a local potential maximum causes resonant backscattering, thereby providing an alternative mechanism for Aharonov-Bohm magnetoconductance oscillations. (From Beenakker and Van Houten [8].)

4.2. Coulomb blockade of the Aharonov-Bohm effect

Single-electron tunneling is governed by the transport of a single quantum of charge e. The Aharonov-Bohm effect is governed by the flux quantum h/e. The present subsection addresses the interplay of these two quanta of nature in the integer quantum Hall effect regime.

In the previous subsection we have discussed how resonant tunneling through circulating edge states can lead to magnetoconductance oscillations in a quantum dot with a well-defined periodicity ΔB, similar to the usual Aharonov-Bohm effect in a ring. There is, however, an essential difference between the two geometries if only a single Landau level is occupied [30]. In each period ΔB the number of states below a given energy increases by one in a dot — but stays constant in a ring. As a result, the Aharonov-Bohm oscillations in the magnetoconductance of a quantum dot are accompanied by an increase of the charge of the dot by one elementary charge per period. That is of no consequence if the Coulomb repulsion of the electrons can be neglected, but becomes important if the dot has a small capacitance C to the reservoirs, since then the electrostatic energy e^2/C associated with the incremental charging by single electrons has to be taken into account.

Following Ref. [30], we analyze this problem by combining the results reviewed in the previous sections. We apply Eq. (12) to the energy spectrum shown in Fig. 16a. We consider here only the edge states from the lowest (spin-split) Landau level, so that the Aharonov-Bohm oscillations have a single periodicity. This corresponds to the strong-magnetic field limit. The magnetic field dependence of the edge states can be described approximately by a sequence of equidistant parallel lines,

$$E_p = \text{constant} - \frac{\Delta E}{\Delta B}(B - p\Delta B), \tag{50}$$

see Fig. 16a. For a circular quantum dot of radius R with a hard-wall confining potential, one can estimate [29] $\Delta B \sim h/eA$ and $\Delta E \sim \hbar\omega_c l_m/2R$. For a smooth confining potential $V(r)$ (with $l_m V' \lesssim \hbar\omega_c$) one has instead the estimates $\Delta B \sim (h/e)[A(B) + BA'(B)]^{-1} \sim (h/eA)[1 - \hbar\omega_c/RV'(R)]^{-1}$ [28], and $\Delta E \sim h/\tau \sim l_m^2 V'(R)/R$, where $A(B)$ is the area enclosed by the equipotential of radius R at the guiding center energy $V(R) = E - \frac{1}{2}\hbar\omega_c$ (cf. Eq. (40) for $n = 1$). [The estimate for ΔE results from the correspondence between the level spacing and the period τ of the classical motion along the equipotential, with guiding-center-drift velocity $V'(R)/eB$.]

On substitution of Eq. (50) into Eq. (12), one finds the condition

$$N\left(\Delta E + \frac{e^2}{C}\right) = \frac{\Delta E}{\Delta B}B_N + E_F + \text{constant} \tag{51}$$

for the magnetic field value B_N of the N-th conductance peak. The B-dependence of the reservoir Fermi energy can be neglected in Eq. (51) in the case of a hard-wall confining potential (since $dE_F/dB \approx \hbar\omega_c/B \ll \Delta E/\Delta B$). The periodicity $\Delta B^* \equiv B_{N+1} - B_N$ of the Aharonov-Bohm oscillations is thus given by

$$\Delta B^* = \Delta B\left(1 + \frac{e^2}{C\Delta E}\right). \tag{52}$$

[In the case of a smooth confining potential, the term ΔB in the enhancement factor of Eq. (52) should be replaced by the term $\Delta B[1 + (\Delta B/\Delta E)(dE_F/dB)]^{-1} \sim h/eA$, under the assumption that the Fermi energy in the reservoir is pinned to the lowest Landau level, i.e. $E_F = \frac{1}{2}\hbar\omega_c$.] We conclude from Eq. (52) that the charging energy enhances the spacing of two subsequent peaks in G versus B by a factor $1 + e^2/C\Delta E$. The periodicity of the magnetoconductance oscillations is lost if ΔB^* becomes so large that the linear approximation (50) for $E_p(B)$ breaks down. Since Eq. (50) holds at most over an energy range of the Landau level separation $\hbar\omega_c$, this suppression of the Aharonov-Bohm effect occurs when $(\Delta E/\Delta B)\Delta B^* \gtrsim \hbar\omega_c$, i.e. when $e^2/C \gtrsim \hbar\omega_c$.

The Aharonov-Bohm oscillations with bare periodicity $\Delta B = h/eA$ are recovered if one makes a hole in the disc, which is sufficiently large that the area S of the conducting region is much smaller than the enclosed area A. The inner perimeter of the resulting ring supports a second set of edge states, which travel around the ring in opposite direction as the first set of edge states at the outer perimeter. We compare in Fig. 16 the energy spectrum for a disc [80] and a ring [81]. The two sets of clockwise and counter-clockwise propagating edge states in a ring are distinguished by the opposite sign of dE_p/dB, i.e. of the magnetic moment. Each set of edge states leads to oscillations in the magnetoconductance of a ring with the same period ΔB, but shifted in phase (and in general with different amplitude, because the edge states at the inner perimeter have a smaller tunneling probability to the reservoir than those at the outer perimeter). The charging energy does not modify ΔB in a ring, because

$$E_p(B) = E_p(B + \Delta B) \text{ (ring)} .$$

In a disc, in contrast, one has according to Eq. (50),

$$E_p(B) = E_{p+1}(B + \Delta B) \text{ (disc)} .$$

To illustrate the difference, we compare in Fig. 20 for disc and ring the renormalized energy levels E_p^* [defined in Eq. (12)]. The effect of the charging energy in a ring is to open an energy gap of magnitude e^2/C in E_p^*. This gap will not affect the conductance oscillations as a function of B (at constant or slowly varying E_F). A controlled experimental demonstration of the influence of Coulomb repulsion on the AB effect may be obtained in a system which can be transformed from a disc into a ring. What we have in mind is a geometry such as shown in Fig. 21, which has an additional gate within the gates shaping the disc. By applying a negative voltage to this additional gate one depletes the central region of the quantum dot, thereby transforming it into a ring. In order to estimate the mutual capacitance C between the undepleted quantum disc and the adjacent 2DEG reservoirs, we note that only a circular strip of width l_m and radius R along the circumference of the disc contributes to C. The central region of the dot is incompressible in the quantum Hall effect regime, and thus behaves as a dielectric as far as the electrostatics is concerned. The capacitance C contains contributions from the self-capacitance of this strip as well as from its capacitance to the gate. (We assume that the gate is electrically connected to the 2DEG reservoirs.) Both contributions are of order ϵR, with a numerical prefactor of order unity which depends only logarithmically on the width of the strip and the separation to the gate [57] (ϵ is the dielectric constant). A dot radius of $1\,\mu m$ yields a charging energy $e^2/C \simeq 1\,meV$ for $\epsilon \simeq 10\epsilon_0$. This exceeds the

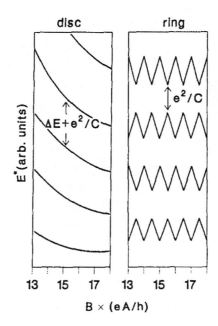

Figure 20. Renormalized energy levels, defined by Eq. (12), corresponding to the bare energy levels shown in Fig. 16. (From Beenakker et al. [30].)

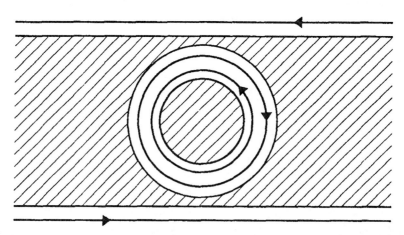

Figure 21. Schematic layout of a semiconductor nanostructure proposed to demonstrate the Coulomb blockade of the Aharonov-Bohm effect in a quantum dot, and its recovery upon transformation of the device into a ring (by applying a negative voltage to the central gate). (From Beenakker et al. [30].)

level separation $\Delta E \simeq \hbar\omega_c l_m/2R \simeq 2 \times 10^{-5}\,\mathrm{eV(T}/B)$ at a field of a few T. A significant increase of the frequency of the AB oscillations should thus be observable on depletion of the central region of the dot, even for a relatively large radius of $1\,\mu$m. To observe a full suppression of the AB effect in a sub-micron disc with $e^2/C \gtrsim \hbar\omega_c$, and its recovery on transformation to a ring, would be an ultimate test of the theory [30] reviewed here.

The difference between a ring and a disc disappears if more than a single Landau level is occupied in the disc. This occurs in the upper-left-hand corner in Fig. 16a. The energy spectrum in a disc now forms a mesh pattern which is essentially equivalent to that in a ring (Fig. 16b). There is no Coulomb-blockade of the Aharonov-Bohm effect in such a case [32], as discussed below.

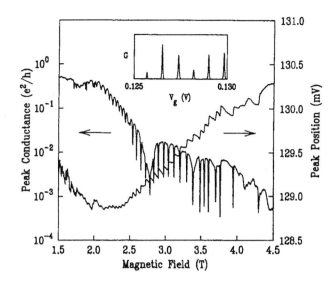

Figure 22. Effect of a magnetic field on the height and position of a conductance peak in a quantum dot in a GaAs-AlGaAs heterostructure, of the design shown in Fig. 2b. The temperature is 100 mK. Inset: Coulomb-blockade oscillations as a function of gate voltage, for $B = 3$ T. (From McEuen et al. [32].)

4.3. Experiments on quantum dots

We propose that the observation in a quantum dot of Aharonov-Bohm magnetoconductance oscillations by Van Wees et al. [82] was made possible by the presence of one or more extended edge channels, as in Fig. 17c (all of the succesful observations were, to our knowledge, made for $G > e^2/h$). In the presence of extended states the charge on the dot varies continuously, so that the Coulomb blockade of the Aharonov-Bohm effect discussed above is not operative. A direct experimental test of this interpretation would be desirable. This could be done by repeating the experiment in different magnetic field regimes, both with and without the presence of an extended edge channel.

Even if the magnetoconductance oscillations are suppressed, it is still possible to observe Coulomb-blockade oscillations in the conductance as a function of gate voltage (at fixed magnetic field). Previous observations of conductance oscillations as a function of gate voltage which were not observed as a function of B have been attributed to the Aharonov-Bohm effect [84, 85], but might well have been Coulomb-blockade oscillations instead.

An extended edge channel is one way to remove the Coulomb blockade of the Aharonov-Bohm effect. A second circulating edge channel in the quantum dot is another way, exploited by McEuen et al. [32] They observed conductance oscillations both as a function of gate voltage and as a function of magnetic field in a quantum dot of the design shown in Fig. 2b. Their main experimental results are reproduced in Fig. 22. The trace of conductance versus gate voltage at $B = 3$ T (Fig. 22, inset) exhibits the Coulomb-blockade oscillations, with an approximately constant periodicity. The main curves in Fig. 22 show that the height and position of a particular peak vary with B in a striking fashion. In the region between 2.5 and 3.5 T the peak height is periodically suppressed by as much as an order of magnitude, while the position of the peak oscillates synchronously around a slowly varying background. In this field regime two Landau levels are occupied in the dot, as in Fig. 17b, the lowest of which is spin-degenerate.

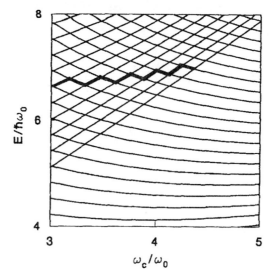

Figure 23. Close-up of the energy spectrum of Fig. 15 (after McEuen et al. [32]). The heavy line indicates the energy of the highest occupied state for a fixed number (23) of electrons in the dot. In each period of the saw-tooth a single electron is transferred from the second Landau level (rising lines) to the first (falling lines).

These observations have been explained by McEuen et al. in terms of the theory of Coulomb-blockade oscillations in the resonant tunneling regime. The one-electron energy spectrum in the range of two occupied Landau levels is shown in Fig. 23 (for the case of a parabolic confining potential, cf. Fig. 15). The experiment is performed at 100 mK, which is presumably in the resonant tunneling regime $k_B T < \Delta E$. Thus, only a single state participates in the conduction through the dot. As indicated in Fig. 23 (heavy line), this state belongs alternatingly to the first and the second Landau level (corresponding, respectively, to the falling and rising line segments of the sawtooth in Fig. 23). Thus, the tunnel rate into this state is alternatingly large and small. The periodic suppression of the peak height seen in Fig. 22 directly reflects this difference in tunnel rates.

According to Eqs. (12) and (14), the gate voltage of the N-th peak shifts with B according to

$$\delta\phi_{\text{gate}} = \frac{1}{\alpha e} \frac{\partial(E_N - E_F)}{\partial B} \delta B ,\qquad(53)$$

with α defined in Eq. (17). McEuen et al. determined α from the temperature dependence of the peak width, and neglected the change in E_F with B, as well as the difference between the electrostatic potential ϕ_{gate} and the measured electrochemical potential V_{gate}. The measured shift of the peak position with B (see Fig. 22) then directly yields the shift in energy of E_N. In this way they were able to map out the one-electron spectrum of the dot (Fig. 24). (To arrive at the bare energy spectrum a constant charging energy e^2/C was subtracted for each consecutive level.) The similarity of Figs. 23 and 24b is quite convincing. An unexplained effect is the gap in the spectrum around 0.2 meV. Also, the level spacing in the first Landau level (the vertical separation between the falling lines in Fig. 24b) appears to be two times smaller than that in the second Landau level (rising lines). Although this might be related to spin-splitting [32], we feel that it is more likely

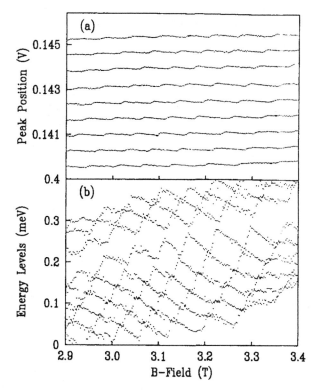

Figure 24. (a) Peak position as a function of magnetic field for a series of consecutive Coulomb-blockade oscillations in a quantum dot with two occupied Landau levels. (b) Energy spectrum of the dot obtained from the data in (a) after subtraction of the charging energy. (From McEuen et al. [32].)

that the assumption of a magnetic-field independent E_F is not justified. If, as should be expected, E_F is pinned to the second Landau level, then a proper correction for the Fermi level shift with B would lead to a clock-wise rotation of the entire level spectrum in Fig. 24b around $(B, E) = (0, 0)$. The agreement with the theoretical spectrum would then improve.

Coulomb-blockade oscillations as a function of gate voltage in the quantum Hall effect regime were studied by Williamson et al. [17] in a quantum dot of the design shown in Fig. 2c. They found that the amplitude of the oscillations was strongly enhanced compared to zero field, whereas the period was not much affected. (A similar enhancement of the amplitude has been seen in disordered quantum wires, and possible explanations are discussed below.) Representative traces of conductance versus gate voltage at zero field and for $B = 3.75$ T are reproduced in Fig. 25. The oscillations in the presence of a field are quite spectacular, of amplitude comparable to e^2/h. These experiments are in the regime where the conductance of the individual barriers approaches e^2/h as well, and virtual tunneling processes may be important. Experimentally, the conductance minima are not exponentially suppressed (see Fig. 25), even though the temperature was low (100 mK). In addition, the conductance maxima in the zero-field trace exceed e^2/h. These observations are also indicative of virtual tunneling processes [53, 54]. Finally, we would like to draw attention to the slow beating seen in the amplitude of the oscillations at zero field, which is suppressed at $B = 3.75$ T. Instead, a weak doublet-like structure becomes visible, reminiscent of that reported by Staring et al. [12] for a disordered quantum wire in a strong magnetic field (see Fig. 26), discussed below. Further experimental and theoretical work is needed to understand these intriguing effects of a magnetic field.

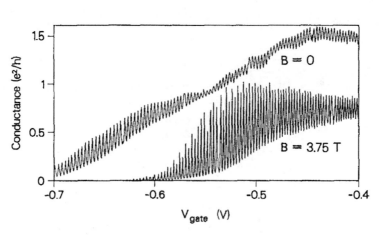

Figure 25. Effect of a magnetic field on the conductance oscillations in a quantum dot in a GaAs-AlGaAs heterostructure, with a geometry as in Fig. 2c. The temperature is 50 mK. This is an effective two-terminal conductance (obtained from a four-terminal conductance measurement, with the voltage measured diagonally across the dot [8, page 183].) (From Williamson et al. [17].)

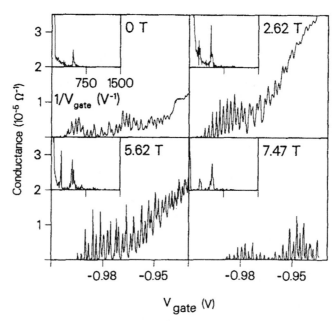

Figure 26. Effect of a magnetic field on the Coulomb-blockade oscillations a disordered quantum wire (as in Fig. 11), at 50 mK. Insets: Fourier transforms of the data, with the vertical axes of the curves at 0 T and 7.47 T magnified by a factor 2.5, relative to the curves at 2.62 T and 5.62 T. (From Staring et al. [12].)

4.4. Experiments on disordered quantum wires

The effect of a parallel and perpendicular magnetic field on the conductance oscillations in a narrow channel in a Si inversion layer has been studied by Field et al. [11]. Staring et al. [12, 13] investigated the effect of a perpendicular field on disordered quantum wires in the 2DEG of a GaAs-AlGaAs heterostructure. Some of the data is reproduced in Fig. 26. The Fourier transforms of the traces of conductance versus gate voltage (insets) demonstrate a B − independent dominant frequency of 450 V^{-1}. Curiously, as the magnetic field is increased a second peak in the Fourier transform emerges at about half the dominant frequency. This second peak corresponds to an amplitude modulation of the peaks, as is most clearly seen in the trace at 5.62 T where high and low peaks alternate in a doublet-like structure. This feature is characteristic of this particular sample. Other channels showed different secondary effects, such as a much more rapid oscillation superposed on the conductance trace for certain values of the magnetic field [12]. It is likely that the presence of additional segments in the wire plays a role. The period $\delta V_{\text{gate}} \sim 2.2$ mV of the dominant conductance oscillations is remarkably insensitive to a strong magnetic field. Spin-splitting of the peaks was not observed, even at the highest fields of 8 T. These qualitative observations agree with our interpretation of the effect as Coulomb-blockade oscillations. In Sec. 3.2 we have already had occasion to show that the temperature dependence of the lineshape of an isolated peak was well accounted for by Eq. (24), for a set of parameter values consistent with zero-field experiments.

The height of the conductance peaks is enhanced by a field of intermediate strength (2 T $< B < 6$ T), followed by a decrease at stronger fields ($B \sim 7.5$ T). Also the width of the peaks is reduced in a strong magnetic field. The largest isolated peaks (found in a different sample [13]) approach a height of e^2/h, measured two-terminally. A similar enhancement of the amplitude of the Coulomb-blockade oscillations by a magnetic field was observed in a quantum dot [17] (see Fig. 25). One explanation is that the inelastic scattering rate is reduced by a magnetic field. In the low-temperature regime $k_{\text{B}}T \lesssim h\Gamma$ this makes the peaks higher and narrower (cf. Sec. 2.2). In a disordered quantum wire the magnetic suppression of backscattering provides another mechanism for an enhancement of the peak height because of the resulting reduction in series resistance [13]. Additionally, the modulation of the Fermi level in the quantum Hall effect regime may lead to a non-monotonic variation with B of the transmission probability $T(E_{\text{F}})$, and thus presumably of the tunnel rates $h\Gamma$. The level degeneracy varies with B, becoming large when the Fermi energy coincides with a bulk Landau level in the dot. This may also give rise to variations in the peak height [34]. These are tentative explanations of the surprising magnetic field dependence of the amplitude of the Coulomb-blockade oscillations, which remains to be elucidated.

We close this subsection by noting that Staring et al. [12] also measured magneto-conductance traces at fixed gate voltage. In contrast to the gate voltage scans, these exhibited irregular structure only, with strong features corresponding to depopulation of Landau levels. The absence of regular oscillations constitutes the first experimental evidence for the predicted [30] Coulomb blockade of the Aharonov-Bohm effect.

ACKNOWLEDGEMENTS. Valuable discussions with S. Colak, L. P. Kouwenhoven, N. C. van der Vaart, J. G. Williamson, and the support of J. Wolter and M. F. H. Schuurmans are gratefully acknowledged. Our experimental work has been made possible by C. T. Foxon who has grown the necessary samples by molecular beam epitaxy, and by C. E. Timmering who took care of the technology. We have benefitted from interactions with the participants of the NATO ASI on Single Charge Tunneling. We thank our colleagues at M.I.T., Delft, and Philips for their permission to reproduce some of their results. This research was partly funded under the ESPRIT basic research action project 3133.

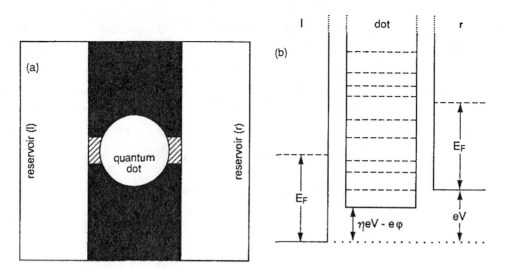

Figure 27. (a) Schematic cross-section of the geometry studied in this appendix, consisting of a confined region ("quantum dot") weakly coupled to two electron reservoirs via tunnel barriers (hatched). (b) Profile of the electrostatic potential energy (solid curve) along a line through the tunnel barriers. The Fermi levels in the left and right reservoirs, and the discrete energy levels in the quantum dot are indicated (dashed lines).

A. Conductance of a quantum dot coupled to two electron reservoirs

Following the treatment by Beenakker [19], we derive in this appendix Eq. (24) for the conductance of a confined region which is weakly coupled via tunnel barriers to two electron reservoirs. The confined region, or "quantum dot", has single-electron energy levels at E_p ($p = 1, 2, \ldots$), labeled in ascending order and measured relative to the bottom of the potential well. Each level contains either one or zero electrons. Spin degeneracy can be included by counting each level twice, and other degeneracies can be included similarly. Each reservoir is taken to be in thermal equilibrium at temperature T and chemical potential E_F. A continuum of states is assumed in the reservoirs, occupied according to the Fermi-Dirac distribution

$$f(E - E_F) = \left[1 + \exp\left(\frac{E - E_F}{kT}\right)\right]^{-1}. \tag{54}$$

In Fig. 27 we show schematically a cross-section of the geometry, and the profile of the electrostatic potential energy along a line through the tunnel barriers.

A current I can be passed through the dot by applying a potential difference V between the two reservoirs. The tunnel rate from level p to the left and right reservoirs in Fig. 27 is denoted by Γ_p^l and Γ_p^r, respectively. We assume that both kT and ΔE are $\gg h(\Gamma^l + \Gamma^r)$ (for all levels participating in the conduction), so that the finite width $h\Gamma = h(\Gamma^l + \Gamma^r)$ of the transmission resonance through the quantum dot can be disregarded. This assumption allows us to characterize the state of the quantum dot by a set of occupation numbers, one for each energy level. (As discussed in Sec. 2.2, the restriction kT, $\Delta E \gg h\Gamma$ results in the conductance being much smaller than the quantum e^2/h.) We also assume conservation of energy in the tunnel process, thus neglecting contributions of higher order in Γ from tunneling via a *virtual* intermediate state in the quantum dot [54, 53] as discussed in Chap. 6. We finally assume that inelastic scattering takes place exclusively in the reservoirs — not in the quantum dot. The effect of inelastic scattering in the quantum dot is considered in Ref. [19].

Energy conservation upon tunneling from an initial state p in the quantum dot (containing N electrons) to a final state in the left reservoir at energy $E^{f,l}$ (in excess of the local electrostatic potential energy), requires that

$$E^{f,l}(N) = E_p + U(N) - U(N-1) + \eta eV. \tag{55}$$

Here η is the fraction of the applied voltage V which drops over the left barrier. (As we will see, this parameter η drops out of the final expression for the conductance in linear response.) The energy conservation condition for tunneling from an initial state $E^{i,l}$ in the left reservoir to a final state p in the quantum dot is

$$E^{i,l}(N) = E_p + U(N+1) - U(N) + \eta eV, \tag{56}$$

where [as in Eq. (55)] N is the number of electrons in the dot *before* the tunneling event. Similarly, for tunneling between the quantum dot and the right reservoir one has the conditions

$$E^{f,r}(N) = E_p + U(N) - U(N-1) - (1-\eta)eV, \tag{57}$$

$$E^{i,r}(N) = E_p + U(N+1) - U(N) - (1-\eta)eV, \tag{58}$$

where $E^{i,r}$ and $E^{f,r}$ are the energies of the initial and final states in the right reservoir.

The stationary current through the left barrier equals that through the right barrier, and is given by

$$I = -e \sum_{p=1}^{\infty} \sum_{\{n_i\}} \Gamma_p^l P(\{n_i\}) \left(\delta_{n_p,0} f(E^{i,l}(N) - E_F) \right.$$

$$\left. - \delta_{n_p,1} \left[1 - f(E^{f,l}(N) - E_F) \right] \right). \tag{59}$$

The second summation is over all realizations of occupation numbers $\{n_1, n_2, \ldots\} \equiv \{n_i\}$ of the energy levels in the quantum dot, each with stationary probability $P(\{n_i\})$. (The numbers n_i can take on only the values 0 and 1.) In equilibrium, this probability distribution is the Gibbs distribution in the grand canonical ensemble:

$$P_{eq}(\{n_i\}) = Z^{-1} \exp\left[-\frac{1}{kT} \left(\sum_{i=1}^{\infty} E_i n_i + U(N) - N E_F \right) \right], \tag{60}$$

where $N \equiv \sum_i n_i$, and Z is the partition function,

$$Z = \sum_{\{n_i\}} \exp\left[-\frac{1}{kT} \left(\sum_{i=1}^{\infty} E_i n_i + U(N) - N E_F \right) \right]. \tag{61}$$

The non-equilibrium probability distribution P is a stationary solution of the kinetic equation

$$\frac{\partial}{\partial t}P(\{n_i\}) = 0 = -\sum_p P(\{n_i\})\delta_{n_p,0}\left(\Gamma_p^l f(E^{i,l}(N) - E_F) + \Gamma_p^r f(E^{i,r}(N) - E_F)\right)$$

$$-\sum_p P(\{n_i\})\delta_{n_p,1}\left(\Gamma_p^l\left[1 - f(E^{f,l}(N) - E_F)\right] + \Gamma_p^r\left[1 - f(E^{f,r}(N) - E_F)\right]\right)$$

$$+\sum_p P(n_1, \ldots n_{p-1}, 1, n_{p+1}, \ldots)\delta_{n_p,0}$$

$$\times \left(\Gamma_p^l\left[1 - f(E^{f,l}(N+1) - E_F)\right] + \Gamma_p^r\left[1 - f(E^{f,r}(N+1) - E_F)\right]\right)$$

$$+\sum_p P(n_1, \ldots n_{p-1}, 0, n_{p+1}, \ldots)\delta_{n_p,1}$$

$$\times \left(\Gamma_p^l f(E^{i,l}(N-1) - E_F) + \Gamma_p^r f(E^{i,r}(N-1) - E_F)\right). \tag{62}$$

The kinetic equation (62) for the stationary distribution function is equivalent to the set of detailed balance equations (one for each $p = 1, 2, \ldots$)

$$P(n_1, \ldots n_{p-1}, 1, n_{p+1}, \ldots)\left(\Gamma_p^l\left[1 - f(E^{f,l}(\tilde{N}+1) - E_F)\right]\right.$$

$$\left. + \Gamma_p^r\left[1 - f(E^{f,r}(\tilde{N}+1) - E_F)\right]\right)$$

$$= P(n_1, \ldots n_{p-1}, 0, n_{p+1}, \ldots)\left(\Gamma_p^l f(E^{i,l}(\tilde{N}) - E_F) + \Gamma_p^r f(E^{i,r}(\tilde{N}) - E_F)\right), \tag{63}$$

with the notation $\tilde{N} \equiv \sum_{i \neq p} n_i$.

A similar set of equations formed the basis for the work of Averin, Korotkov, and Likharev on the Coulomb staircase in the non-linear I–V characteristic of a quantum dot [34]. To simplify the solution of the kinetic equation, they assumed that the charging energy e^2/C is much greater than the average level spacing ΔE. In this chapter we restrict ourselves to the regime of linear response, appropriate for the Coulomb-blockade oscillations. Then the conductance can be calculated exactly and analytically.

The (two-terminal) linear response conductance G of the quantum dot is defined as $G = I/V$ in the limit $V \to 0$. To solve the linear response problem we substitute

$$P(\{n_i\}) \equiv P_{eq}(\{n_i\})\left(1 + \frac{eV}{kT}\Psi(\{n_i\})\right) \tag{64}$$

into the detailed balance equation (63), and linearize with respect to V. One finds

$$P_{eq}(n_1, \ldots n_{p-1}, 1, n_{p+1}, \ldots)$$

$$\times \left(\Psi(n_1, \ldots n_{p-1}, 1, n_{p+1}, \ldots)(\Gamma_p^l + \Gamma_p^r)[1 - f(\epsilon)] - [\Gamma_p^l\eta - \Gamma_p^r(1-\eta)]kTf'(\epsilon)\right)$$

$$= P_{eq}(n_1, \ldots n_{p-1}, 0, n_{p+1}, \ldots)$$

$$\times \left(\Psi(n_1, \ldots n_{p-1}, 0, n_{p+1}, \ldots)(\Gamma_p^l + \Gamma_p^r)f(\epsilon) + [\Gamma_p^l\eta - \Gamma_p^r(1-\eta)]kTf'(\epsilon)\right), \tag{65}$$

where $f'(\epsilon) \equiv df(\epsilon)/d\epsilon$, and we have abbreviated $\epsilon \equiv E_p + U(\tilde{N}+1) - U(\tilde{N}) - E_F$.

Equation (65) can be simplified by making subsequently the substitutions

$$1 - f(\epsilon) = f(\epsilon)e^{\epsilon/kT}, \tag{66}$$

$$P_{eq}(n_1, \ldots n_{p-1}, 1, n_{p+1}, \ldots) = P_{eq}(n_1, \ldots n_{p-1}, 0, n_{p+1}, \ldots)e^{-\epsilon/kT}, \tag{67}$$

$$kT f'(\epsilon)\left(1 + e^{-\epsilon/kT}\right) = -f(\epsilon). \tag{68}$$

The factors P_{eq} and f cancel, and one is left with the simple equation

$$\Psi(n_1, \ldots n_{p-1}, 1, n_{p+1}, \ldots) = \Psi(n_1, \ldots n_{p-1}, 0, n_{p+1}, \ldots) + \frac{\Gamma_p^r}{\Gamma_p^l + \Gamma_p^r} - \eta. \tag{69}$$

The solution is

$$\Psi(\{n_i\}) = \text{constant} + \sum_{i=1}^{\infty} n_i \left(\frac{\Gamma_i^r}{\Gamma_i^l + \Gamma_i^r} - \eta\right). \tag{70}$$

The constant first term in Eq. (70) takes care of the normalization of P to first order in V, and need not be determined explicitly. Notice that the first order non-equilibrium correction Ψ to P_{eq} is *zero* if $\eta = \Gamma_i^r/(\Gamma_i^l + \Gamma_i^r)$ for all i. This will happen in particular for two identical tunnel barriers (when $\eta = \frac{1}{2}$, $\Gamma_i^l = \Gamma_i^r$). Because of the symmetry of the system, the distribution function then contains only terms of *even* order in V.

Now we are ready to calculate the current I through the quantum dot to first order in V. Linearization of Eq. (59), after substitution of Eq. (64) for P, gives

$$I = -e\frac{eV}{kT}\sum_p \sum_{\{n_i\}} \Gamma_p^l P_{eq}(\{n_i\})\left(\delta_{n_p,0}\eta kT f'(\epsilon) + \delta_{n_p,1}\eta kT f'(\epsilon)\right.$$

$$\left. + \Psi(\{n_i\})\delta_{n_p,0}f(\epsilon) - \Psi(\{n_i\})\delta_{n_p,1}[1 - f(\epsilon)]\right)$$

$$= \frac{e^2 V}{kT}\sum_p \sum_{\{n_i\}} \Gamma_p^l P_{eq}(\{n_i\})\delta_{n_p,0}f(E_p + U(N+1) - U(N) - E_F)$$

$$\times [\eta + \Psi(n_1, \ldots n_{p-1}, 1, n_{p+1}, \ldots) - \Psi(n_1, \ldots n_{p-1}, 0, n_{p+1}, \ldots)]$$

$$= \frac{e^2 V}{kT}\sum_p \sum_{\{n_i\}} \frac{\Gamma_p^l \Gamma_p^r}{\Gamma_p^l + \Gamma_p^r} P_{eq}(\{n_i\})\delta_{n_p,0}f(E_p + U(N+1) - U(N) - E_F). \tag{71}$$

In the second equality we have again made use of the identities (66)–(68), and in the third equality we have substituted Eq. (69). Notice that the parameter η has dropped out of the final expression for I.

We define the equilibrium probability distributions

$$P_{eq}(N) = \sum_{\{n_i\}} P_{eq}(\{n_i\})\delta_{N,\sum_i n_i} = \frac{\exp(-\Omega(N)/kT)}{\sum_N \exp(-\Omega(N)/kT)}, \tag{72}$$

$$F_{\text{eq}}(E_p \mid N) = \frac{1}{P_{\text{eq}}(N)} \sum_{\{n_i\}} P_{\text{eq}}(\{n_i\}) \delta_{n_p,1} \delta_{N,\sum_i n_i}$$

$$= \exp(\mathcal{F}(N)/kT) \sum_{\{n_i\}} \exp\left(-\frac{1}{kT} \sum_{i=1}^{\infty} E_i n_i\right) \delta_{n_p,1} \delta_{N,\sum_i n_i}. \tag{73}$$

Here $\Omega(N)$ is the thermodynamic potential of the quantum dot, and $\mathcal{F}(N)$ is the free energy of the internal degrees of freedom:

$$\Omega(N) = \mathcal{F}(N) + U(N) - N E_{\text{F}}, \tag{74}$$

$$\mathcal{F}(N) = -kT \ln\left[\sum_{\{n_i\}} \exp\left(-\frac{1}{kT} \sum_{i=1}^{\infty} E_i n_i\right) \delta_{N,\sum_i n_i}\right]. \tag{75}$$

The function $P_{\text{eq}}(N)$ is the probability that the quantum dot contains N electrons in equilibrium; The function $F_{\text{eq}}(E_p \mid N)$ is the conditional probability in equilibrium that level p is occupied given that the quantum dot contains N electrons. In terms of these distribution functions, the conductance $G = I/V$ resulting from Eq. (71) equals

$$G = \frac{e^2}{kT} \sum_{p=1}^{\infty} \sum_{N=0}^{\infty} \frac{\Gamma_p^l \Gamma_p^r}{\Gamma_p^l + \Gamma_p^r} P_{\text{eq}}(N)[1 - F_{\text{eq}}(E_p \mid N)]$$

$$\times f(E_p + U(N+1) - U(N) - E_{\text{F}}). \tag{76}$$

In view of Eqs. (66) and (67), Eq. (76) can equivalently be written in the form

$$G = \frac{e^2}{kT} \sum_{p=1}^{\infty} \sum_{N=1}^{\infty} \frac{\Gamma_p^l \Gamma_p^r}{\Gamma_p^l + \Gamma_p^r} P_{\text{eq}}(N) F_{\text{eq}}(E_p \mid N)$$

$$\times [1 - f(E_p + U(N) - U(N-1) - E_{\text{F}})]. \tag{77}$$

Redefining $P_{\text{eq}}(N) F_{\text{eq}}(E_p \mid N) = P_{\text{eq}}(N, n_p = 1)$ we find Eq. (24) as it appears in Sec. 2.2.

References

[1] L. S. Kuz'min and K. K. Likharev, Pis'ma Zh. Eksp. Teor. Fiz. **45**, 389 (1987) [JETP Lett. **45**, 495 (1987)].

[2] T. A. Fulton and G. J. Dolan, Phys. Rev. Lett. **59**, 109 (1987).

[3] K. Mullen, E. Ben-Jacob, R. C. Jaklevic, and Z. Schuss, Phys. Rev. B **37**, 98 (1988); M. Amman, K. Mullen, and E. Ben-Jacob, J. Appl. Phys. **65**, 339 (1989).

[4] K. K. Likharev, IBM J. Res. Dev. **32**, 144 (1988); D. V. Averin and K. K. Likharev, in: *Mesoscopic Phenomena in Solids*, B. L. Al'tshuler, ed. by P. A. Lee, and R. A. Webb, (Elsevier, Amsterdam, 1991). This is a comprehensive review of single-electron tunneling in metals.

[5] L. I. Glazman and R. I. Shekhter, J. Phys. Condens. Matter **1**, 5811 (1989).

[6] R. I. Shekhter, Zh. Eksp. Teor. Fiz. **63**, 1410 (1972) [Sov. Phys. JETP **36**, 747 (1973)].

[7] I. O. Kulik and R. I. Shekhter, Zh. Eksp. Teor. Fiz. **68**, 623 (1975) [Sov. Phys. JETP **41**, 308 (1975)].

[8] For a review of theoretical and experimental aspects of quantum transport in semiconductor nanostructures, see: C. W. J. Beenakker and H. van Houten, Solid State Physics **44**, 1 (1991).

[9] J. H. F. Scott-Thomas, S. B. Field, M. A. Kastner, H. I. Smith, and D. A. Antoniadis, Phys. Rev. Lett. **62**, 583 (1989).

[10] U. Meirav, M. A. Kastner, M. Heiblum, and S. J. Wind, Phys. Rev. B **40**, 5871 (1989).

[11] S. B. Field, M. A. Kastner, U. Meirav, J. H. F. Scott-Thomas, D. A. Antoniadis, H. I. Smith, and S. J. Wind, Phys. Rev. B **42**, 3523 (1990).

[12] A. A. M. Staring, H. van Houten, C. W. J. Beenakker, and C. T. Foxon, in: *High Magnetic Fields in Semiconductor Physics III*, ed. by G. Landwehr, (Springer, Berlin, 1991).

[13] A. A. M. Staring, H. van Houten, C. W. J. Beenakker, and C. T. Foxon, Phys. Rev. B. to be published.

[14] C. de Graaf, J. Caro, S. Radelaar, V. Lauer, and K. Heyers, submitted to Phys. Rev. B.

[15] U. Meirav, M. A. Kastner, and S. J. Wind, Phys. Rev. Lett. **65**, 771 (1990).

[16] L. P. Kouwenhoven, N. C. van der Vaart, A. T. Johnson, C. J. P. M. Harmans, J. G. Williamson, A. A. M. Staring, and C. T. Foxon, *Festkörperprobleme / Advances in Solid State Physics* **31** (Vieweg, Braunschweig, 1991)

[17] J. G. Williamson, A. A. M. Staring, H. van Houten, L. P. Kouwenhoven, and C. T. Foxon, to be published; A. A. M. Staring, J. G. Williamson, H. van Houten, C. W. J. Beenakker, L. P. Kouwenhoven, and C. T. Foxon, Physica B, to be published.

[18] A. L. Efros and B. I. Shklovskii, in: *Electron-Electron Interactions in Disordered Systems*, ed. by A. L. Efros and M. Pollak, (North-Holland, Amsterdam, 1985).

[19] C. W. J. Beenakker, Phys. Rev. B. **44**, 1646 (1991).

[20] Y. Meir, N. S. Wingreen, and P. A. Lee, Phys. Rev. Lett. **66**, 3048 (1991).

[21] M.A. Kastner, S. B. Field, U. Meirav, J. H. F. Scott-Thomas, D. A. Antoniadis, and H. I. Smith, Phys. Rev. Lett. **63**, 1894 (1989).

[22] A. I. Larkin and P. A. Lee, Phys. Rev. B **17**, 1596 (1978).

[23] P. A. Lee and T. M. Rice, Phys. Rev. B **19**, 3970 (1979).

[24] D. V. Averin and K. K. Likharev, unpublished.

[25] H. van Houten and C. W. J. Beenakker, Phys. Rev. Lett. **63**, 1893 (1989).

[26] F. M. de Aguiar and D. A. Wharam, Phys. Rev. B **43**, 15 April (1991).

[27] J. Mašek and B. Kramer, submitted to Europhys. Lett.

[28] B. J. van Wees, L. P. Kouwenhoven, C. J. P. M. Harmans, J. G. Williamson, C. E. Timmering, M. E. I. Broekaart, C. T. Foxon, and J. J. Harris, Phys. Rev. Lett. **62**, 2523 (1989).

[29] U. Sivan and Y. Imry, Phys. Rev. Lett. **61**, 1001 (1988); U. Sivan, Y. Imry, and C. Hartzstein, Phys. Rev. B **39**, 1242 (1989).

[30] C. W. J. Beenakker, H. van Houten, and A. A. M. Staring, Phys. Rev. B. **44**, 657 (1991).

[31] C. W. J. Beenakker, H. van Houten, and A. A. M. Staring, in: *Granular Nano-electronics*, ed. by D. K. Ferry, J. Barker, and C. Jacoboni, (Plenum, New York, 1991).

[32] P. L. McEuen, E. B. Foxman, U. Meirav, M. A. Kastner, Y. Meir, N. S. Wingreen, and S. J. Wind, Phys. Rev. Lett. **66**, 1926 (1991).

[33] P. A. Lee, Phys. Rev. Lett. **65**, 2206 (1990).

[34] D. V. Averin and A. N. Korotkov, Zh. Eksp. Teor. Fiz. **97**, 1661 (1990) [Sov. Phys. JETP **70**, 937 (1990)]; A. N. Korotkov, D. V. Averin, and K. K. Likharev, Physica B **165** & **166**, 927 (1990); D. V. Averin, A. N. Korotkov, and K. K. Likharev, submitted to Phys. Rev. B.

[35] A. Groshev, Phys. Rev. B **42**, 5895 (1990); A. Groshev, T. Ivanov, and V. Valtchinov, Phys. Rev. Lett. **66**, 1082 (1991).

[36] L. P. Kouwenhoven, A. T. Johnson, N. C. van der Vaart, C. J. P. M. Harmans, and C. T. Foxon, unpublished.

[37] L. J. Geerligs, V. F. Anderegg, P. A. M. Holweg, J. E. Mooij, H. Pothier, D. Estève, C. Urbina, M. H. Devoret, Phys. Rev. Lett. **64**, 2691 (1990).

[38] H. Pothier, P. Lafarge, C. Urbina, D. Estève, and M. H. Devoret, Europhys. Lett. **17**, 249 (1992).

[39] M. A. Reed, J. N. Randall, R. J. Aggarwal, R. J. Matyi, T. M. Moore, and A. E. Wetsel, Phys. Rev. Lett. **60**, 535 (1988).

[40] B. Su, V. J. Goldman, and J. E. Cunningham, preprint.

[41] W. B. Kinard, M. H. Weichold, G. F. Spencer, and W. P. Kirk, in: *Nanostructure Physics and Fabrication*, ed. by M. A. Reed and W. P. Kirk, (Academic, New York, 1989).

[42] H. van Houten, C. W. J. Beenakker, and B. J. van Wees, in: *Nanostructured Systems*, ed. by M. A. Reed, (a volume of Semiconductors and Semimetals, Academic, New York, 1991).

[43] B. J. van Wees, H. van Houten, C. W. J. Beenakker, J. G. Williamson, L. P. Kouwenhoven, D. van der Marel, and C. T. Foxon, Phys. Rev. Lett. **60**, 848 (1988).

[44] D. A. Wharam, T. J. Thornton, R. Newbury, M. Pepper, H. Ahmed, J. E. F. Frost, D. G. Hasko, D. C. Peacock, D. A. Ritchie, and G. A. C. Jones, J. Phys. C **21**, L209 (1988).

[45] J. A. Nixon and J. H. Davies, Phys. Rev. B. **41**, 7929 (1990).

[46] G. W. Bryant, Phys. Rev. Lett. **59**, 1140 (1987).

[47] P. A. Maksym and T. Chakraborty, Phys. Rev. Lett. **65**, 108 (1990).

[48] A. Kumar, S. E. Laux, and F. Stern, Phys. Rev. B **42**, 5166 (1990).

[49] M. Büttiker, Phys. Rev. B **33**, 3020 (1986).

[50] M. Büttiker, IBM J. Res. Dev. **32**, 63 (1988).

[51] A. D. Stone and P. A. Lee, Phys. Rev. Lett. **54**, 1196 (1985).

[52] R. Landauer, in: *Localization, Interaction, and Transport Phenomena*, ed. by B. Kramer, G. Bergmann, and Y. Bruynseraede, (Springer, Berlin, 1985).

[53] D. V. Averin and A. A. Odintsov, Phys. Lett. A **140**, 251 (1989); D. V. Averin and Yu. V. Nazarov, Phys. Rev. Lett. **65**, 2446 (1990).

[54] L. I. Glazman and K. A. Matveev, Pis'ma Zh. Eksp. Teor. Fiz. **51**, 425 (1990) [JETP Lett. **51**, 484 (1990)].

[55] Yu. V. Nazarov, Solid State Comm. **75**, 669 (1990); A. N. Korotkov and Yu. V. Nazarov, Physica B, to be published.

[56] S. E. Laux, D. E. Frank, and F. Stern, Surf. Sci. **196**, 101 (1988).

[57] L. D. Landau and E. M. Lifshitz, *Electrodynamics of Continuous Media* (Pergamon, New York, 1960).

[58] K. K. Likharev, N. S. Bakhvalov, G. S. Kazacha, and S. I. Serdyukova, IEEE Trans. Magn. **25**, 1436 (1989).

[59] M. Amman, E. Ben-Jacob, and K. Mullen, Physics Letters A **142**, 431 (1989).

[60] A. B. Fowler, A. Hartstein, and R. A. Webb, Phys. Rev. Lett. **48**, 196 (1982); R. A. Webb, A. Hartstein, J. J. Wainer, and A. B. Fowler, Phys. Rev. Lett. **54**, 1577 (1985).

[61] R. F. Kwasnick, M. A. Kastner, J. Melngailis, and P. A. Lee, Phys. Rev. Lett. **52**, 224 (1984); M. A. Kastner, R. F. Kwasnick, J. C. Licini, and D. J. Bishop, Phys. Rev. B **36**, 8015 (1987).

[62] P. A. Lee, Phys. Rev. Lett. **53**, 2042 (1984).

[63] R. K. Kalia, W. Xue, and P. A. Lee, Phys. Rev. Lett. **57**, 1615 (1986).

[64] C. J. Adkins, in: *Hopping and Related Phenomena*, ed. by H. Fritsche and M. Pollak, (World Scientific, New Jersey, 1990).

[65] B. L. Al'tshuler, Pis'ma Zh. Eksp. Teor. Fiz. **41**, 530 (1985) [JETP Lett. **41**, 648 (1985)].

[66] P. A. Lee and A. D. Stone, Phys. Rev. Lett. **55**, 1622 (1985).

[67] A. B. Fowler, G. L. Timp, J. J. Wainer, and R. A. Webb, Phys. Rev. Lett. **57**, 138 (1986); A. B. Fowler, J. J. Wainer, and R. A. Webb, IBM J. Res. Dev. **32**, 372 (1988).

[68] M. Green and M. Pollak, in: *Hopping and Related Phenomena*, ed. by H. Fritsche and M. Pollak, (World Scientific, New Jersey, 1990).

[69] J. J. Wainer, A. B. Fowler, and R. A. Webb, Surf. Sci. **196**, 134 (1988).

[70] T. E. Kopley, P. L. McEuen, and R. G. Wheeler, Phys. Rev. Lett. **61**, 1654 (1988).

[71] T. K. Ng and P. A. Lee, Phys. Rev. Lett. **61**, 1768 (1988).

[72] Y. Aharonov and D. Bohm, Phys. Rev. **115**, 485 (1959); **123**, 1511 (1961).

[73] S. Washburn and R. A. Webb, Adv. Phys. **35**, 375 (1986).

[74] A. G. Aronov and Yu. V. Sharvin, Rev. Mod. Phys. **59**, 755 (1987).

[75] P. H. M. van Loosdrecht, C. W. J. Beenakker, H. van Houten, J. G. Williamson, B. J. van Wees, J. E. Mooij, C. T. Foxon, and J. J. Harris, Phys. Rev. B **38**, 10162 (1988).

[76] T. Ando, A. B. Fowler, and F. Stern, Rev. Mod. Phys. **54**, 437 (1982).

[77] M. Büttiker, Phys. Rev. B **38**, 9375 (1988).

[78] V. Fock, Z. Phys. **47**, 446 (1928).

[79] C. G. Darwin, Proc. Camb. Phil. Soc. **27**, 86 (1930).

[80] F. Geerinckx, F. M. Peeters, and J. T. Devreese, J. Appl. Phys. **68**, 3435 (1990).

[81] M. Büttiker, Y. Imry, and R. Landauer, Phys. Lett. A **96**, 365 (1983).

[82] B. J. van Wees, E. M. M. Willems, C. J. P. M. Harmans, C. W. J. Beenakker, H. van Houten, J. G. Williamson, C. T. Foxon, and J. J. Harris, Phys. Rev. Lett. **62**, 1181 (1989).

[83] J. K. Jain, Phys. Rev. Lett. **60**, 2074 (1988).

[84] R. J. Brown, C. G. Smith, M. Pepper, M. J. Kelly, R. Newbury, H. Ahmed, D. G. Hasko, J. E. F. Frost, D. C. Peacock, D. A. Ritchie, and G. A. C. Jones, J. Phys. Condens. Matter **1**, 6291 (1989).

[85] C. G. Smith, M. Pepper, H. Ahmed, J. E. F. Frost, D. G. Hasko, D. C. Peacock, D. A. Ritchie, and G. A. C. Jones, J. Phys. C **21**, L893 (1989).

Chapter 6

Macroscopic Quantum Tunneling of Charge and Co-Tunneling

D. V. AVERIN

*Department of Physics, State University of New York, Stony Brook
NY 11794, USA,
and Department of Physics, Moscow State University
Moscow 119899 GSP, USSR*

and

YU. V. NAZAROV

*Nuclear Physics Institute, Moscow State University
Moscow 119899 GSP, USSR*

1. Introduction

As it is well known from the high-energy physics, the most basic property of quantum mechanical particles is that they obey only fundamental symmetries, and always can find the way around the lower-level restrictions. It means that any process that is not forbidden by fundamental symmetries should take place, although the rate of some processes can be small.

As illustration of this general "freedom" principle, let us consider an example when the relevant fundamental symmetry is energy conservation, and the particle, say, electron, can occupy two localized states (which we will call "jail" and "top") with unequal energies (E_{jail} and E_{top}) and continuous band of states ("ocean") - see Fig. 1. Wave functions of

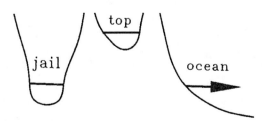

Figure 1. The potential profile that provide escape from the "jail" via virtual intermediate "top" state.

Single Charge Tunneling, Edited by H. Grabert and
M.H. Devoret, Plenum Press, New York, 1992

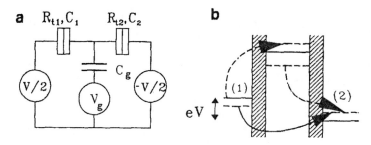

Figure 2. (a) Equivalent circuit of a symmetrically biased double junction system. (b) Energy diagram illustrating tunneling processes in two junctions involved in one act of inelastic q-mqt. An electron tunnels to the central island through the first junction (1), and resulting virtual state decays by a different electron tunneling across the second junction (2) (or vice versa). b_1 and b_2 indicate the tunnel barriers, and solid lines represent the Fermi levels in the electrodes.

the states overlap slightly providing possibility for the electron to hop between jail and top, and top and ocean, but one restriction is imposed on electron: there is no way from the jail to the ocean. In other words, there is no matrix elements of the Hamiltonian $\hat{H}$ of the system between these states.

Suppose one puts electron in jail and tries to keep it there. For a moment it looks like one really can do this, since the energy conservation makes it impossible for the electron to hop first to the top and then from the top to the ocean. Nevertheless it escapes to the ocean! It solves the problem with energy by making virtual its presence on the top during this two-step transition "jail→top→ocean", so that only the transition "jail→ocean" as a whole should conserve energy. The amplitude A and the rate Γ of such a complex transition are [1] :

$$A = \frac{\langle \mathrm{jail}|\hat{H}|\mathrm{top}\rangle\langle \mathrm{top}|\hat{H}|\mathrm{ocean}\rangle}{E_{\mathrm{jail}} - E_{\mathrm{top}}}, \tag{1}$$

$$\Gamma = \frac{2\pi}{\hbar} \sum_{\mathrm{ocean}} |A|^2 \delta(E_{\mathrm{jail}} - E_{\mathrm{ocean}}). \tag{2}$$

In the present chapter we discuss complex transitions of this kind in systems of small tunnel junctions. To explain their relevance, let us consider one of such systems, a small metallic island connected by tunnel junctions to bulk external electrodes, forming two tunnel junctions in series (see Fig. 2). As it was explained in Chap. 2 (see also [2]), if the tunnel resistance R_t of these junctions is much larger than the quantum resistance: $R_t \gg R_K$, where $R_K \equiv h/e^2 \cong 25.8$ kΩ, Coulomb blockade of tunneling takes place in this system at low temperatures and small bias voltages V across the two junctions. It means that the energy conservation makes it impossible for an electron to tunnel through either of the junctions, because such a tunneling would charge the island and increase electrostatic energy of the system by $\Delta E \simeq E_c$, where $E_c \equiv e^2/C_\Sigma$, and C_Σ is the total electric capacitance of the island. It means that the double junction system in the Coulomb blockade regime is quite similar to the "jail" system considered above. One of the external electrodes can be identified with the jail, another one with the ocean, and the charged state of the island with the top. Thus, the discussion above tells us that although Coulomb blockade suppresses sequential tunneling of electrons in the two junctions, there should exist more complex quantum tunneling process that goes via virtual charged state of the island. This process causes a finite current through the junctions, and hence, limits the Coulomb blockade of tunneling.

Below we consider small tunnel junctions that are still large on atomic scale, so that electrodes of these junctions contains a large number of free electrons and have quasicontinuous energy spectra. In such junctions the main contribution to quantum tunneling process is due to transfer of different electrons in different junctions. For instance, while one electron enters the island of the double junction system through one of the junctions, another electron leaves it through another junction. Hence, one can call this process *co-tunneling* of two electrons.

On the other hand, the process as a whole can be viewed as a quantum tunneling through the Coulomb energy barrier between the different charge states. The barrier arises from electrostatic interaction of macroscopically large number of electrons. Hence, one can call this tunneling process (which causes decay of the Coulomb blockade) *macroscopic quantum tunneling of charge* (abbreviated to "q-mqt") in analogy to macroscopic quantum tunneling of Josephson phase difference φ ("φ-mqt") in conventional Josephson junctions (see, e.g., [3]).[1]

In this chapter we will use both terms "q-mqt" and "co-tunneling". They will denote, respectively, *phenomenon* of the Coulomb blockade decay, and *elementary process* of simultaneous tunneling of two or several electrons, which typically underlies this phenomenon. It would be probably easier to use one term. However, the authors of this chapter did not come to an agreement, which term is better. Moreover, it can be useful to keep two terms because, first, the q-mqt can be associated not only with the co-tunneling (see Sec. 3), and, second, the co-tunneling can be encountered not only in the situation of the Coulomb blockade decay (see Sec. 4).

How one can estimate the rate $\Gamma^{(2)}$ of two-electron co-tunneling? Let single electron tunneling be forbidden by the energy conservation law. Nevertheless electron is allowed by uncertainty relation to spend some time in the forbidden state. This time can be estimated as $\hbar/\Delta E$, where ΔE is the energy deficit associated with single electron tunneling, $\Delta E \simeq E_c$. If an electron tunnels through another junction while the first electron is in the forbidden state, the process of charge transfer through the whole system will be successfully completed. Hence, the rate can be estimated as follows

$$\Gamma^{(2)} \simeq \Gamma^{(1)}\Gamma^{(1)}\hbar/E_c \simeq \Gamma^{(1)}R_K/R_t, \tag{3}$$

where $\Gamma^{(1)} \simeq 1/R_t C_\Sigma$ is the characteristic rate of single electron tunneling. One can see that in the limit under consideration, $R_t \gg R_K$, the co-tunneling rate is much smaller than the characteristic rate of single electron tunneling.

The co-tunneling can involve several electrons. Moreover, in accordance with the freedom principle discussed above *any* electron tunneling process that conserves energy can take place in a system of junctions. Although the rate of several-electron processes decreases rapidly with increasing number of involved electron transitions (cf. the estimate (3) of two-electron co-tunneling rate), it can be high enough to modify considerably dynamics of electron tunneling.

---

[1] This analogy is not superficial, but stems from the general duality between electric charge and magnetic flux. In a single Josephson junctions this analogy is quantitative, and in this case there is precise quantitative correspondence between q-mqt and φ-mqt problems [4].

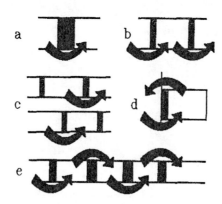

Figure 3. The various types of co-tunneling processes. The electron transitions involved in the co-tunneling are denoted by arrows. The shaded areas represent tunnel barriers between metallic electrodes.

Since electrons involved in the co-tunneling are transferred from the energy states below the Fermi levels of junction electrodes to the states above Fermi level, the co-tunneling creates electron-hole excitations in the electrodes, and hence, transfers electrostatic energy of the system into the energy of these excitations. In this respect the co-tunneling is inelastic process, and one can call q-mqt associated with the co-tunneling inelastic q-mqt. Inelastic q-mqt in the double junction system, where the q-mqt is due to the two-electron co-tunneling (Fig. 3b), and in the 1D arrays of junctions, where it is due to co-tunneling of larger number of electrons (Fig. 3d), is discussed in Sec. 2.

Besides the co-tunneling, q-mqt can be caused by the one-electron tunneling through two or several junctions. The process of this kind (which is called elastic q-mqt) in the double junction system is discussed in Sec. 3. In Sec. 4 we consider the system in which the co-tunneling can take place in the junctions that are not coupled galvanically (Fig. 3c). Such a co-tunneling provides an interesting possibility to manipulate with electrons, based on the fact that electrons involved in co-tunneling are transferred in different junctions in the completely correlated manner. We conclude in Sec. 5 by the discussion of unsolved problems and perspectives of the field.

It should be noted that the present chapter does not cover all aspects of the q-mqt and co-tunneling. In particular, limitations of the practical single electron tunneling devices related to q-mqt are discussed in Chap. 9. Besides this, we do not discuss the co-tunneling in a single junction. One can show [5] that at low temperatures, when one-electron tunneling (Fig. 3a) is suppressed, co-tunneling of two electrons in one junction (Fig. 3d) determines the real part of ac junction conductance at low frequencies $\hbar\omega < E_c$. In particular, at $T = 0$ and $\hbar\omega \ll E_c$:

$$\mathrm{Re}[G(\omega)] \simeq (R_K/R_t)^2 (\hbar\omega/E_c)^6. \tag{4}$$

More detailed discussion of this co-tunneling can be found in the original publication [5].

2. Inelastic q-mqt

2.1. Inelastic q-mqt in the double junction system

We begin quantitative discussion of the q-mqt with the simplest example of the system of two junctions connected in series, which is the basis of the single electron tunneling (SET) transistor [6] (see Chap. 2, Sec. 6). The rate of electron tunneling in this system depends, as usual, on the change E of electrostatic energy caused by this tunneling. From equivalent circuit shown in Fig. 2a, one can find that the energy change associated with the tunneling in the first and the second junction is

$$E_1 = \frac{e}{C_\Sigma}[e(n + \frac{1}{2}) - V(C_2 + \frac{C_g}{2}) + C_g V_g], \tag{5}$$

and

$$E_2 = \frac{e}{C_\Sigma}[-e(n - \frac{1}{2}) - V(C_1 + \frac{C_g}{2}) - C_g V_g], \tag{6}$$

respectively. Here n is the number of excess electrons on the central electrode of the system before tunneling, C_g is the capacitance between this electrode and the gate electrode, C_1, C_2 are capacitances of the junctions, and $C_\Sigma \equiv C_1 + C_2 + C_g$.

At large voltages V across the junction array, $V > V_t$, where V_t is some threshold voltage dependent of the gate voltage V_g, both energies E_1, E_2 are negative and electrons can tunnel classically in both junctions. At small voltages, $V < V_t$, there is a number n_0 of electrons on the central electrode which provide the minimum of electrostatic energy, i.e. for which $E_1, E_2 > 0$. Hence, the tunneling in either of two junctions of the array in this state could only increase electrostatic energy of the system, and it is suppressed at low temperatures, $k_B T \ll E_c$, where $E_c \equiv e^2/2C_\Sigma$. Thus, the system is trapped in this state and exhibits the Coulomb blockade of tunneling.

However, as it was discussed in the introduction, the Coulomb blockade in this system suppresses only the sequential tunneling of electrons through the junctions. There is more complex tunneling process that transfers electrons through the whole system. Such a process is energetically favorable at any non-vanishing driving voltage V, since $E_1(n) + E_2(n + 1) = E_2(n) + E_1(n + 1) = -eV < 0$. It is essentially quantum because only quantum mechanics can transfer the charge through both junction without transferring it through either of them separately. In other words, this process goes via virtual intermediate state with increased electrostatic energy, and due to the virtual character of the intermediate state the tunneling processes in two junctions that are involved in it can not be separated from one another.

Since this macroscopic quantum tunneling of the charge (q-mqt) is due to the co-tunneling of two electrons, it is of the second order in small tunnel conductance of the junctions $R_t^{-1} \ll R_K^{-1}$ (see estimate (3)). Hence, the rate of the q-mqt is small compared to rate of the single electron tunneling, but nevertheless, q-mqt is dominant in the Coulomb blockade regime, when the single electron tunneling is suppressed.

One can find the q-mqt rate applying the standard "Golden Rule" for the higher-order transitions [1] (similar to Eqs. (1), (2) of the introduction). The transition amplitude of the underlying two-electron tunneling process consists, first, of matrix elements

$T^{(1)}$ and $T^{(2)}$ describing the tunneling in the first and the second junction and, second, of energy denominator $1/\Delta E$ that reflects the fact that the process goes via virtual intermediate state (ΔE is the energy of this state relative to the initial state). In fact, it can go via two intermediate states: one with extra electron on the central electrode of the array (when electron tunnels first in the first junction), and another with a hole on this electrode (when electron tunnels first in the second junction). If electron is transferred in the first junction from the energy state ε_1 below the Fermi level of external electrode ($\varepsilon_1 < 0$) to the energy state ε_2 above the Fermi level of the central electrode ($\varepsilon_2 > 0$), see Fig. 2b, then the energy of the intermediate state of the first type is $\Delta E = \varepsilon_2 - \varepsilon_1 + E_1$. For the second intermediate state $\Delta E = \varepsilon_4 - \varepsilon_3 + E_2$, where ε_3, ε_4 are also internal electron energies relative to the Fermi levels of the electrodes.

The processes with $\varepsilon_2 \neq \varepsilon_3$ not only transfer the charge through the array, but also create electron-hole excitation on the central electrode: electron is taken from the state ε_3 below the Fermi level and transferred to the state ε_2 above the Fermi level (Fig. 2b). Hence, these processes of two-electron co-tunneling transfer energy to the internal degrees of freedom of the central electrode and one can call the associated charge tunneling *inelastic q-mqt*.[2]

The different sets of energies $\varepsilon_1,\ldots,\varepsilon_4$ characterizes the different *final* states of this tunneling process, so that processes with different $\varepsilon_1,\ldots,\varepsilon_4$ should be summed up incoherently. It means that one should sum up the tunneling *rates* and not the tunneling *amplitudes* corresponding to different energies $\varepsilon_1,\ldots,\varepsilon_4$. Thus, the amplitude of elementary two-electron co-tunneling process is the sum of the two terms corresponding to the two intermediate states with different charging energies:

$$\langle i|M|f\rangle = T^{(1)}T^{(2)}\left(\frac{1}{\varepsilon_2 - \varepsilon_1 + E_1} + \frac{1}{\varepsilon_4 - \varepsilon_3 + E_2}\right). \qquad (7)$$

The total rate γ of the inelastic q-mqt is given by the sum over $\varepsilon_1,\ldots,\varepsilon_4$ of the "partial" tunneling rates Γ:

$$\Gamma = \frac{2\pi}{\hbar}|\langle i|M|f\rangle|^2\delta(\varepsilon_i - \varepsilon_f), \qquad (8)$$

multiplied by the appropriate probabilities to find an electron in the state ε_1, $f(\varepsilon_1)$, to find an unoccupied state ε_2, $1 - f(\varepsilon_2)$, and so forth. Since electrodes are in equilibrium, $f(\varepsilon)$ is the Fermi distribution function.

The absolute value of the tunnel matrix elements $T^{(i)}$ in Eqs. (7), (8) is directly related to the tunnel conductances of the junctions:

$$R_{ti}^{-1} = \frac{4\pi e^2}{\hbar}|T^{(i)}|^2\nu_0\nu_i, \qquad (9)$$

where ν_0 and ν_i are, respectively, the densities of states of the central electrode and the ith external electrode at the Fermi level.[3] Thus, summing the tunneling rates (8) and

---

[2] As it is clear from this arguments the term "inelastic" has nothing to do with inelastic scattering of electrons in the electrodes.

[3] As usual, we neglect the variations of both $T^{(i)}$ and ν_0, ν_i with energy. This approximation is sufficient as long as the charging energies E_1, E_2 are assumed to be much smaller than the heights of the tunnel barriers and the Fermi energies of the electrodes.

taking into account Eq. (9) we can write down the rate of the inelastic q-mqt as follows:

$$\gamma = \frac{\hbar}{2\pi e^4 R_{t1} R_{t2}} \int d\varepsilon_1 d\varepsilon_2 d\varepsilon_3 d\varepsilon_4 \, f(\varepsilon_1)(1 - f(\varepsilon_2)) f(\varepsilon_3)(1 - f(\varepsilon_4))$$

$$\times \left(\frac{1}{\varepsilon_2 - \varepsilon_1 + E_1} + \frac{1}{\varepsilon_4 - \varepsilon_3 + E_2} \right)^2 \delta(eV + \varepsilon_1 - \varepsilon_2 + \varepsilon_3 - \varepsilon_4). \quad (10)$$

At vanishing temperatures, $T = 0$, integrals in Eq. (10) can be calculated explicitly and one gets [4]:

$$\gamma = \frac{\hbar}{2\pi e^3 R_{t1} R_{t2}} \left[\left(1 + \frac{2}{eV} \frac{E_1 E_2}{E_1 + E_2 + eV} \right) \left(\sum_{i=1,2} \ln(1 + eV/E_i) \right) - 2 \right] V. \quad (11)$$

Since different acts of q-mqt are uncorrelated, the dc current I flowing through the double-junction array in the Coulomb blockade regime is proportional to the tunneling rate, $I = e\gamma(V)$. One can see from Eq. (11) that at small voltages, $eV \ll E_1, E_2$, the current varies as the third power of the voltage V:

$$I(V) = e\gamma(V) = \frac{\hbar}{12\pi R_{t1} R_{t2}} \left(\frac{1}{E_1} + \frac{1}{E_2} \right)^2 V^3. \quad (12)$$

For the non-vanishing temperatures the inelastic q-mqt rate can be calculated for small voltages, $eV \ll E_1, E_2$, when the energies $\varepsilon_1, ..., \varepsilon_4$ are small compared to charging energies E_1, E_2 and Eq. (10) reduces to

$$\gamma = \frac{\hbar}{2\pi e^4 R_{t1} R_{t2}} \left(\frac{1}{E_1} + \frac{1}{E_2} \right)^2 \int \prod_{i=1}^{4} (d\varepsilon_i f(\varepsilon_i)) \delta\left(eV + \sum_{i=1}^{4} \varepsilon_i\right). \quad (13)$$

Here we have used the identity $f(-\varepsilon) = 1 - f(\varepsilon)$.

The q-mqt rate (13) can be directly related to the current I through the system even at $T \neq 0$ when the current is determined by the balance between forward and backward tunneling rates:

$$I(V) = e[\gamma(V) - \gamma(-V)]. \quad (14)$$

From Eq. (13) one can see that the forward and backward tunneling rates are related in a simple way: $\gamma(-V) = \exp(-eV/k_B T)\gamma(V)$. Substituting this relation into Eq. (14) we get

$$\gamma(V) = \frac{I(V)}{e}[1 - \exp(-eV/k_B T)]^{-1}. \quad (15)$$

Making use of Eq. (14) one can calculate the integrals in Eq. (13) and find the current I explicitly [7, 8]:

$$I(V) = \frac{\hbar}{12\pi e^2 R_{t1} R_{t2}} \left(\frac{1}{E_1} + \frac{1}{E_2} \right)^2 [(eV)^2 + (2\pi k_B T)^2] V. \quad (16)$$

Equations (15), (16) determine the finite-temperature rate of inelastic q-mqt.

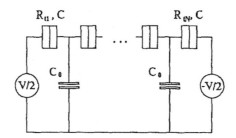

Figure 4. Equivalent circuit of the 1D array of small tunnel junctions. C_0 is the self-capacitance of the junction electrodes.

Equation (16) implies that the zero-bias conductance of the double junction system associated with the inelastic q-mqt vanishes at vanishing temperatures. This result can be easily understood, since under the conditions $V \to 0$, $T \to 0$, there is no energy available for creation of electron-hole excitations associated with the inelastic q-mqt process. However, there is another q-mqt process that does not involve creation of these excitations and leads to finite zero-temperature linear conductance of the system. We will consider this *elastic* q-mqt later in Sec. 3 after the discussion of more practically important inelastic q-mqt in the multijunction circuits.

2.2. Multijunction circuits

We begin the discussion of q-mqt in more complex circuits with the 1D uniform array of N junctions shown in Fig. 4. Similarly to the double-junction system, the array is in the Coulomb blockade state at small voltages V across the array, $V < V_t$, and the q-mqt causes the finite current through the array in this regime. When the voltage V is sufficiently small, $V \ll V_t$, only the transfer of an electron through the whole array is energetically favorable with respect to electrostatic energy of the system, so that one act of q-mqt includes transfer of electrons in N junctions, and $N - 1$ virtual intermediate states are involved in the tunneling. Hence, in this case the q-mqt is due to co-tunneling of N electrons. Similarly to the amplitude of the second-order tunneling in the double junction system (7) one can write down the amplitude of this Nth-order tunneling as follows:

$$\langle i|M|f\rangle = T^{(1)}...T^{(N)} \sum_{\{k_i\}} \left(\prod_{i=1}^{N} \frac{1}{\varepsilon_{2i} - \varepsilon_{2i-1} + E_i(\{k_i\})} \right). \tag{17}$$

Here $T^{(i)}$ is the amplitude of electron tunneling in the ith junction, and the sum is taken over different "paths" of the tunneling. Each path is characterized by the set $\{k_i\}$ that determines the sequence of electron tunneling through the array: k_i is the junction number through which the ith tunneling takes place. The electrostatic energies E_i of the intermediate states in general are different for different paths $\{k_i\}$.

Summing up partial tunneling rates with different internal energies $\{\varepsilon_i\}$ in the same way as it was done for the double junction system we get the rate of the Nth-order inelastic q-mqt:

$$\gamma = \frac{2\pi}{\hbar}(\prod_{i=1}^{N}\frac{R_K}{4\pi^2 R_{ti}})\int d\varepsilon_1...d\varepsilon_{2N}[\prod_{i=1}^{N}f(\varepsilon_{2i})(1-f(\varepsilon_{2i-1}))]$$

$$\times S^2\delta(eV-\sum_{i=1}^{N}[\varepsilon_{2i}-\varepsilon_{2i-1}]), \qquad (18)$$

where

$$S \equiv \sum_{k_i}(\prod_{i=1}^{N-1}\frac{1}{\varepsilon_{2i}-\varepsilon_{2i-1}+E_i(\{k_i\})}). \qquad (19)$$

At small voltages, $eV \ll E_i$, one can neglect the energies ε_i in Eq. (19). In this case the dc current I flowing through the array is related to the q-mqt rate γ via the same Eqs. (14), (15) as in the double-junction array. Making use of these equations one can calculate the integrals in Eq. (18) and get [9]

$$\gamma = \frac{2\pi}{\hbar}(\prod_{i=1}^{N}\frac{R_K}{4\pi^2 R_{ti}})S^2[1-\exp(-\frac{eV}{k_BT})]^{-1}\sum_{k=0}^{N-1}c_k^{(N)}(k_BT)^{2k}(eV)^{2(N-k)-1}. \qquad (20)$$

The coefficients $c_k^{(N)}$ can be calculated explicitly for several N. In particular, at $T=0$

$$\gamma = \frac{2\pi}{\hbar}(\prod_{i=1}^{N}\frac{R_K}{4\pi^2 R_{ti}})\frac{S^2}{(2N-1)!}(eV)^{2N-1} \qquad (21)$$

for any N.

The sum S of the energy denominators also can be calculated explicitly for $eV \ll E_i$ in several simple cases. Let us consider, for example, the array with negligible self-capacitances C_0 of the junction electrodes.[4] From the equivalent circuit of the 1D array (Fig. 4) it is straightforward to check that in this case the energies E_i of the virtual intermediate states are independent of the sequence $\{k_i\}$ of electron tunneling through the array. Hence,

$$S = N!\left(\prod_{i=1}^{N-1}E_i\right)^{-1}, \qquad (22)$$

where E_i is the charging energy of electron on the ith internal electrode of the array at $V=0$. This energy is

$$E_i = \frac{e^2}{2C_i}, \qquad (23)$$

where C_i is the capacitance between this electrode and external electrodes of the array:

$$C_i = C\left(\frac{1}{i}+\frac{1}{N-i}\right) = C\frac{N}{i(N-i)}. \qquad (24)$$

---

[4] One can neglect C_0 in comparison to the capacitance C of the junction when $C_0 \ll C$, and the array is not very long, $N \ll (C/C_0)^{1/2}$.

Substituting Eqs. (23), (24) into Eq. (22) we get finally

$$S = \left(\frac{e^2}{2C}\right)^{-(N-1)} \frac{N^N}{(N-1)!}. \tag{25}$$

Equations (18), (20) show that the q-mqt rate depends exponentially on the order N of the q-mqt transition. Each virtual intermediate state involved in this tunneling adds a small factor $R_K/4\pi^2 R_{ti}$ to the tunneling rate, and at small voltage it adds also another small factor of the order of $(eV/E_c)^2$ or $(k_B T/E_c)^2$, depending on $eV/k_B T$ ratio. The physical reason for the latter factor is the necessity to create additional electron-hole excitation, which at small voltages and temperatures limits the available phase volume for tunneling electrons, since the excitation can get the energy only from external voltage or thermal fluctuations. Thus, the q-mqt rate decreases rapidly with increasing N.

Equations (18)-(20) enable one, in principle, to calculate the q-mqt rate not only in the linear 1D arrays of junctions, but also in arbitrary multijunction circuit. In respect of the q-mqt these systems differ only in specific values of the energies E_i of the virtual intermediate states. However, there is one important new element introduced by the systems that are more complex than linear arrays. Linear array is either in the Coulomb blockade state at small driving voltages, or above the Coulomb blockade at large voltages. In the Coulomb blockade regime the classical tunneling is suppressed, and all the energies E_i of the intermediate states of the q-mqt process that determines the current through the array are *positive*. It means that each junction of the array involved in the q-mqt process provides the Coulomb barrier for electron tunneling. Above the Coulomb blockade electron can tunnel classically along the whole array and the q-mqt is irrelevant. In contrast to that, in complex circuits part of the junctions can be in the Coulomb blockade state, while another part is above the Coulomb blockade. In this case there are q-mqt processes that go via intermediate states with *negative* energies [9]. The physical meaning of this processes is that the energy favorable classical tunneling in the "open" junctions induces the tunneling in the junctions in which the classical tunneling is energy unfavorable. The rate of this induced tunneling is given by the same Eqs. (18), (19) as the rate of the usual q-mqt with $E_i > 0$. One can see that these equations are well defined for $E_i < 0$, as long as $|E_i| > eV$, where eV is the total decrease of the electrostatic energy of the circuit due to the q-mqt transition. This condition has the simple physical meaning that at least part of the junctions should provide the barrier for electron tunneling, so that the tunneling in this junctions "absorb" the energy supplied by the classical tunneling in the open junctions.

Thus, the basic principles of the inelastic q-mqt in the multijunction circuits seem to be well understood. However, specific calculations of dynamics of electron tunneling in this circuits including higher-order q-mqt transitions present a considerable technical problem, since the number of possible transitions increases very rapidly with increase of both the number of junctions in the circuit and transition order. This problem is very important for a serious design of practical single electron devices, because q-mqt decreases reliability of operation of these devices (see Chap. 3).

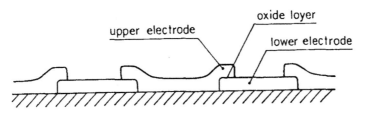

Figure 5. Cross-section of the overlapping metal islands on insulating substrate, which form 1D array of small tunnel junctions (schematically).

2.3. Experimental observation of inelastic q-mqt

There are several experimental evidences of the inelastic q-mqt in systems of small tunnel junctions. The first one was obtained in experiment [10] with linear arrays of small number ($N = 2, 3$) of normal-metal junctions. The junctions in this experiment were fabricated in the standard suspended-mask technology. This technology enables one to create a series of thin and narrow aluminum strips (electrodes) on the insulating substrate in such a way that they overlap at the edges, as shown schematically in Fig. 5. The lower electrodes are oxidized, so that they are covered with the thin layer of the natural oxide, and tunnel junctions are formed in the regions of overlap.

The thickness d of the oxide layer that determines junction resistance is controlled by varying the oxidation pressure, and typically is about 1 nm. The junction capacitance C can be estimated as $\varepsilon\varepsilon_0 S/d$, where S is the area of the overlapping regions, and $\varepsilon \approx 12$ is the dielectric constant of aluminum oxide. In the experiment discussed here the electrodes were 60 nm wide, 20 or 40 nm thick, and about 1 μm long, with the area S of the overlapping regions about $(60\,\text{nm})^2$. In the double junction system that was studied more thoroughly, the gate electrode was evaporated at 1.5 μm distance from the central electrode, with the coupling capacitance C_g of about 0.07 fF. Hence, the total capacitance C_Σ of the central electrode of this system can be estimated as 1 fF, which gives the characteristic charging energy E_c about 0.1 meV. The samples were cooled down to 30 mK, and the junction resistance R_t was of the order of 100 kΩ. Hence, conditions $k_B T \ll E_c$ and $R_t \gg R_K$ were satisfied, so that both thermal and quantum fluctuations of the charge on the central electrode should be small, and the charging effects should be well pronounced.

In accordance with these estimates, the observed $I - V$ curves of the double junction clearly exhibit the Coulomb blockade on the large current scale, $e/R_t C_\Sigma$, given by the "classical" electron tunneling rate (one of these curves is shown in Fig. 6). These $I - V$ curves can be well described within the simplest approach, assuming both junctions in the array to be equal and taking into account only the classical tunneling events, which are of the first order in small junction conductances $1/R_t$.

The threshold V_t of the Coulomb blockade is modulated periodically by the gate voltage V_g and can be maximized by adjustment of V_g.[5] From Eqs. (5), (6) one can readily

---

[5] At some gate voltage between those which maximize V_g the Coulomb blockade is completely suppressed (as shown in the insert of Fig. 6) and the zero-bias conductance reaches its maximum value $1/4R_t$. After the first observation [11] in metal tunnel junctions similar behavior was observed [12] in the double tunnel junctions formed on semiconductor heterostructures.

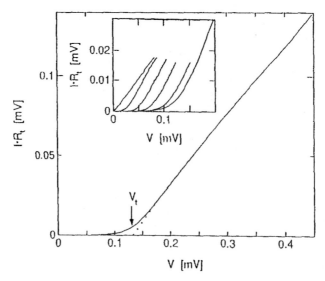

Figure 6. $I - V$ curve of a double junction with $R_t = 78$ kΩ. The maximum Coulomb blockade threshold voltage V_t is obtained by adjusting V_g. The insert shows that the Coulomb blockade can be continuously controlled by the gate voltage (curves are given for, from right to left, $V_g C_g \approx 0$, $0.1e$, $0.2e$, $0.3e$, $0.4e$, and finally, $0.5e$, when the Coulomb blockade is completely suppressed).

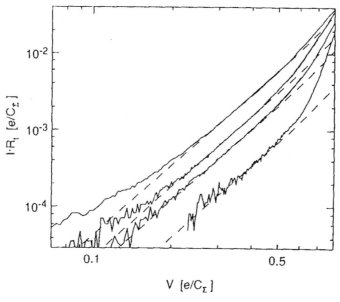

Figure 7. Comparison of $I - V$ curves below the Coulomb blockade threshold with theory, for four double junctions of different resistance. The measurements (solid curves) have been scaled with R_t and C_Σ, the latter being a fit parameter. The dashed curves give the predictions for inelastic q-mqt.

obtain that the maximum threshold voltage V_t depends only on the total capacitance C_Σ of the central electrode, $\max\{V_t\} = e/C_\Sigma$, and one can use experimental value of maximized V_t to determine C_Σ.

However, a considerable current is flowing through the double junction below the Coulomb blockade threshold even at low temperatures $T \ll E_c/k_B$. Figure 7 shows $\log(I)$-$\log(V)$ curves for the four double junctions with different resistances at voltages below V_t. Classical tunneling current (that could flow below V_t at finite temperatures) would increase exponentially with voltage. In contrast, the experimental $\log(I)$-$\log(V)$ curves yield straight lines with a slope equal to 3, as expected for inelastic q-mqt, see

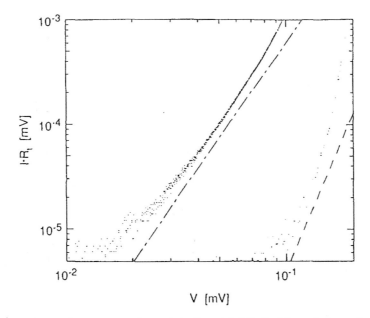

Figure 8. $I - V$ curves for a double junction with $R_t = 78$ kΩ (left) and three junctions in series with $R_t = 84$ kΩ (right) below the Coulomb blockade threshold. As a guide to the eye, lines for which $I \propto V^5$ (dashed) and $I \propto V^3$ (dash-dotted) have been drawn, representing the inelastic q-mqt predictions for the slope of the log(I)-log(V) curves.

Eq. (12), except for high or very low voltages. All curves can be very well fitted by calculations for q-mqt from Eq. (11) in a broad voltage range inside the Coulomb blockade. The smaller slope of the curves at $V \to 0$ is probably caused by the transition to the linear $I - V$ curves, either due to finite temperature (see Eq. (16)), or due to elastic q-mqt.[6] The larger slope for high voltages is probably due to the crossover to the thermally assisted classical tunneling.

There is small quantitative discrepancy in the experimental result described above, that have no satisfactory explanation at present. The best fit values of the capacitance C_Σ obtained from the q-mqt $I - V$ curves differs by about 30% from C_Σ determined from the Coulomb blockade threshold V_t. This difference in principle could arise due to asymmetry of the two junctions in the arrays. However, the degree of such asymmetry that is necessary to account for the difference in C_Σ is much larger than the possible asymmetry characteristic for the technology used to fabricate the arrays.

The measured $I - V$ curve of a three-junction array below the Coulomb blockade threshold also provides support of the inelastic q-mqt. The log(I)-log(V) curves of this array and of a double junction of almost equal R_t are compared in Fig. 8. The prediction from Eq. (21) is that for the double junction $I \propto V^3$, whereas for three junctions $I \propto V^5$. As a guide to the eye, two lines in Fig. 8 gives the expected slopes of the curves, which are in fair agreement with the results. Because of the higher exponent of I, the voltage

---

[6] As it will be shown in the next section, elastic q-mqt rate is smaller than inelastic rate by the factor Δ/E_c, where Δ is the energy level spacing of the central electrode. For junction electrodes with the dimensions cited above, Δ is about 10^{-7} eV, so that the ratio Δ/E_c is of the order of 10^{-3} and transition to elastic q-mqt should occur (see the next section) when both k_BT and eV are less than $(\Delta E_c)^{1/2} \simeq 0.03E_c$. This is not very far from the parameter range corresponding to $I - V$ curves shown in Fig. 7.

range in which the current is neither unobservably small nor for a significant amount due to thermally assisted tunneling is smaller for the longer array.

Another experimental evidence of the inelastic charge mqt in metal tunnel junctions is provided by the recent experiments with single electron turnstile and pump, which are described in Chap. 3. In the classical approximation dc current I in these devices is determined by the frequency f of applied external rf voltage, $I = ef$. Observed small deviations δI from this quantized current can be well explained by the inelastic q-mqt.

Recently, the inelastic q-mqt also was observed [13] in the double junction formed on the semiconductor heterostructure. In this experiment the lateral transport through a relatively large ($\simeq 1.2\mu m$ in diameter) quantum dot was proved to be governed by the charging effects. In particular, it was shown that at low temperatures the zero-bias conductance of the quantum dot in the Coulomb blockade regime increases with temperature as T^2, in accordance with Eq. (16) for inelastic q-mqt.

3. Elastic q-mqt and virtual electron diffusion

The inelastic q-mqt considered in the previous section is associated with the co-tunneling of different electrons in different junctions that take part in the q-mqt. This allows one to calculate the rate of this tunneling neglecting a coherence between wave functions of electrons tunneling in different junctions. However, this approximation (employed in the previous section) does not account for elastic contribution to the q-mqt rate [8, 14], for which the coherence of the wave functions is important.

As it will be shown below, in the most realistic case when the electron density of states Δ^{-1} of the junction electrodes is high, $\Delta \ll E_c$, the elastic q-mqt rate is smaller than inelastic rate by a factor of Δ/E_c. Nevertheless, elastic q-mqt can be important at small voltages and low temperatures, $eV, k_B T \ll E_c$, when the current associated with inelastic q-mqt is small (see Eqs. (16), (20)).

In this section we derive a general expression describing both elastic and inelastic contributions to the q-mqt in the double-junction system, and calculate the current associated with elastic tunneling. In the relevant limit of small junction conductances, $G_i \ll R_K^{-1}$, one can do this applying the standard perturbation theory in tunnel Hamiltonians H_{Ti} that describe electron tunneling in the junctions.

The total Hamiltonian H of the double-junction system is the sum of the tunnel Hamiltonians H_{T1}, H_{T2}, Hamiltonians H_c, H_1, H_2 of the central electrode and external electrodes, and electrostatic energy U of the system:

$$H = H_0 + H_T, \quad H_0 = H_c + H_1 + H_2 + U, \quad H_T = H_{T1} + H_{T2}. \tag{26}$$

Electrostatic energy U in general can be written in the following form (see, e.g., [2]):

$$U = \frac{Q^2}{2C_\Sigma} - \frac{eV}{C_\Sigma}(C_1' n_2 + C_2' n_1), \tag{27}$$

$$Q = e(n_1 - n_2) + Q_0, \quad C_\Sigma = C_1' + C_2',$$

where n_i is the number of electrons that have tunneled through the ith junction, Q_0 describes the potential difference between central and external electrodes at vanishing bias voltage V, and C_1', C_2' are effective junction capacitances. For example, for symmetrically biased system shown in Fig. 2 $C_1' = C_1 + C_g/2$, $C_2' = C_2 + C_g/2$, and $Q_0 = V_g C_g$. The tunnel Hamiltonians have the standard form:

$$H_{Ti} = H_i^+ + H_i^-, \quad H_i^+ = \sum_{n,m} T_{mn}^{(i)} c_m^\dagger c_n, \quad H_i^- = (H_i^+)^\dagger. \tag{28}$$

In the Coulomb blockade regime the current associated with the one-electron tunneling (which is of the second order in H_{Ti}) is suppressed, and the finite current I flowing through the system is due to the q-mqt. Since one act of q-mqt includes two virtual electron transitions, the current associated with this tunneling is of the fourth order in H_T. Treating H_T as a perturbation one can get the following expression of the fourth (lowest nonvanishing) order for the current I:

$$\begin{aligned}
I = \frac{2e}{\hbar^4} \mathrm{Re} \Bigg(&\int_{-\infty}^t d\tau \int_{-\infty}^\tau d\tau' \int_{-\infty}^{\tau'} d\tau'' \left[\langle H_1^+(t) H_T(\tau) H_T(\tau') H_T(\tau'') \rangle \right. \\
&\left. - \langle H_T(\tau'') H_T(\tau') H_T(\tau) H_1^+(t) \rangle \right] \\
&+ \int_{-\infty}^t d\tau \int_{-\infty}^\tau d\tau' \int_{-\infty}^{\tau'} d\tau'' \left[\langle H_T(\tau'') H_T(\tau') H_1^+(t) H_T(\tau) \rangle \right. \\
&\left. - \langle H_T(\tau) H_1^+(t) H_T(\tau') H_T(\tau'') \rangle \right] \Bigg).
\end{aligned} \tag{29}$$

Here the time dependence of H_T is determined by H_0, and the average $\langle \ldots \rangle$ is taken over the equilibrium density matrix corresponding to H_0.

Equation (29) contains the terms which describe various two-electron tunneling processes, and the terms which correspond to one-electron tunneling. In the Coulomb blockade regime all these terms vanish except for those which describe q-mqt, and Eq. (29) can be reduced to the following form:

$$I = e[\gamma^{(+)} - \gamma^{(-)}], \tag{30}$$

$$\begin{aligned}
\gamma^{(+)} = \frac{2}{\hbar^4} \sum_{\substack{p,q=1,2 \\ p \neq q}} \mathrm{Re} \Bigg(&\int_{-\infty}^t d\tau \int_{-\infty}^\tau d\tau' \int_{-\infty}^{\tau'} d\tau'' \exp\left\{ -\frac{i}{\hbar} [E_2(t-\tau) + eV(t-\tau') - E_p(\tau' - \tau'')] \right\} \\
&\times \langle H_p^-(\tau'') H_q^-(\tau') H_1^+(t) H_2^+(\tau) \rangle \\
&- \int_{-\infty}^t d\tau \int_{-\infty}^\tau d\tau' \int_{-\infty}^{\tau'} d\tau'' \exp\left\{ \frac{i}{\hbar} [E_1(t-\tau) - eV(\tau-\tau') + E_p(\tau' - \tau'')] \right\} \\
&\times \langle H_p^-(\tau'') H_q^-(\tau') H_2^+(\tau) H_1^+(t) \rangle \Bigg),
\end{aligned} \tag{31}$$

$$\gamma^{(-)} = \frac{2}{\hbar^4} \sum_{\substack{p,q=1,2 \\ p \neq q}} \mathrm{Re}\left(\int_{-\infty}^{t} d\tau \int_{-\infty}^{\tau} d\tau' \int_{-\infty}^{\tau'} d\tau'' \exp\left\{ \frac{i}{\hbar}[E_1(t-\tau) - eV(\tau - \tau') - E_p(\tau' - \tau'')] \right\} \right.$$

$$\times \langle H_2^+(\tau) H_1^+(t) H_p^-(\tau') H_q^-(\tau'') \rangle$$

$$- \int_{-\infty}^{t} d\tau \int_{-\infty}^{\tau} d\tau' \int_{-\infty}^{\tau'} d\tau'' \exp\left\{ -\frac{i}{\hbar}[E_2(t-\tau) + eV(t - \tau'') + E_p(\tau' - \tau'')] \right\}$$

$$\left. \times \langle H_1^+(t) H_2^+(\tau) H_p^-(\tau') H_q^-(\tau'') \rangle \right), \tag{32}$$

In Eqs. (31), (32) we have written down explicitly the time dependent phase factors of operators $H_i^\pm$ related to electrostatic energy (27):

$$E_i = U(n_i + 1) - U(n_i).$$

Applying Wick theorem to the averages in Eqs. (31), (32) we get, for instance:

$$T_{km}^{(1)} T_{k'm}^{(1)*} T_{nl}^{(2)} T_{nl'}^{(2)*} \langle c_{l'}^\dagger c_k c_l c_{k'}^\dagger \rangle$$

$$\longrightarrow -|T_{km}^{(1)}|^2 |T_{nl}^{(2)}|^2 \langle c_l^\dagger c_l \rangle \langle c_k c_k^\dagger \rangle + T_{km}^{(1)} T_{lm}^{(1)*} T_{nl}^{(2)} T_{nk}^{(2)*} \langle c_k^\dagger c_k \rangle \langle c_l c_l^\dagger \rangle. \tag{33}$$

(Here and below in this section the indices m, n refer to the energy eigenstates of the external electrodes, while k, l to those of the central electrode.)

Carrying out the transformation (33) in Eqs. (31), (32) we get:

$$\gamma^\pm = \gamma_{\mathrm{in}}^\pm + \gamma_{\mathrm{el}}^\pm, \tag{34}$$

where $\gamma_{\mathrm{in}}^\pm$ and $\gamma_{\mathrm{el}}^\pm$ contain those terms in Eqs. (31), (32) which are similar, respectively, to the first and the second one in the right hand side of relation (33). This relation shows that $\gamma_{\mathrm{in}}^\pm$ is only dependent on the absolute value of the tunneling amplitudes $T^{(i)}$ and thus describe the tunneling process without any coherence between tunneling events in two junctions. One can check that this contribution to the q-mqt is the inelastic q-mqt considered in the previous section. In particular, the tunneling rate γ_{in}^+ coincides with that given by Eq. (10) of the Sec. 2.1, and the rate γ_{in}^- is given by the same equation with eV changed to $-eV$.

However, besides the terms corresponding to inelastic q-mqt, Eqs. (31), (32) contain the terms similar to the second one in the right hand side of relation (33). The contribution of these term can be written down as follows:

$$\gamma_{\mathrm{el}}^{(+)} = \frac{2\pi}{\hbar} \sum_{m,n,k,l} T_{km}^{(1)} T_{lm}^{(1)*} T_{nk}^{(2)*} T_{nl}^{(2)} f(\varepsilon_m)[1 - f(\varepsilon_n)]$$

$$\times F(\varepsilon_l, \varepsilon_m, \varepsilon_n) F(\varepsilon_k, \varepsilon_m, \varepsilon_n) \delta(\varepsilon_m - \varepsilon_n + eV), \tag{35}$$

$$F(\varepsilon, \varepsilon_m, \varepsilon_n) = \frac{1 - f(\varepsilon)}{E_1 + \varepsilon - \varepsilon_m} - \frac{f(\varepsilon)}{E_2 - \varepsilon + \varepsilon_n}.$$

The backward tunneling rate $\gamma_{\mathrm{el}}^{(-)}$ is given by Eq. (35) with $V \to -V$, $E_i \to E_i + eV$, and $\varepsilon_m \leftrightarrow \varepsilon_n$.

Comparing the arguments of the δ-functions in Eq. (35) and Eq. (10) one can see that in contrast to inelastic q-mqt the tunneling process described by Eq. (35) does not create electron-hole excitations on the central electrode of the system. In a sense, this implies that such an *elastic q-mqt* is due to the tunneling of one electron through both of the junctions.[7] Hence, the rate of this tunneling should be very sensitive to the electron motion inside the central electrode. In accordance with this interpretation, $\gamma_{\mathrm{el}}^{\pm}$ depends not only on the absolute values of the tunneling amplitudes $T^{(i)}$, but also on the phase factors of these amplitudes, which contain the information about electron motion in the electrode. Thus, elastic q-mqt presents the example when results of calculations with the standard tunnel Hamiltonian (28) depend crucially on phases of the tunnel matrix elements $T^{(i)}$. It means, in particular, that elastic q-mqt rate depends not only on junction conductances, but also on the specific geometry of the junctions.

In order to illustrate the importance of the phase factors of $T^{(i)}$ let us assume for a moment that the central electrode is one dimensional box. The electron eigenfuctions in the box are either even or odd, and the neighboring energy levels are of different parity. The tunneling matrix elements are determined by overlap of the wave functions in the box and wave functions in external electrodes. Hence, the matrix element of tunneling from the even state are of the same sign for tunneling to the left and right external electrode, whereas they are of different signs for the odd state. The first term in (33) does not feel these signs. In contrast to this, the second term changes its sign every time when the parity of levels k, l is changed. If we sum up this term over the electron states in the central electrode, the contributions of the neighboring levels tend to cancel each other. As a result, the total contribution of this term should be much smaller than the contribution from the first term, which is independent of the phase factors of $T^{(i)}$.

As it will be shown below, in accordance with these considerations the elastic tunneling rate is smaller than the inelastic rate at least by a factor of Δ/E_c. Thus, the elastic tunneling can be essential only for $eV, k_B T \ll E_c$, when the current due to inelastic tunneling is small (see Eq. (16)). At these voltages the elastic current, $I_{\mathrm{el}} = e(\gamma_{\mathrm{el}}^{(+)} - \gamma_{\mathrm{el}}^{(-)})$, depends linearly on the voltage, and it is sufficient to calculate the corresponding conductance $G_{\mathrm{el}} = (dI_{\mathrm{el}}/dV)|_{V=0}$. In order to take into account phases of the tunneling

---

[7] In particular, one can note that Eq. (35) has the usual form of one-electron tunneling rate with the amplitude $\langle m | M_{\mathrm{el}} | n \rangle$ of tunneling through the two junctions:

$$\langle m | M_{\mathrm{el}} | n \rangle = \sum_k T_{km}^{(1)} T_{nk}^{(2)*} F(\varepsilon_k, \varepsilon_m, \varepsilon_n).$$

amplitudes it is convenient to write down the amplitudes in the coordinate representation:

$$T_{km} = \int d^3y \int d^3z \; T(y,z) \psi_k^*(y) \psi_m(z). \tag{36}$$

Substituting (36) into (35) we get for the elastic conductance

$$G_{\rm el} = \frac{2\pi e^2}{\hbar} \int d\varepsilon \int d\varepsilon' \; F(\varepsilon) F(\varepsilon') R(\varepsilon, \varepsilon'), \tag{37}$$

$$R(\varepsilon, \varepsilon') = \int d^3z_1 \, d^3z_2 \, d^3z_3 \, d^3z_4 \, d^3y_1 \, d^3y_2 \, d^3y_3 \, d^3y_4 \; T^{(1)}(y_1, z_1) T^{(2)}(y_3, z_3)$$

$$\times T^{(1)*}(y_2, z_2) T^{(2)*}(y_4, z_4) K_0(z_2, z_1) K_0(z_3, z_4) K_\varepsilon(y_1, y_3) K_{\varepsilon'}(y_4, y_2). \tag{38}$$

where the points z_j and y_j are located in the external electrodes and the central electrode, respectively; $F(\varepsilon) \equiv F(\varepsilon, 0, 0)$, and

$$K_\varepsilon(x, x') = \sum_q \psi_q^*(x) \psi_q(x') \delta(\varepsilon - \varepsilon_q). \tag{39}$$

All quantum mechanical properties of electron propagation in the central electrode are taken into account in Eqs. (37), (38). However, if the size of this electrode is much larger than electron wavelength, it is much more natural to treat electrons as classical particles. (This situation is similar to light propagation in an optical system. Characteristic dimensions of the system determine whether geometrical optics is sufficient or one should take into account diffraction and interference phenomena.)

Powerful quasiclassical methods have been developed [15, 16], in particular for the case of electron tunneling [17], which enable one to reduce quantum expressions like Eq. (38) to classical description of electron motion. We are not able to give here a comprehensive review of these methods. What follows is the outline of the basic ideas underlying the methods in application to Eq. (38).

For the sake of simplicity we assume that the Fermi surfaces of the electrodes of the junctions are isotropic, $\varepsilon(p) \equiv \varepsilon(|p|)$, and Hamiltonian H_0 is symmetric under inversion of time, which allows us to choose real wavefunctions of electrons. Following [16] we introduce the quasiclassical probability $\Pi(x, p, \varepsilon; x', p', t)$ for the electron to be in a certain point x with a certain momentum p and energy ε at time 0, and to be in another point x' with momentum p' at time t. One can express it in terms of quantum mechanical operators $\hat{x}$, $\hat{p}$:

$$\Pi(x, p, \varepsilon; x', p', t) = {\rm Tr}[\delta(\hat{x} - x)\delta(\hat{p} - p)\delta(\hat{H} - \varepsilon)\delta(\hat{x}(t) - x')\delta(\hat{p}(t) - p')]. \tag{40}$$

We consider quasiclassical quantity, and the operator ordering in Eq. (40) is not essential. Since operator of momentum acts on electron wavefunction as follows:

$$\delta(\hat{p} - p)\psi(x) = \frac{1}{2\pi} \int d^3y \, e^{ip(x-y)} \psi(y), \tag{41}$$

we can rewrite Eq. (40) in terms of eigenfunctions and eigenenergies:

$$\Pi = \frac{1}{(2\pi)^2} \int d^3x_1 \, d^3x_2 \, e^{ip(x-x_1)} e^{ip'(x'-x_2)}$$

$$\times \sum_{n,m} \psi_m(x)\psi_n(x_1)\delta(\varepsilon_n - \varepsilon)\psi_m(x')\psi_n(x_2)e^{i(\varepsilon_n - \varepsilon_m)t/\hbar}. \tag{42}$$

Thus,

$$\int dt \, e^{i(\varepsilon'-\varepsilon)t/\hbar} \Pi(x, p, \varepsilon; x', p', t)$$

$$= \frac{1}{2\pi} \int d^3x_1 \, d^3x_2 \, e^{ip(x-x_1)} e^{ip'(x'-x_2)} K_\varepsilon(x, x') K_{\varepsilon'}(x_1, x_2), \tag{43}$$

where K_ε is defined by Eq. (39).

As long as we are interested in electrons near the Fermi surface we can neglect dependence of the probability Π on the absolute value of momentum, and consider only directions n, n' of the momenta p, p' on the Fermi surface. The absolute value of momentum is conserved in the processes of electron motion: $|p| = |p'|$, i.e. $\varepsilon(p) = \varepsilon(p')$. Besides this, one can introduce *conditional* probability $P(x, n, 0; x', n', t)$ for the electron to have coordinate x' and momentum direction n' at time t, if it had coordinate x and momentum direction n at time $t = 0$. It is natural to define this probability only for positive t. Such a conditional probability differs from Π by the probability to have coordinate x, momentum direction n, and energy ε at *any* time. The latter probability does not depend on the space coordinates, and is concentrated in momentum space on the surface corresponding to the energy ε. Thus, we can write down Π in the following form:

$$\Pi(x, p, \varepsilon; x', p', t)$$

$$= \frac{\delta(\varepsilon(p) - \varepsilon)}{(2\pi\hbar)^3}[P(x, n, 0; x', n', t)\Theta(t) + P(x', n', 0; x, n, t)\Theta(-t)]\frac{\delta(|p| - |p'|)}{4\pi|p|^2}. \tag{44}$$

It is convenient to transform the last term in Eq. (44) introducing the electron density of states ν, and to interchange the variables in $P(x', n', 0; x, n, t)$ using the time inversion symmetry. In this way we get more symmetric expression:

$$\Pi(x, p, \varepsilon; x', p', t) = \frac{1}{\nu} \frac{\delta(\varepsilon(p) - \varepsilon)}{(2\pi\hbar)^3} P(x, n, 0; x', n', |t|) \frac{\delta(\varepsilon(p') - \varepsilon)}{(2\pi\hbar)^3}. \tag{45}$$

Now let us express the junction conductances in terms of similar quasiclassical probabilities. At fixed voltage, the zero-bias conductance of, say, the first junction is given by the following Golden-Rule-type formula:

$$G_1 = \frac{4\pi e^2}{\hbar} \sum_{k,m} \delta(\varepsilon_m)\delta(\varepsilon_k)|T_{km}|^2. \tag{46}$$

This formula can be rewritten in coordinate representation similarly to Eq. (38) above:

$$G_1 = \frac{4\pi e^2}{\hbar} \int d^3z_1\, d^3z_2\, d^3y_1\, d^3y_2\, K_0(z_2, z_1) K_0(y_1, y_2) T^{(1)}(y_1, z_1) T^{(1)}(y_2, z_2). \quad (47)$$

At this point we make quasiclassical approximation, assuming that the points z_1, z_2 are separated only by few electron wavelengths, so that one can use the following expression for K_0:

$$K_0(z_1, z_2) = \frac{1}{(2\pi\hbar)^3} \int d^3p\, e^{ip(z_1 - z_2)} \delta(\varepsilon(p)). \quad (48)$$

This enables one to transform Eq. (47) as follows:

$$G_1 = \int d^2n\, d^3z\, g_1(z, n), \quad (49)$$

where

$$g_1(z, n)$$
$$= \frac{4\pi e^2}{\hbar} \int d^3u\, d^3y_1\, d^3y_2 \int \frac{|p|^2 d|p|}{(2\pi\hbar)^3} \delta(\varepsilon(p)) e^{ipu} K_0(y_1, y_2) T^{(1)}(y_1, z_1) T^{(1)}(y_2, z_2). \quad (50)$$

(This approximation corresponds to the idea that electron tunnels to a certain point $z = (z_1 + z_2)/2$.)

The tunneling matrix elements in coordinate representation should be non-vanishing only if the points z, y are close to the metal surfaces. Hence, we can integrate $g_1(z, n)$ over the coordinate orthogonal to the metal surface. In the result we obtain the functions $g_i(x, n)$, which can be viewed as quasiclassical probabilities of electron tunneling from ith external electrode to the state n near the point x, where x lies on the junction surface. As one can see from Eq. (50), these probabilities are normalized in such a way that the junction conductances per unit area $g_i(x)$ and their total conductances G_i are:

$$g_i(x) = \int d^2n\, g_i(x, n), \quad G_i = \int d^2x\, g_i(x). \quad (51)$$

Here the integration d^2x is carried out over the ith junction area.

Combining Eqs. (43), (45), and (50) one can show that $R(\varepsilon, \varepsilon')$ (38) in the quasi-classical approximation is

$$R(\varepsilon, \varepsilon') = R(\varepsilon - \varepsilon') = \frac{\hbar}{8\pi^3 e^4 \nu} \int d^2x_{1,2}\, d^2n_{1,2}\, g_1(x_1, n_1) g_2(x_2, n_2)$$

$$\times \int dt\, \exp[i(\varepsilon - \varepsilon')t/\hbar] P(x_1, n_1, 0; x_2, n_2, |t|). \quad (52)$$

Elastic conductance G_{el} can be presented in the following compact form:

$$G_{\text{el}} = \frac{2}{e^2 \nu} \int_0^\infty dt\, |F(t)|^2 \int d^2x_{1,2} d^2n_{1,2}\, g_1(x_1, n_1) g_2(x_2, n_2) P(x_1, n_1, 0; x_2, n_2, t), \quad (53)$$

where $F(t)$ is Fourier transform of $F(\varepsilon)$. However, for practical calculations it is more convenient to use Eqs. (37), (38), and (52).

Thus, despite the fact that the intermediate electron state on the central electrode of the double junction system during the tunneling is virtual, the rate of elastic q-mqt is expressed via Eqs. (37), (52), and (53) in terms of classical electron motion through the central electrode. The only fact that reminds us of the virtual character of electron motion in the tunneling process is that the tunneling rate depends on the characteristics of classical motion on the time scale $\hbar/E_c$. Thus, $\hbar/E_c$ plays the role of the time which the virtual electron can spend to propagate from one junction to another.

The rate of the elastic tunneling and the corresponding conductance G_{el} depend, in general, on the geometry of the junctions. If the characteristic dimension L of the central electrode is larger than electron elastic mean free path l ($L \gg l$), one can use the usual diffusion equation to describe the electron motion inside the electrode. In this case the probability P in Eq. (53) does not depend on n, so that

$$R(\varepsilon) = \frac{\hbar}{8\pi^3 e^4 \nu} \int d^2 x_{1,2}\, g_1(x_1) g_2(x_2) \int dt\, e^{i\varepsilon t/\hbar} P(x_1, 0; x_2, |t|), \tag{54}$$

and the tunneling rate is determined by the process of "virtual diffusion" of electrons on the time scale of $\hbar/E_c$.

We consider two limiting cases. If the characteristic time $\hbar/E_c$ is much larger than the classical time L^2/D of electron diffusion through the electrode, (where D is diffusion coefficient, i.e., $E_c \ll E_{th} \equiv \hbar D/L^2$), the probability P is constant on the large time scale, $P = 1/V$, where V is the volume of the electrode. It follows from Eqs. (37), (54), that in such a case, irrespective of the shape of the central electrode,

$$G_{el} = \frac{\hbar G_1 G_2 \Delta}{4\pi e^2} \left(\frac{1}{E_1} + \frac{1}{E_2} \right). \tag{55}$$

Thus, we can estimate the elastic current as

$$I_{el} \simeq V R_K G_1 G_2 \frac{\Delta}{E_c}. \tag{56}$$

Comparing Eq. (56) with the inelastic current (16) we see that the elastic current should dominate if voltage and temperature are much less than the crossover values V_{cr}, T_{cr}:

$$eV_{cr}, k_B T_{cr} \simeq \sqrt{\Delta E_c} \ll E_c. \tag{57}$$

In the opposite limit, $E_c \gg E_{th}$, the conductance G_{el} depends on the specific form of the central electrode and non-uniformity of conductances $g_i(x_i)$ along the junction areas. Solving the diffusion equation for the simplest case of a rectangular electrode and $g_i(x_i) = \text{const}$ we get for low temperatures ($k_B T \ll E_{th}$):

$$G_{el} = \frac{\hbar^2 G_1 G_2 \Delta}{16 e^2} \left(\frac{1}{E_1} + \frac{1}{E_2} \right)^2 \frac{\pi^{1/2} D}{L^2}, \tag{58}$$

where L is the length of the electrode (the distance between parallel planes of the junctions). Hence, elastic conductance for $E_c \gg E_{th}$ is even smaller than conductance (55) by a factor of (E_{th}/E_c), so that the crossover voltage is also smaller than the value given by Eq. (57):

$$eV_{\text{cr}} \simeq \sqrt{\Delta E_{\text{th}}}. \tag{59}$$

At larger temperatures, $E_c \gg k_B T \gg E_{\text{th}}$, the elastic conductance rapidly decreases and becomes exponentially small:

$$G_{\text{el}} \simeq \exp\{-(4\pi k_B T/E_{\text{th}})^{1/2}\}. \tag{60}$$

Equation (60) implies that there should exist a temperature region where the total linear conductance of the system, $G_{\text{el}} + G_{\text{in}}$, also decreases with temperature.

When the central electrode of the junctions is comparable to electron elastic mean free path, the probability P depends essentially on momentum direction n_i, so that the n_i-dependence of the tunneling probabilities $g_i(x_i, n_i)$ becomes of importance. In order to find $P(t)$ in this case one should solve a kinetic equation for the specific shape of the electrode with the specific boundary conditions describing the surface scattering. In the most realistic case when the surface scattering is diffusive and $E_c \ll E_{\text{th}}$ ($E_{\text{th}} = \hbar v_F/L$), the elastic conductance is again given by the Eq. (55).

4. Transport of electron-hole pairs in coupled arrays of small tunnel junctions

4.1. General picture of electron-hole transport

Electron tunneling transitions involved in co-tunneling are bound together by the virtual character of intermediate states. Hence, in the circuits that are more complex than 1D arrays of junctions, co-tunneling can give rise to correlations between acts of electron transfer through the junctions that are not coupled galvanically. In general, such correlations of the currents in galvanically uncoupled junctions can take place even if the junctions are not isolated from electrodynamic environment, so that the quantum fluctuations of the charge on the junction are not small [18]. In this section we consider an example of the system, in which such correlations are absolute and give rise to non-trivial charge transport behavior.

The system consists [19] of the two 1D arrays of metal islands connected by small tunnel junctions and capacitively coupled as shown in Fig. 9. We assume that the tunneling between the arrays is negligible, so that we have two circuits that are independent galvanically but are coupled electrostatically. Electrostatics of the system is characterized by the two capacitances: C_0 between the adjacent metal islands in the different arrays, and C between the islands connected by the junction (see Fig. 9). We are interested in the case when the coupling between the arrays is dominant, $C_0 \gg C$.

In this case particular charge configuration on the metal islands which consists of an electron in one of the islands and a hole in the adjacent island of another array is energetically favorable. Electrostatic energy of such a pair (that can be called an "exciton") is of the order e^2/C_0 and much smaller than the energy of the unpaired

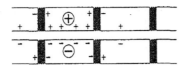

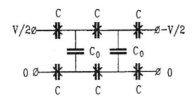

Figure 9. Schematic drawing and equivalent circuit of two electrostatically coupled 1D arrays of metal tunnel junctions. The dashed region denotes the insulator layers through which electrons can tunnel. Also shown is the charge distribution associated with electron-hole pair ("exciton") on the adjacent electrodes of the two arrays. On the equivalent circuit the tunnel junctions are represented by crossed capacitances.

electron ($\simeq e^2/C$). Thus, significant energy is needed to separate the electron and the hole and destroy the exciton.

One can see that these excitons can enter the junction arrays and move along them due to the two-electron *co-tunneling* even when the driving voltages applied to the system are much smaller than e/C and the arrays are in the Coulomb blockade state with respect to the *single electron* tunneling. At first an electron tunnels virtually through the junction in one of the arrays, and increases electrostatic energy. From the equivalent circuit of our system (shown in Fig. 9 for the specific case of the three-junction arrays) one can get that at $C_0 \gg C$ the energy ε of resulting charge configuration is

$$\varepsilon = \frac{e^2}{4C}\left(1 - \frac{1}{N}\right),\qquad(61)$$

where N is the number of the junctions in one array. It can be viewed as the binding energy of the exciton. The energy is decreased when the second electron tunnels through the coupled junction of another array in the direction opposite to that of the first electron. (In other words, one can say that the hole tunnels in the same direction.) This tunneling restores the exciton charge configuration that is already shifted one site along the arrays. The rate γ of such a two-electron tunneling process is given by the Eqs. (15), (16) derived in Sec. 2.1. In these equations E_1, E_2 are now the energies of the intermediate states arising after electron transfer through the junction of the first or the second array, respectively, (for $C_0 \gg C$ both energies are equal to ε (61)), and eV is the energy gain due to the one-site shift of the exciton.

Since the different acts of the exciton hopping between the sites of the system are uncorrelated, the dynamics of exciton motion can be described quantitatively by the conventional master equation for probabilities $\sigma(n_1, ..., n_{N-1}) \equiv \sigma\{n\}$ to find n_i excitons on the i-th site:

$$\dot{\sigma}\{n\} = \sum_{k=1}^{N}\left[\gamma_k^+\{n\}_k^+ \sigma\{n\}_k^+ + \gamma_k^-\{n\}_k^- \sigma\{n\}_k^- - (\gamma_k^+\{n\} + \gamma_k^-\{n\})\sigma\{n\}\right].\qquad(62)$$

Here N is the number of the junctions in each array, $\gamma_k^\pm$ are the rates of forward and backward exciton tunneling in the k-th junction, and by $\{n\}_k^\pm$ we denote the exciton configurations which differs from the $\{n\}$ configuration by the forward or backward tunneling of one exciton in the k-th junction: $\{n\}_k^\pm \equiv \{n_1, ..., n_k \pm 1, n_{k+1} \mp 1, ..., n_{N-1}\}$. Equation (62) is solved in the following sections for several specific cases.

Excitons in our system of coupled metal islands are qualitatively similar to the solid state microscopic excitons. In this respect the system under consideration is the next one in a series of mesoscopic manmade devices that allow one to model the concepts of microscopic solid state physics. Another system of this kind is the "artificial crystal" of Kouwenhoven et al. [20], that exhibited the 1D band structure.

However, in contrast to excitons in a real semiconductor crystal, in our "artificial crystal" the charges forming the "exciton" are well separated spatially, and exciton transport leads to charge transport. Exciton hopping shifts one electron along the upper junction array and one electron along the lower array in the opposite direction. Thus, the currents in the circuits that contains the arrays are exactly equal in magnitude and opposite in direction. Moreover, the driving voltage for the exciton is the difference between the voltages across the two arrays. Hence, there is a possibility to establish the current *against* the voltage in one of the circuits. The energy needed to do this is supplied by the voltage in the other circuit. It means that from the point of view of device application this system can be considered as dc current transformer.

An accuracy of the equality of the currents in the two arrays is limited by the q-mqt process that transfers an electron along one of the arrays without corresponding tunneling of the hole in another array. This "leakage" current is proportional to $(R_K/R_t)^N$. Since the rate of the exciton tunneling is proportional to $(R_K/R_t)^2$, the current I flowing in the arrays is mainly due to the exciton transport only when the arrays contain at least three junctions. In this case the current carried by excitons is larger than the leakage current by a factor of R_K/R_t.

So far we discussed the main properties of the exciton transport which do not depend on the number of the junctions in the arrays, and for driving voltages smaller than e/C are independent of the voltage. Now we turn to calculations of the current I carried by excitons, that is sensitive to the specific dynamics of exciton tunneling and depends on both factors.

4.2. Three-junction arrays

We begin our discussion of exciton tunneling dynamics with the simplest case of the three-junction arrays biased as shown in Fig. 9. At small driving voltages $V \simeq e/C_0$ each exciton creates voltage drop between the sites of the system comparable with the driving voltage, so that there are essential correlations between excitons. At low temperatures, $k_B T \ll e^2/C_0$, for an exciton to enter the system the voltage should be larger than the threshold voltage $V_e \equiv e/C_0$.[8] For $C_0 \gg C$ the driving voltage for the excitons drops

---

[8] As usual, we assume that uncontrolled residual potential differences between metallic islands of the system are negligible. Experiments [21] with long 1D junction arrays seem to indicate that there is some relaxation mechanism for these residual voltages, so that they are small at least for aluminum junctions (see a discussion of this problem in Chap. 9). However, dynamics of these voltages in systems of small tunnel junctions is not understood completely at present.

completely across the edge junctions and at voltages V above V_e, $V_e < V < 3V_e$, the picture of the exciton motion through the system is as follows.

When an exciton jumps into the system through one of the edges it is trapped on the edge site, since there is no voltage drop that would drive the exciton through the middle pair of junctions. At small driving voltages, $V < 3V_e$, exciton on the edge site makes it impossible for another exciton to enter the array through the same edge. Hence, the tunneling is blocked until the exciton of the opposite sign (with reversed position of electron and hole) enters the system through another edge. Then one of the excitons hops across the middle junctions, they annihilate, and the process recurs.

The master equation (62) that corresponds to this qualitative picture of the exciton tunneling can be written down as follows:

$$\dot{\sigma}(-1,0) = \gamma[\sigma(0,0) - \sigma(-1,0)],$$

$$\dot{\sigma}(0,1) = \gamma[\sigma(0,0) - \sigma(0,1)],$$

$$\dot{\sigma}(0,0) = \gamma_a\sigma(-1,1) - 2\gamma\sigma(0,0),$$

$$\dot{\sigma}(-1,1) = \gamma[\sigma(-1,0) + \sigma(0,1)] - \gamma_a\sigma(-1,1),$$

where γ and γ_a are the rates of tunneling through the edge and the middle junctions, respectively. These rates are given by Eqs. (15), (16), in which E_1, E_2 are equal to ε (61), junction resistances R_{ti} are assumed to be equal for all junctions, $R_{ti} \equiv R$, and the energy gain eV is $e(V - V_e)/2$ for γ and e^2/C_0 for γ_a.

From the steady-state solution of this set of equations one can find the dc current I carried by excitons at small driving voltages

$$I = \frac{2e\gamma\gamma_a}{2\gamma + 3\gamma_a}. \tag{63}$$

Near the threshold, $V - V_e \ll V_e$, the tunneling through the middle junctions is much faster than the tunneling in the edge junctions, $\gamma \ll \gamma_a$, so that

$$I = \frac{2}{3}e\gamma = \frac{\hbar C^2(V - V_e)}{\pi R^2 e^4}\frac{(V - V_e)^2 + (4\pi k_B T/e)^2}{1 - \exp[-e(V - V_e)/2k_B T]}. \tag{64}$$

When the voltage and/or temperature are relatively large, $e^2/C_0 \ll eV, k_B T \ll \varepsilon$, the correlations between excitons are suppressed. As follows from the results of the next section, in this case the exciton current I through the three-junction array is

$$I = \frac{4\hbar C^2 V}{\pi R^2 e^4}\left[(V/3)^2 + (2\pi k_B T/e)^2\right]. \tag{65}$$

In order to find the dc $I-V$ curves the three-junction array for arbitrary voltages one can solve the master equation (62) numerically. The result of this calculations for $T = 0$ are shown in Fig. 10. At small and large voltages the $I - V$ curve reaches asymptotes (64) and (65) respectively. Deviations from the small-voltage asymptote at very small currents, which are visible on the log-log scale, result from the fact that C/C_0 ratio is finite. Note, that the large-voltage asymptote is reached only on this scale, since at

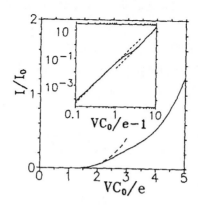

Figure 10. dc $I-V$ curve of the three-junction arrays in the regime of exciton transport for $C/C_0 = 0.01$ and $T = 0$. The insert shows the same curve plotted on a log-log scale. Dashed lines are the small-voltage and large-voltage asymptotes given by Eqs. (64), (65). The current is measured in units I_0, $I_0 \equiv eR_K C^2/R^2 C_0^3$.

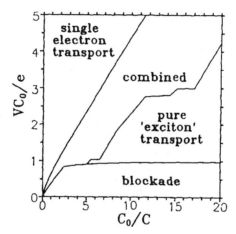

Figure 11. Parameter regions with different charge transport mechanisms at $T = 0$ for the three-junction arrays.

$k_B T \ll eV$ Eq. (65) gives correctly only the leading (in voltage) term that is proportional to V^3.

At finite C/C_0 ratios the exciton transport can coexist with the single electron tunneling. The zero-temperature boundaries between different regimes of the charge transport are displayed in Fig. 11. At small driving voltages ($V < e/C_0$ for $C \ll C_0$) both single electron and exciton transport are suppressed. At somewhat larger voltages only exciton transport becomes possible. For $C \ll C_0$ the boundaries of the parameter region with pure exciton transport are $V = e/C$ and $V = e/4C$. At larger voltages ($e/4C < V < e/2C$ for $C \ll C_0$) single electrons can not tunnel through the array, but the exciton motion can create the charge configuration that makes the single electron tunneling possible. Since at $T = 0$ number n of excitons on the sites of the array changes discretely with increasing voltage, the boundary between the pure exciton and combined transport regions exhibits steps at $VC_0/e \cong 1, 3, 5, \ldots$, where n is increased by 1. Finally, at $V > e/2C$ single electrons can tunnel along the array and charge transport is governed by the single electron tunneling.

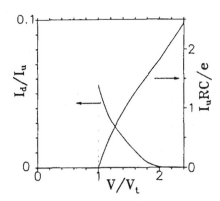

Figure 12. The ratio I_d/I_u of the currents in the lower and upper junction arrays and the current I_u in the upper array at $T = 0$ and $C/C_0 = 0.01$ in the regime of the single electron tunneling. The induced current I_u flows in the opposite direction with respect to I_d. The threshold voltage V_t is $e/2C$.

In principle, the correlated single electron tunneling can simulate to some extent the co-tunneling responsible for the exciton transport. Figure 12 shows numerically calculated $I - V$ curves of the the three-junction arrays at voltages above the threshold of the single electron tunneling. One can see that the current I_u in the upper circuit induces a finite reverse current $-I_d$ in the lower circuit, which implies that the tunneling of single electron along one of the junction arrays drives the hole along the coupled array.[9] However, in this case the electron and the hole are only weakly bound together, so that the relative magnitude of induced current, I_d/I_u, is rather small and rapidly decreases with increasing voltage (see Fig. 12). It means that in the regime of exciton transport single electron tunneling events activated by finite temperature or large external voltage presents another source of inaccuracy of the equality of the currents in the two arrays in addition to the q-mqt leakage current discussed in the previous section.

4.3. Multijunction arrays

Similarly to the three-junction arrays, in arrays with larger number N of junction correlations between excitons are important at small temperatures, $k_B T \ll e^2/C_0$, and small driving voltages, $V \simeq V_e$, where V_e is the threshold voltage for exciton transport. The threshold V_e is of the order of Ne/C_0 and depends on voltage U between two junction arrays.

When the driving voltage and/or temperature are large, $(eV/N), k_B T \gg e^2/C_0$, characteristic number of excitons on each site of the system is also large, $|n_i| \gg 1$, and the correlations between excitons are suppressed. Besides this, for large $|n_i| \gg 1$ the shot-noise fluctuations of n_i are relatively small, so that one can neglect fluctuations of

---

[9] It is interesting to note that the similar effect of the *opposite* sign takes place in the coupled 2DEG, i.e. an electron moving in one of the 2DEG pushes *electrons* in another 2DEG (see, e.g., [22, 23] and references therein). It would be interesting to study the crossover between the two regimes, where the effect changes the sign.

n_i and reduce the master equation for probability density $\sigma\{n\}$ (62) to the equations for the average numbers $\langle n_i \rangle$ of excitons on the sites:

$$\langle \dot{n_i} \rangle = (I_{i-1} - I_i)/e\,, \qquad I_i = e(\gamma_{i,i+1} - \gamma_{i+1,i}). \tag{66}$$

Here $\gamma_{i,i+1}$ is the rate of the exciton hopping from the site i to the site $i+1$. The energy gain in this processes (that determines $\gamma_{i,i+1}$) is $e(\varphi_i - \varphi_{i+1})$, where φ_i is the average potential difference between two electrodes of i-th site. At $C_0 \gg C$ the voltages φ_i for internal sites are directly related to the average numbers $\langle n_i \rangle$ of excitons on the sites, $\varphi_i = e\langle n_i \rangle/C_0$, $(i = 1, ..., N-1)$, and φ_0, φ_N are the potential differences between respectively left and right external electrodes. When the temperature is larger than the voltage drop between the sites, $k_B T \gg e(\varphi_i - \varphi_{i+1})$, Eq. (66) gives the discrete version of the diffusion equation for the voltages φ_i with the diffusion constant D:

$$\dot{\varphi_i} = D(\varphi_{i+1} + \varphi_{i-1} - 2\varphi_i), \quad D \equiv \frac{4\pi\hbar}{3e^2 R^2 C_0}(T/\varepsilon)^2. \tag{67}$$

Equation (66) implies that in the stationary state the voltages φ_i change linearly along the arrays, $e(\varphi_i - \varphi_{i+1}) = \text{const} = (V_u - V_d)/N \equiv \Delta V$, where V_u, V_d are the voltages across the upper and lower arrays. Thus, the excitons are driven by the difference of the voltages across the arrays, and the current I carried by excitons is given by Eq. (16) with the driving voltage equal to ΔV:

$$I = \frac{\hbar \Delta V}{3\pi R^2 \varepsilon}[\Delta V^2 + (\frac{2\pi k_B T}{e})^2]. \tag{68}$$

In particular, for the three-junction arrays of Fig. 9, Eq. (68) gives Eq. (65) of the previous section.

Discussion of coexistence of exciton and single electron transport for the three-junction arrays (Sec. 4.2) can be qualitatively applied to arrays with larger number of junctions. Up to now there are no quantitative results on this coexistence.

Exciton transport in electrostatically coupled arrays of small tunnel junctions so far has not been observed experimentally. One of the new requirements for such an observation in comparison to the "conventional" experiments with submicron tunnel junctions (discussed, e.g., in Sec. 2.3) is the large inter-array capacitances $C_0 \gg C$. Possible way to fabricate these capacitances is to connect the adjacent electrodes of the arrays by additional metal strips evaporated over them. In this way one can make the ratio C/C_0 as small as 0.01, while keeping the junction capacitance C (and the self-capacitances of the electrodes) close to the typical values 10^{-15}–10^{-16} F. It means that slight modification of the existing fabrication technique should provide the possibility to observe the "exciton" charge transport.

5. Unsolved problems and perspectives

In conclusions we would like to outline shortly our view on the main unsolved problems of the field considered in this chapter. First, let us note that the single electronics in general belongs to large extent to applied physics (or at least potentially applicable one). In this respect, it can be considered as an attempt to make use of inter-electron Coulomb forces by fabricating artificial structures in which one is able to control these forces. On the other hand, well defined artificial structures provide sometimes unique possibilities to study various fundamental properties of electron transport. These facts should determine the development of the subject.

From the point of view of the applications the most important problem associated specifically with the q-mqt and co-tunneling is calculation of operation accuracy of single electron devices such as, e.g., a current standard or a logic cell. To calculate this accuracy, one should be able to describe dynamics of electron tunneling taking into account both single electron tunneling and multi-electron co-tunneling processes. This seems to be purely technical problem for $R_t \gg R_K$, nevertheless it has not been solved yet in general. The reason for this is that in multi-junction systems the number of higher-order electron transitions is very large, and dynamics of electron tunneling becomes extremely complex.

Quantum fluctuations of electric charge responsible for q-mqt and co-tunneling become large at $R_t \simeq R_K$. The crossover between the controllable change of charge states due to electron tunneling events and complex quantum mixture of these states where tunneling events can not be separated from each other should take place in this region. Quantitative understanding of this crossover presents another important problem. In particular, the simplest but fundamental problem is to understand behavior of a single metallic island connected with external bulk electrode by one junction. Despite several attempts to analyze this problem (see, e.g., [24, 25]) we are far from real understanding. Recent result obtained in the limit $R_t \ll R_K$ [26] indicates that there should exist some new order in tunnel junction behavior in this limit.

The next interesting problem is the coexistence of elastic and inelastic q-mqt in situation when they are of the same order of magnitude. Such a situation can be encountered in semiconductor heterostructures where the carrier wavelength is large, so that probability of elastic motion through the structure is rather high. One can consider, for example, lateral electron transport through 1D array of quantum dots. The array can be described by a generalized Hubbard model with a large number of states per site. Evidently, the charged excitations in this system should originate from the localized charge state of nearly isolated dots, but on the other hand, they can move coherently from site to site as a single carrier. It would be interesting to understand the nature of the charged excitations in such a system.

Another set of problems is related to systems of superconducting tunnel junction. Treated naively, Cooper pair in the metallic island is just a double electron charge and one can do with pairs the same tricks as with electrons. In particular, one can study the superconducting analog of the system discussed in Sec. 4: capacitively coupled 1D arrays of superconducting islands. In a certain range of parameters the co-tunneling of Cooper pairs in different arrays should dominate over the single Cooper pair tunneling in one array. In this case superconducting zero-voltage state of a finite array is mainly due to coherent tunneling of compound objects, a Cooper pair and "anti Cooper pair" placed in adjacent island of the two arrays. The critical currents of the arrays should be maximum if the bias currents in the circuits are equal in magnitude and opposite in direction. Superconducting zero-voltage state of the infinite arrays should be stable *only* if this condition on the bias currents is fulfilled. Another possibility is to couple capacitively normal and superconducting arrays. In this case one can play with the compound objects consisting of hole in one array and Cooper pair in another array.

Even the above list of the most obvious problems concerning macroscopic quantum tunneling of charge and electron co-tunneling shows that there is enough space to keep the field growing in the nearest future. However, the history of single electronics provides grounds to believe that attempts to solve these obvious problems will lead to exciting unexpected developments.

References

[1] R. P. Feynman and A. R. Hibbs, *Quantum Mechanics and Path Integrals* (McGraw-Hill, N.Y. 1965), Chap. 6.

[2] D. V. Averin and K. K. Likharev, in: *Mesoscopic Phenomena in Solids*, ed. by B. L. Altshuler, P. A. Lee, and R. A. Webb (Elsevier, Amsterdam, 1991), p. 173.

[3] A. O. Caldeira and A. J. Leggett, Ann. Phys. (N.Y.) **149**, 374 (1983).

[4] D. V. Averin and A. A. Odintsov, Phys. Lett. A **140**, 251 (1989).

[5] Yu. V. Nazarov, Fiz. Nizk. Temp. **16**, 718 (1990) [Sov. J. Low Temp. Phys. **16**, 422 (1990)].

[6] K. K. Likharev, IEEE Trans. Magn. **23**, 1142.

[7] D. V. Averin and G. Schön, in: *Quantum Coherence in Mesoscopic Systems*, ed. by B. Kramer (Plenum, New York, 1991), p. 531.

[8] D. V. Averin and Yu. V. Nazarov, Phys. Rev. Lett. **65**, 2446 (1990).

[9] D. V. Averin, A. A. Odintsov, and S. V. Vyshensky, to be published.

[10] L. J. Geerligs, D. V. Averin, and J. E. Mooij, Phys. Rev. Lett. **65**, 3037 (1990).

[11] L. J. Geerligs, V. F. Anderegg, and J. E. Mooij, Physica B **165/166**, 973 (1990).

[12] U. Meirav, M. A. Kastner, and S. J. Wind, Phys. Rev. Lett. **65**, 771 (1990).

[13] D. C. Glattli, C. Pasquier, U. Meirav, F. I. B. Williams, Y. Jin, and B. Etienne, Z. Phys. B **85**, 375 (1991).

[14] L. I. Glazman and K. A. Matveev, Pis'ma Zh. Eksp. Teor. Fiz. **51**, 425 (1990) [JETP Lett. **51** 484 (1990)].

[15] E. A. Shapoval, Zh. Eksp. Teor. Fiz. **49**, 930 (1965) [Sov. Phys. JETP **22**, 647 (1966)].

[16] L. P. Gor'kov and G. A. Eliashberg, Zh. Eksp. Teor. Fiz. **48**, 1407 (1965) [Sov. Phys. JETP **21**, 940 (1965)].

[17] Yu. V. Nazarov, Zh. Eksp. Teor. Fiz. **96**, 240 (1989) [Sov. Phys. JETP **68**, 561 (1989)]; Yu. V. Nazarov, Zh. Eksp. Teor. Fiz. **98**, 306 (1990) [Sov. Phys. JETP **71**, to be published].

[18] U. Geigenmüller and Yu. V. Nazarov, Phys. Rev. B **44**, 10953 (1991).

[19] D. V. Averin, A. N. Korotkov, and Yu. V. Nazarov, Phys. Rev. Lett. **66**, 2818 (1991).

[20] L. P. Kouwenhoven, F. W. J. Hekking, B. J. van Wees, C. J. P. M. Harmans, C. E. Timmering, and C. T. Foxon, Phys. Rev. Lett. **65**, 361 (1990).

[21] P. Delsing, K. K. Likharev, L. S. Kuzmin, and T. Claeson, Phys. Rev. Lett. **63**, 1861 (1989).

[22] P. M. Solomon, P. J. Price, D. J. Frank, and D. C. La Tulipe, Phys. Rev. Lett. **63**, 2508 (1989).

[23] T. J. Gramila, J. P. Eisenstein, A. H. MacDonald, L. N. Pfeiffer, and K. W. West, Phys. Rev. Lett. **66**, 1216 (1991).

[24] G. Schön and A. D. Zaikin, Phys. Rep. **198**, 237 (1990).

[25] K. A. Matveev, Zh. Eksp. Teor. Fiz. **99**, 1598 (1991) [Sov. Phys. JETP **72**, 892 (1991)].

[26] S. V. Panyukov and A. D. Zaikin, Phys. Rev. Lett. **67**, 3168 (1991).

Chapter 7

One-Dimensional Arrays of Small Tunnel Junctions

P. DELSING

Department of Physics, Chalmers University of Technology
S-41296 Göteborg, Sweden

1. Introduction

Several of the characteristic and interesting single charge tunneling effects have been observed in 1D arrays of tunnel junctions, i.e., in series coupled arrays of many, closely spaced, ultrasmall junctions. For example, the time correlation of tunnel events was first observed in such an array. The interpretation of the results obtained with an array may be complicated as compared to a single junction, but the additional degrees of freedom for an array, like the formation of charge "solitons", may be of advantage in developing the single charge effects and, in particular, for applications of these effects. One of the advantages of series coupled junctions is that it is easier to fabricate high resistance tunnel junctions closely connected to the junction under study than high resistance conventional resistors.

The electromagnetic environment strongly influences the charging properties of single tunnel junctions. This is discussed in detail in Chap. 2. A stray capacitance may add to the capacitance of a small junction, make it large, and swamp the single electron tunneling (SET) effects. Usually, the impedance of the measurement leads connected to a junction is much smaller than the junction resistance, R. This means that the junction will be voltage biased and it will be very hard to observe any charging effects like charge (or voltage) oscillations. The junction needs to be "protected" from the environment by high resistance resistors located as close to the junction as possible to avoid stray capacitances.

The situation is drastically improved when two, or more, junctions are coupled in series. Each junction is disconnected from the low impedance load by its high resistive neighbor(s). The individual junctions can be quasi-current-biased in a 1D array, even though the array itself is voltage biased. There are properties that are characteristic of arrays of junctions and which are not found in a few junction system. For example, time correlation of tunnel events may arise in voltage biased arrays of junctions but not in voltage biased double junctions. Space correlation of tunnel events, on the other hand, occurs in both systems. The tunneling rates depend upon the charge on the interconnecting island(s) (or electrodes) between the tunnel junctions and a tunnel event in one junction may cause another tunnel event in a neighboring junction.

As other chapters of this book treat properties of double junctions or a few junctions coupled in series to form SET transistors, turnstiles, or pumps (see Chaps. 2–5), we will concentrate on properties specific to 1D arrays of many junctions. 2D arrays are treated separately in Chap. 8.

We will restrict the treatment to the case of weak Josephson coupling ($E_J \ll E_c$, where E_J is the Josephson coupling energy and E_c the Coulomb energy, $e^2/2C$ (for charging the capacitance C), and large normal state resistance, $R > R_K$ ($R_K = h/e^2 \approx 25.8$ kΩ)). In other words, we operate in the lower left corner of the Schmid diagram with E_J/E_c and R/R_K as parameters [1]. The influence of superconductivity and Josephson coupling effects will only be commented upon shortly in this chapter.

A specific property of 1D arrays is the occurrence of charge "solitons" and "anti-solitons". If a single electron is added (or subtracted) to (from) an intermediate island inside the array, the surrounding tunnel junction capacitances are charged as well. If the self capacitances (stray capacitances) of the islands were zero, each junction would be charged an equal amount. With non-zero self capacitances (as in a real situation), the charge has to be shared between the junction capacitance and the self capacitance. Therefore the charge on the junctions will decay with the distance from the initially charged electrode so that the polarization and the charge distribution will be localized in space over a few junctions. The whole charge distribution moves as a single electron is tunneled between islands. The reason for calling this object a charge "soliton" is that it does not change its form as it moves, provided it is far from the edges of the array. We can compare with the situation in long Josephson junctions where the magnetic field may tread the tunnel barrier and form flux solitons.

Tunneling in 1D arrays of junctions has been treated theoretically in Refs. [2]-[5]. We give a short theoretical background in Sec. 2. The discrete model of Refs. [2]-[4] is described and the most important features are derived. $I - V$ characteristics for arrays are presented and discussed in Sec. 3 as well as experiments that indicate time and space correlation of single electron tunneling events. Time correlation is discussed further in Sec. 4 and some future experiments are suggested. Descriptions of the fabrication procedure and measurement methods can be found in Refs. [6]-[9].

2. Theoretical background

To describe a 1D array of N junctions, we need to specify the capacitance C and resistance R of each junction and the self capacitance C_0 of each of the interconnecting electrodes. Here we will follow the model developed in Ref. [2]. For simplicity the array is assumed to be homogeneous. By neglecting capacitive coupling between next nearest neighbors and higher order terms we get the coupling capacitance between different electrodes i and j :

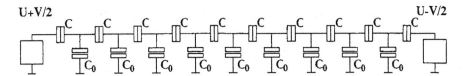

Figure 1. Schematic drawing of a 1D array. The array is biased with a differential voltage V and a common mode voltage U. C denotes the junction capacitance and C_0 denotes the self capacitance of each electrode. The array is assumed to be homogeneous so that all junction capacitances are equal and all self capacitances are equal.

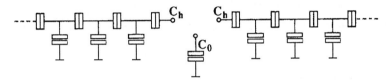

Figure 2. The components of the effective capacitance $C_{\mathrm{eff}} = C_0 + 2C_h$. C_h is the capacitance of a half infinite array.

$$C_{ij} = \begin{cases} C_0 & \text{if } i = j \\ C & \text{if } |i - j| = 1 \\ 0 & \text{if } |i - j| > 1 . \end{cases} \qquad (1)$$

The influence of Josephson coupling is also neglected, i.e., $E_J \ll E_c$. The effect of the superconducting energy gap can however be taken into account by assuming a voltage-dependent quasi-particle resistance. The array is assumed to be voltage biased at both ends with a differential voltage V and a common mode voltage U. Using these approximations some properties of the 1D arrays and of the solitons can be derived analytically. A schematic picture of the array is shown in Fig. 1.

If we place a single electron on the kth electrode of an infinitely long array, a potential φ_k will be generated. This way an effective capacitance is defined, such that $\varphi_k = -e/C_{\mathrm{eff}}$. This effective capacitance is $C_{\mathrm{eff}} = C_0 + 2C_h$, C_h being the capacitance seen from the edge of a half infinite array (see Fig. 2). It is easy to see that $C_h^{-1} = C^{-1} + (C_0 + C_h)^{-1}$ so that for C_h and C_{eff} we get

$$C_h = \frac{1}{2}\left(\sqrt{C_0^2 + 4CC_0} - C_0\right) \qquad (2)$$

and

$$C_{\mathrm{eff}} = \sqrt{C_0^2 + 4CC_0} . \qquad (3)$$

In the experiments which will be described in this chapter the self capacitance of each electrode is small ($C_0 \ll C$) so that the effective capacitance is about $C_{\mathrm{eff}} \approx \sqrt{4CC_0}$.

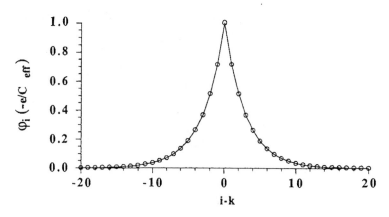

Figure 3. The potential distribution created by a single electron placed on an electrode k inside an infinite 1D array, i.e., a charge soliton. The "soliton length" M is chosen to 3. The rings represent the potentials of the electrodes, the solid line is just a guide for the eye.

The potential of an arbitrary electrode i as a function of the distance from the charged k^{th} electrode can be calculated. We get

$$\varphi_i = -\frac{e}{C_{\text{eff}}} \left(\frac{C}{C + C_0 + C_h} \right)^{|i-k|} \tag{4}$$

which can be written in an exponential form

$$\varphi_i = -\frac{e}{C_{\text{eff}}} e^{-|i-k|/M} . \tag{5}$$

In other words the potential falls of exponentially with a characteristic fall off length M which can be expressed as

$$M^{-1} = \ln \left(\frac{C_{\text{eff}} + C_0}{C_{\text{eff}} - C_0} \right) . \tag{6}$$

If the junction capacitance is of the order of or larger than the self capacitance of the electrodes, we can use the much simpler expression

$$M \approx \sqrt{\frac{C}{C_0}} \quad \text{if } C > C_0 . \tag{7}$$

For the case of a single electron on the kth electrode there is a potential distribution with a maximum value of $\varphi_k = -e/C_{\text{eff}}$. The potential falls off exponentially on both sides as illustrated in Fig. 3. If an electron tunnels from electrode k to electrode $k + 1$, this potential distribution will move one step to the right. Since this potential distribution does not change its form as it moves we may call it a soliton. The soliton extends over approximately $2M$ junctions. Typically in the experiments which will be described M is of the order of 3-5. If we remove an electron from an electrode, we will in the same way create an antisoliton with the same form but with opposite sign. The energy E_S of a soliton or antisoliton is

$$E_S = \frac{e^2}{2C_{\text{eff}}} . \tag{8}$$

Solitons of equal sign repel each other. A soliton and an antisoliton attract each other and may, if they meet, annihilate. As a soliton approaches the edge of the array the form changes. An unbiased edge of the array can be described as a mirror, in the sense that a soliton at electrode k near the edge acts as if the array was infinite and an antisoliton was sitting at electrode $-k$. This means that a soliton (or an antisoliton) is attracted to an unbiased edge. The interaction energies between solitons (antisolitons) and edges, biased or unbiased, have been worked out in Ref. [2]. If a voltage is applied to the edge of a long array $(N \gg M)$ the electric field will penetrate the distance M into the array. If a sufficiently large voltage is applied a single electron will eventually tunnel into the first electrode. In other words a soliton is injected into the array. We can easily use the interaction energies to calculate the voltage needed to inject a soliton. This threshold voltage is

$$V_t = \frac{e}{2C_{\text{eff}}}(1 + e^{-1/M}) = \frac{e}{C_{\text{eff}} + C_0} \ . \tag{9}$$

As V_t is exceeded, solitons start entering the array and since they repel each other they push each other down the array giving rise to a current. If the array is biased symmetrically, solitons are injected from one side and antisolitons from the other and they annihilate in the middle of the array. In our case when the array is biased symmetrically and $C \gg C_0$ the current switches on at

$$V_S = 2V_t \approx \frac{e}{\sqrt{CC_0}} \ . \tag{10}$$

If the array is short $(N < 2M)$ the tails of the electric field from the two edges will interact thereby lowering the V_S.

This situation is very similar to the situation in a current biased Josephson junction. Here the current creates a magnetic field which penetrates into the junction the distance λ_J from both sides (λ_J being Josephson's penetration depth). If the junction width is of the order of $2\lambda_J$, i.e., if the tails of the magnetic field are non-vanishing in the center, a flux quantum can more easily be transferred. Thus the critical current is lowered.

In the 1D array the threshold voltage can be suppressed if a common mode voltage U is applied, since the solitons start to enter the array as soon as either $|U+V/2|$ or $|U-V/2|$ gets larger than V_t. It was shown in Refs. [2] and [4] that also the $I - V$ characteristic above the threshold is affected by a common mode voltage. This can be interpreted as C-SET transistor behavior, i.e., space correlation. In the case of asymmetric bias [4] the average potential U changes continuously as V is changed and a Coulomb-staircase-like feature appears in the $I - V$ characteristic.

The dynamics of a 1D array cannot be calculated analytically but has to be simulated numerically. The methods are described in Refs. [2]-[4]. A very important result of these simulations is that the tunnel events are time correlated inside the array despite the voltage bias. This can be understood as follows: As one soliton has entered the array the probability for an other soliton to enter gets very low. The repelling force between the biased edge and the soliton drives the soliton into the array and then, the next soliton can enter the array. The solitons line up in a moving, quasi 1D Wigner lattice inside the array, pushing each other further into the array.

The frequency spectra of the potential of one of the electrodes was obtained in Ref. [2]. These calculations clearly show peaks at the frequency I/e, with a line width of about 30 %. As the current is increased above e/RC, the time correlation is destroyed, i.e., the peak in the frequency spectra disappears.

It was also shown [2] that if the array was irradiated with microwaves phase locking between the SET-oscillations and the external microwave frequency should occur. This should manifest itself as steps in the $I-V$ characteristics at currents $I = nef$, (where $n = \pm 1, \pm 2, \ldots$). The numerical simulations show that it is important to keep the current low compared to the characteristic current $I_0 = e/RC$ of the individual junction in order to observe steps in the $I-V$ characteristic, i.e., $\omega RC \ll 1$.

To observe these steps the solitons should be "long", i.e., $2M \gg 1$ so that they interact with each other at a fairly long distance. Furthermore the array should be long compared to the soliton. Thus we get, $N > 2M \gg 1$.

For high voltages the usual offset voltage should be observed. The current reaches the asymptotic value $I = (V - V_{\text{off}})/NR$. The offset voltage scales linearly with the number of junctions so that

$$V_{\text{off}} \approx N\frac{e}{2C} .$$

(11)

3. Experiments

This experimental section is divided into five different subsections. In each subsection experimental data for 1D arrays are shown and the results are compared to numerical simulations. In the first subsection, general properties of the $I-V$ characteristics and its dependences on temperature, magnetic field, and microwave irradiation are described. In the other four subsections, four separate experiments related to the time and space correlation of the tunnel events, the influence of environment, and an application in the form of charge controlled transistor, are presented. The figures shown in this section are all from measurements on arrays of 15 to 53 aluminum junctions. They were fabricated using the Dolan "hanging bridge" technology. The size of the junctions were typically $0.006 - 0.01\ \mu\text{m}^2$ and the superconducting transition temperature of the aluminium electrodes was between $1.20\,\text{K}$ and $1.25\,\text{K}$. The superconductivity (and the Josephson coupling) could be suppressed in some of the experiments by a magnetic field.

3.1. General properties, I-V curves

Several experiments have been performed to test the properties of 1D arrays of ultrasmall tunnel junctions [6]-[9]. Mainly aluminum has been used as junction material but also arrays of tin junctions have been measured [9].

The $I-V$ characteristics of all arrays show the offset voltage V_{off} which is typical to the Coulomb blockade. The experimental values of V_{off} agree well with those estimated from the capacitances of the junctions. This was deduced from the estimated area using a specific capacitance of 40-45 fF/μm^2. Furthermore, there is a very sharp threshold voltage V_S below which virtually no tunneling occurs. Both these characteristic features are shown for a 15 junction array and a 53 junction array in Fig. 4. Note that the

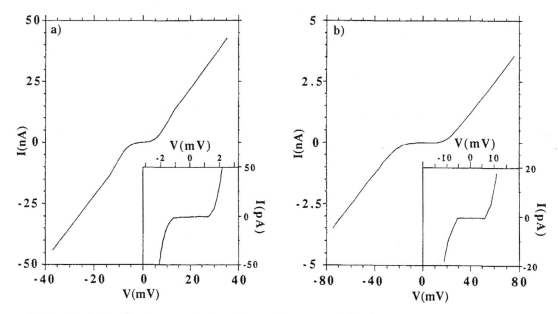

Figure 4. Typical $I - V$ characteristics of two 1D arrays. The inserted figures show magnifications of the low bias regions. $T \leq 90\,\mathrm{mK}$. (a) A 15 junction array with the individual junction resistance $R \approx 53\,\mathrm{k\Omega}$ and $V_\mathrm{off} \approx 280\,\mu\mathrm{V}$. Note the smeared superconducting gap structure, which is seen as a change of slope at about $12\,\mathrm{mV}$. (b) A 53 junction array with the individual junction resistance $R \approx 280\,\mathrm{k\Omega}$ and $V_\mathrm{off} \approx 415\,\mu\mathrm{V}$. From Ref. [8].

current below the threshold is less than $50\,\mathrm{fA}$ and that the differential resistance below V_S is greater than $200\,\mathrm{G\Omega}$. These numbers are limited by the leakage resistance and the noise of the measurement system. Some of the arrays that have been investigated showed a larger V_S than expected from theory (Eq. (10)). However, this might be explained by inhomogeneities in the array or by tunneling through localized states in the barrier (see the discussion in Ref. [8]).

It seems as if the offset voltage decreases when the individual junction resistance decreases and becomes close to or lower than R_K [8, 11]. The reason for this is probably that the quantum fluctuations due to the junction resistance (which are of the order of $\hbar/RC$) become larger than the charging energy, thereby smearing the state of definite charge. Furthermore the junctions become less well isolated from the environment, since the neighbors are more low resistive. Therefore the quantum fluctuations due to the environment also become larger.

The superconductivity of the electrodes affects the system in two ways. Firstly the Josephson coupling gives a finite E_J/E_c value. However, the arrays which will be treated here, all have a low E_J/E_c ratio and therefore this effect is small. Secondly the superconducting energy gap 2Δ, gives rise to a voltage-dependent resistance. When an electron tunnels the junction voltage is of the order of $e/2C$ (which in these junctions is close to $2\Delta/e$), or higher. The voltage at which an electron tunnels will therefore sometimes be below and sometimes be above $2\Delta/e$. Thus, the superconducting gap structure in the $I - V$ characteristic of the whole array will be smeared, and the gap will not be visible if $e/2C > 2\Delta/e$. This can be seen by comparing the two arrays in Fig. 4, where a smeared superconducting gap is observable in Fig. 4a ($e/2C = 280\mu\mathrm{V}$) but not in Fig. 4b ($e/2C = 415\mu\mathrm{V}$).

The temperature dependence of a 19 junction array is shown in Fig. 5. The thresh-

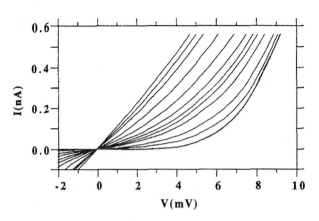

Figure 5. The temperature dependence of the $I - V$ characteristics of a 19 junction array. Curves are shown for the temperatures 0.06, 0.12, 0.25, 0.50, 0.71, 0.82, 0.94, 1.02, 1.07, 1.13, 1.22, 1.30, 1.43, 1.52, and 1.83 K. The Coulomb blockade feature seen at low voltages is gradually smeared as the temperature is increased. Note that the curves for $T \leq 0.5\,\text{K}$ coincide, giving one thick curve. $R \approx 210\,\text{k}\Omega$ per junction, $V_{\text{off}} \approx 370\,\mu\text{V}$ per junction. From Ref. [8].

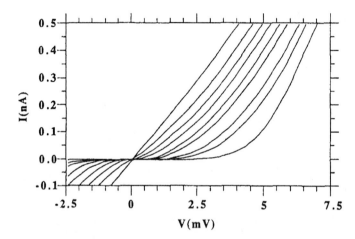

Figure 6. dc $I - V$ curves for the same 19 junction array as in Fig. 5 under microwave irradiation. Curves are given for no microwave power and for increasing power: -20, -17, -15, -14, -13, -13, -12, -11, and -10 dBm. The power level is referred to the room temperature end of the coaxial cable. T=50 mK, f=687 MHz. From Ref. [8].

old is washed out already at $\sim 0.5\,\text{K}$, but the large scale $I - V$ characteristic is unaffected by temperature below $\sim 0.5\,\text{K}$. As the temperature is increased further the $I - V$ characteristic is gradually smeared. At temperatures well above $e^2/2k_BC$ the $I - V$ curve is practically linear with the normal resistance slope. Traces of the offset voltage are observable up to fairly high temperatures, $\sim 10\,\text{K}$.

The $I - V$ characteristics can be smeared, not only by temperature, but also by applying microwaves to the array. In Fig. 6 the $I - V$ characteristic of the 19 junction array is shown for different values of the microwave amplitude. The reduced current at low voltages is gradually increased as the amplitude of the microwaves is increased.

The cause of the smearing is the same in both cases. As an electron tunnels it can absorb energy from the thermal fluctuations or from the microwave field to overcome the Coulomb blockade. The blockade of tunneling for low voltages is therefore no longer

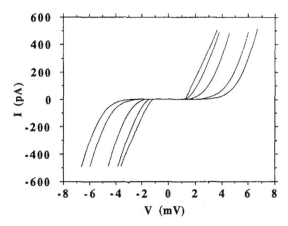

Figure 7. dc $I - V$ curves for a 25 junction array at different magnetic fields. Curves are given for fields of 0, 0.11, 0.19, 0.26, and 0.33 T. $T = 50$ mK, $R \approx 60$ kΩ per junction, $V_{\text{off}} \approx 440\,\mu$V per junction.

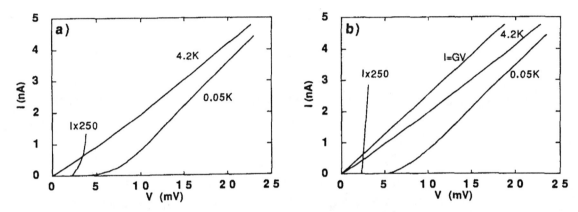

Figure 8. Experimental dc $I - V$ curves for the 19 junction array in Fig. 5 compared with numerical simulations. Curves are given for both 50 mK and 4.2 K. (a) Experimental curves. A blow-up of the current scale for the 50 mK curve shows that the current is practically zero at low bias below $V_t \approx 2$ mV. (b) Numerical simulations with the parameters: $C_0 = 1.2 \times 10^{-17}$ F, $C = 2.4 \times 10^{-16}$ F, $R = 210$ kΩ per junction, and the superconducting gap $2\Delta = 0.38$ meV. From Ref. [10].

effective if the amplitude of the thermal fluctuations or the microwave field is large enough.

As a magnetic field is applied to the array, superconductivity is suppressed. This suppression decreases Josephson coupling energy E_J. However, this fact does not alter the properties of the arrays shown here since we are already in the weak coupling limit ($E_J \ll E_c$). Array properties of samples in the strong coupling limit have only been investigated briefly [9, 11]. Furthermore, the superconducting gap is decreased and eventually it vanishes, which leads to the disappearance of the smeared gap structure in the $I - V$ characteristic, which is seen for some of the arrays. The largest effect of the magnetic field on the $I - V$ characteristics is observed close to the threshold voltage. In Fig. 7 the $I - V$ characteristics of a 25 junction array is shown for several different values of the magnetic field. As Δ/e is suppressed below $e/2C$ the tunneling probability is increased and therefore the differential resistance above the threshold is increased.

The experimental $I - V$ characteristic of a 19 junction array is compared to numerical simulations for two different temperatures in Fig. 8. The numerical simulations were performed by Likharev et al. as described in Ref. [2]. The parameters used for this simulation were: normal state resistance $R = 210$ kΩ (taken from the large voltage asymptote of the $I - V$ curve); estimated values for capacitance $C = 0.24$ fF and stray capacitance $C_0 = 0.012$ fF; the superconducting gap $2\Delta = 0.38$ meV and the quasi particle resistance $R/R_{qp} = 0.15$. The simulations of this array show good agreement with the experimental data as can be seen in Fig. 8. The large scale $I - V$ curves show very good agreement for both 50 mK and for 4.2 K. There is a small discrepancy between experiment and simulation just above the threshold voltage of the 50 mK curve. This discrepancy is probably due to the crude assumption of a voltage independent and fairly low quasiparticle resistance below the superconducting gap. From more recent measurements on single junctions we know that R/R_{qp} is substantially smaller than 0.15 (rather 0.005) and voltage dependent. This is in agreement with the magnetic field dependence shown in Fig. 7. The onset of current is smooth with no magnetic field but gets sharper as the magnetic field is increased and hence the R/R_{qp} ratio is increased. However, the assumption of a too low R/R_{qp} ratio does not make a large difference for the large scale $I - V$ curve, since most of the tunneling events takes place at voltages larger than $2\Delta/e$.

3.2. Time correlation of tunnel events

As was discussed in Sec. 2, it should be possible to get time correlation of the tunnel events in a 1D array even if the array is voltage biased. It should also be possible to phase-lock the resulting SET-oscillations to external microwaves giving flat voltage steps in the $I - V$ characteristic. In several experiments [8, 10, 12] a number of 1D arrays with different number of junctions were measured to test these predictions.

It turns out that the way in which the microwaves are coupled to the array is essential for the outcome of the experiments. Therefore we start by discussing the microwave coupling. In the first experiments, the microwaves were coupled inductively to the arrays. A three turn coil terminated the microwave coaxial cable close to the substrate (see Fig. 9a). The diameter of the coil was about 3 mm. The size of the arrays were of the order of 2 μm, i.e., much smaller than both the coil and the wavelength of the radiation. This kind of coupling was satisfactory only at certain resonance frequencies for which radiation was coupled into an array without heating up the whole substrate holder. On the other hand, the poor coupling of other frequencies effectively prevented broad band noise from reaching the arrays.

In more recent experiments a capacitive coupling via very small chip capacitors was used. Small chip capacitors were soldered between the microwave coaxial cable and the spring-loaded contact pins which were used to make electrical connection to the substrate. This should give a microwave modulation of the differential voltage V (see Fig. 9b).

However, there will also be a modulation of the common mode voltage (see Fig. 9c). This is due to stray capacitances on the chip. In one of the measurements, one of the contact pins did not make contact to the substrate so that the array had to be biased through another junction on the chip. To summarize, with the capacitive coupling there is a mixture of differential mode (V) and common mode (U) modulation. Since there are many sources to the asymmetry, the ratio between the U and the V modulation is hard to estimate.

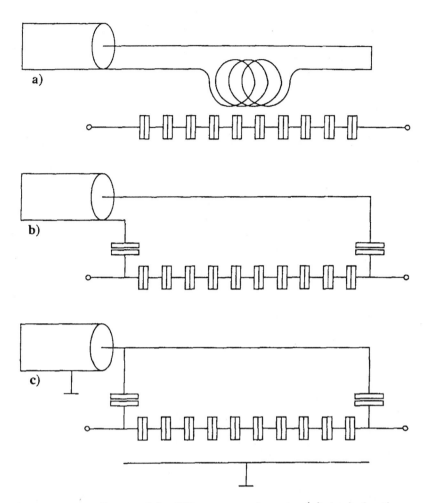

Figure 9. The microwave coupling used in different experiments. (a) An inductive coupling consisting of a three turn coil, giving a differential mode modulation. (b) A capacitive coupling giving a differential mode modulation. (c) A capacitive coupling giving a common mode modulation. The coupling capacitances used were $\sim 100\,\text{pF}$.

The first experimental evidence of SET-oscillations was found [8, 10] in two arrays of 15 and 19 junctions, where an inductive coupling was used. The differential resistance was measured as a function of microwave irradiation. The phase-locking of the SET-oscillations to the external microwaves was observed as small peak structures in the differential resistance. These peaks occurred at currents corresponding to multiples of ef_{ext}. First and second order peaks were observed for both positive and negative currents as can be seen for the 19 junction array in Fig. 10. The peaks were observed for frequencies in the range of 0.7–5 GHz. The peaks were more pronounced at lower frequency. No data could be extracted at frequencies below 0.6 GHz, since the microwave coupling became very poor at low frequency. All peaks showed good agreement with the predicted $I = nef_{\text{ext}}$, with $n = \pm1, \pm2$. The peaks did not change their locations in current as the amplitude of the microwaves was altered. However the location in voltage changed as a function of microwave amplitude. This clearly shows that the peaks were not related to a Josephson effect or photon assisted tunneling. In these experiments the samples were in the superconducting state. Numerical simulations of the microwave irradiated 19 junction array (by Likharev et al.) are shown in Fig. 10b. The parameters

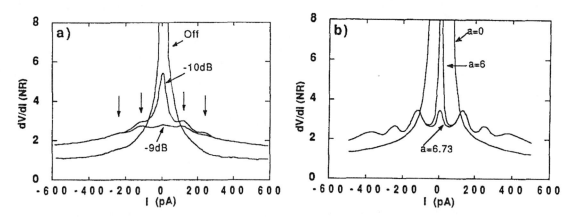

Figure 10. The differential resistance as a function of current for the same array as in Fig. 5 for three values of the microwave power. It is normalized to the array resistance NR. The microwaves were inductively coupled to the array, $T = 50$ mK, $f = 750$ MHz. (a) Experimental curves. (b) Calculated ones. The same junction parameters as in Fig. 8 were used. The difference in the microwave amplitude, a, between the two pumped curves in (b) corresponds to 1 dB in power, i.e., a similar difference as in (a). Arrows show nominal positions of the resistance peaks as expected from $I = nef$. From Ref. [10].

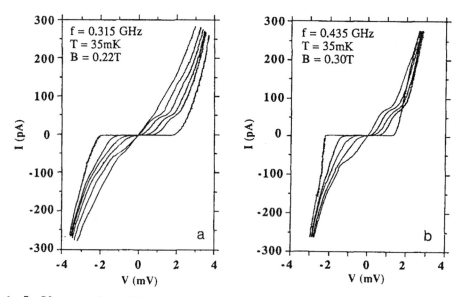

Figure 11. dc $I - V$ curves for a 25 junction array under microwave irradiation at 35 mK. The different curves in each figure are for different microwave amplitudes. The microwaves were capacitively coupled to the array. The steps are clearly seen directly in the $I-V$ curve. $R \approx 64$ kΩ per junction, $V_{\text{off}} \approx 360\,\mu$V per junction. (a) $f = 315$ MHz, $B = 0.22$ T. (b) $f = 435$ MHz, $B = 0.30$ T.

used were the same as for the simulation shown in Fig. 8b. The experimental peaks are somewhat more smeared than in the simulated curve.

A number of changes were made in order to investigate the effect further. As mentioned above, the microwave coupling was changed from an inductive to a capacitive one. A small magnet was also mounted in the cryostat to be able to suppress superconductivity in the arrays. The direction of the magnetic field was perpendicular to the substrate. The new measurements gave substantially sharper steps in the $I - V$ characteristics when microwaves were applied. Figure 11 shows the $I - V$ characteristic of a microwave irradiated 25 junction array with junction parameters $R \approx 50$ kΩ, $C \approx 0.25$ fF. Figures 11a and 11b are for different magnetic field and different frequencies. The available

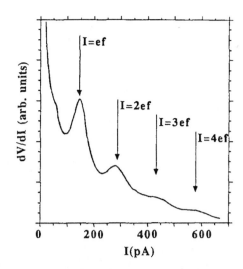

Figure 12. The differential resistance as a function of current for the 25 junction array in Fig. 7. The array is irradiated by microwaves at 900 MHz. Up to fourth order peaks are observable. The peaks fall close to their expected values $I = nef_{\mathrm{ext}}$. Note that there is even a sign of a subharmonic peak at ~ 70 pA. $T = 50$ mK, $B = 0.26$ T.

magnetic field (0.3 T) was not quite enough to suppress superconductivity completely. The critical magnetic field of the superconducting thin films was much higher than the bulk value since the dimensions of the array were of the order of London's penetration depth. The different curves in each figure are for different microwave amplitudes. Steps occur directly in the $I - V$ characteristic at currents corresponding to $I = nef_{\mathrm{ext}}$. The curves were not always symmetric. This was caused by a non-symmetric measurement. The array was in this case measured through a single junction so that the bias circuit was not symmetric.

Much sharper steps have been reported for turnstiles [13] and pumps [14] discussed in Chap. 3. It should be noted, however, that the frequency in the array experiments was of the order of 10-100 times higher than for the turnstiles and the pumps. The steps are expected to be smeared at higher frequency.

In the derivative measurements for an array, up to fourth order peaks could be observed, as can be seen in Fig. 12. The stability against thermal fluctuations was very good. The steps could be clearly observed even up to 0.5 K although they were a bit smeared compared to the low temperature curves (see Fig. 13). The stability against thermal fluctuations is an important feature for future applications. Much would be gained if the operating temperature of a future current standard could be raised above 1 K.

A gate electrode was incorporated on the chip a few micrometers from the array. The $I - V$ characteristic of the array could be altered by changing the voltage on this gate electrode, this will be discussed in the next subsection. The microwave induced steps could also be affected by the gate voltage. A set of $I - V$ curves for an irradiated array is shown in Fig. 14. The different curves are for different gate voltages. The sharpness of the different steps change with changing gate voltage.

The steps are observed directly in the $I - V$ characteristic in the new measurements. This is in contrast to the first experiments where the effect was observed in the derivative curves only. There are several reasons for the improved results. One reason is that the applied magnetic field suppresses superconductivity in the electrodes, thereby giving a

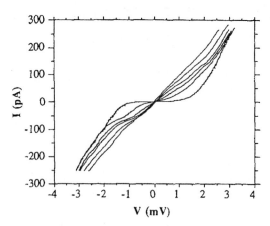

Figure 13. dc $I - V$ curves of the same 25 junction array as in Fig. 11 at 0.5 K. The array is irradiated with microwaves at the frequency 427 MHz. The different curves are for different microwave amplitudes. The steps are clearly observable even at this high temperature. However, they are smeared compared to the steps at low temperature. $B = 0.22$ T.

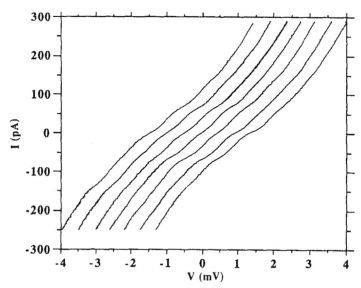

Figure 14. dc $I - V$ curve of the same 25 junction array as in Fig. 11 irradiated with microwaves at the frequency 435 MHz. The different curves are for different voltages applied to an electrode placed $\sim 5\,\mu m$ from the array. Curves are given (from the left) for $V_g = 0, 1, 2, 3, 4, 5$, and 6 mV. The sharpness of the different steps change as a function of the gate voltage. $T = 50$ mK, $B = 0.22$ T.

much higher tunnel conductance at voltages below $2\Delta/e$ (compare Fig. 7). The effective resistance seen by the tunneling electrons is therefore lower. This gives a lower ωRC product which is favorable for observing the steps. If no magnetic field was applied, the steps were only observable in the derivative measurements.

Another reason for the improvements is that a mixture of common mode and differential mode modulation was used for the microwaves. Preliminary numerical simulations [12] show that the relatively sharp steps which were observed could not be explained if only differential modulation was assumed. If, on the other hand, a mixture of the two modulation types were assumed the simulations could be made to agree with the observations. Furthermore, slightly longer arrays with somewhat lower resistances were

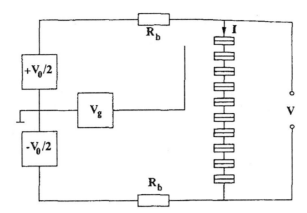

Figure 15. A schematic drawing of the measurement circuit for the capacitively coupled array. The gate was not placed symmetrically with respect to the array.

used in the second set of experiments. This should also give a moderate increase of the step sharpness.

3.3. Space correlation of tunnel events

By space correlation we mean the correlation of tunnel events in different junctions; in other words that a tunnel event in one junction causes a tunnel event in a neighboring junction. The charge on the electrode between the two junctions changes due to the first tunnel event, and if the stray capacitance of the electrode is not too large that change in charge is sufficient to make tunneling in the next junction favorable. This has been described in Chap. 2 for double junctions. In a 1D array there is a kind of avalanche effect the tunnel event in the second junction triggers a tunnel event in the third junction and so on. In the soliton picture this can be thought of as a soliton just having entered the array being pushed further into the array by the interaction with the biased edge. When the soliton has entered far ($k \gg M$) into the array, the soliton stops (since the interaction with the soliton weakens) unless a new soliton enters the array.

The average charge on the electrodes between the junctions in a 1D array changes as the bias voltage is changed. Therefore there is a Coulomb staircase like structure in the $I - V$ characteristics of these arrays, if any kind of asymmetry is present [3, 15]. This asymmetry can be due either to the bias circuit or to inhomogeneities in the array parameters. If the system is symmetric and the array is homogeneous the $I - V$ characteristic should be smooth. The situation is the same as for a double junction where a Coulomb staircase is observed if the two junctions are different.

The charge on the electrodes of an array can be modulated by an external gate electrode in the same way as in a double junction system. This means that SET transistors can be realized where the drain-to-source path is an array of many junctions instead of only two as for a double junction SET transistor. This kind of transistor was realized in Ref. [15] where a capacitively coupled gate electrode was used to modulate a 13 junction array. A schematic picture of the circuit is shown in Fig. 15.

The $I - V$ characteristic showed a clear Coulomb staircase which could be shifted by changing the gate voltage. The $I - V$ curve of the array is shown for two different values of gate voltage in Fig. 16. The temperature in this case was 1.3 K. The array

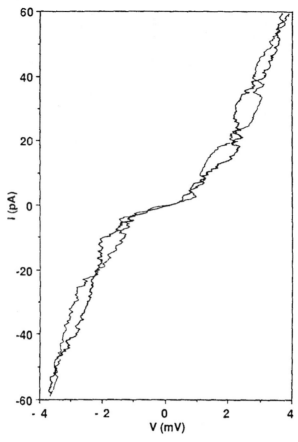

Figure 16. dc $I - V$ curves of a 13 junction array at 1.35 K. The thick and thin curves are given for $V_g = 0$ and 12 mV respectively. This figure illustrates the sensitivity to changes in the background charge caused by mechanical vibrations. The total resistance of the array was about 30 MΩ and the offset voltage of the whole array was about 1.7 mV. From Ref. [15].

voltage could be measured as a function of the gate voltage, by fixing the current. The array voltage changed periodically as a function of gate voltage. This is shown in Fig. 17. The periodicity in gate voltage was about 27 mV. This corresponds to a change of charge on an intermediate electrode of one electron charge. A voltage gain of 0.2 was determined. Biasing at the point of maximum gain and measuring the noise density at 10 Hz, a charge sensitivity as high as $2 \times 10^{-4} e/\sqrt{\text{Hz}}$ could be deduced. A fair agreement between experiment and numerical simulations was found assuming that the array was homogeneous and that the Coulomb staircase was due to asymmetry in the bias circuit. An even better agreement was found if one junction was assumed to be much smaller than the others. This interpretation is also in agreement with the unusually high resistance of that sample.

Recent experiments have shown that it is possible to alter the $I - V$ characteristics substantially also in a homogeneous array by changing the gate voltage. The circuit was the same as for the previous array. Figure 18 shows the $I - V$ characteristics of a homogeneous array (one of the arrays that were used in the time correlation experiments). The different curves are for different gate voltages. The $I - V$ curves are not symmetric around the origin. This is due to two reasons. The gate electrode was not placed symmetrically with respect to the array. Furthermore, the array had to be biased through a single high resistive junction giving an asymmetric bias.

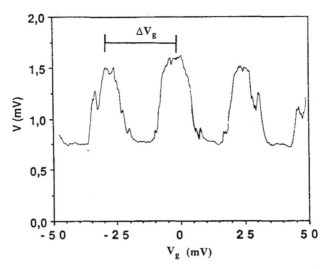

Figure 17. The response to the gate voltage is illustrated for the 13 junction array shown in Fig. 16 at $T = 1.3\,\text{K}$. The bias current was fixed to about $5\,\text{pA}$ and the voltage over the array is plotted as a function of applied gate voltage. A periodic response is seen, the period corresponds to a change of one electron charge on an intermediate electrode. By measuring the noise level at $10\,\text{Hz}$, biased at the steepest part of the response curve, a sensitivity for change in charge of $2 \times 10^{-4}\,e/\sqrt{\text{Hz}}$ was deduced. The transistor "gain", i.e. the relation between changes in the output and gate voltages, is of the order of 0.2 for this curve. From Ref. [15].

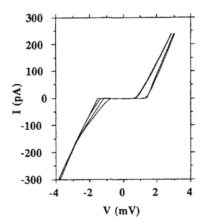

Figure 18. dc $I - V$ curves for the 25 junction array in Fig. 7. The different curves are for different gate voltages. Curves are given for $V_g = 0, 2.5, 8.75$, and $10\,\text{mV}$. The threshold voltage is dependent on the gate voltage. $T = 44\,\text{mK}$, $B = 0.33\,\text{T}$.

3.4. An R-SET transistor using an array as gate resistor

We just discussed how the $I - V$ characteristics of two junctions in series can be altered by changing the charge on the middle electrode between the two junctions. This was done by inducing an extra charge via a gate capacitance in a C-SET transistor configuration. However, there is also another possibility. In an R-SET transistor [16] charge is injected through a resistor to the middle electrode.

In our next experiment to be described [17], a number of R-SET transistors were investigated. Being easier to fabricate than high resistance resistors, arrays of junctions were used as gate resistors. Each device consisted of two nominally equal tunnel junctions

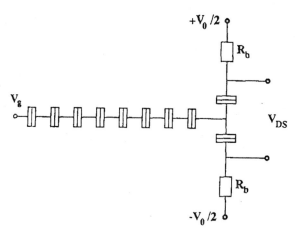

Figure 19. A schematic drawing of the measurement circuit for the resistively coupled SET-transistor. A 1D array of 11 or 13 tunnel junctions is used as gate resistor.

(the drain and source junctions) and a 1D array of 11 or 13 junctions connected to the common electrode between the drain and source junctions, see Fig. 19. The design parameters were the same for all the junctions of each device.

The measured current $I=(I_D+I_S)/2$ was the average of the drain and source currents, and the voltage V_{DS} was the sum of source and drain junction voltages. Three different devices A, B, and C were investigated at temperatures down to 50 mK. Results will be presented mainly for the samples A and B. The data for sample C was very similar to that of sample A.

A set of $I-V$ curves for different gate voltages is shown in Fig. 20 for samples A and B. There are pronounced Coulomb blockades at zero gate voltage for both samples. The current at low bias is practically zero. The offset voltage, V_{off}, at large bias currents is 2–3 times larger in sample B than in A. The gate voltage behavior of sample B (to be discussed below) and a comparison of V_{off} for sample B with V_{off} for other arrays on the chip tell us that the resistances (and to some extent the capacitances) of the drain and source junctions are considerably different. The resistance of one junction is higher than the one of the second junction of the pair and those of the gate junctions. This is in contrast to sample A where the source and drain junctions are nearly equal. Device parameters for the three samples are listed in Table I.

Changes in the gate voltage affect the $I-V$ curves considerably. The Coulomb blockade is gradually smeared and the differential resistance at the origin approaches R_D+R_S, as the gate voltage is increased. For device B, the whole $I-V$ curve is shifted in voltage as the gate voltage is altered. This is due to the fact that the center electrode is not at zero potential, which, in turn, is due to unequal R_D and R_S.

The drain-to-source voltage, V_{DS}, could also be measured as a function of V_g at constant bias voltage, V_o. The response dV_{DS}/dV_g gives the gain of the transistor. A set of curves (for both A and B) is shown in Fig. 21 for a number of bias voltages. For sample A, a maximum voltage gain G_V (slope) of 0.15 is obtained, whereas sample B shows a maximum voltage gain of 0.85($\pm$0.15). There is an additional slope, $dV_{DS}/dV_g \approx 0.06$, superimposed on the response curve for sample B. This is again due to the asymmetry in the device. A real power gain G_P of about 4 was deduced for all the samples.

Numerical simulations of these R-SET transistors [17] show that the non-linearity of the elements due to superconductivity gives $I-V$ characteristics which show similarities

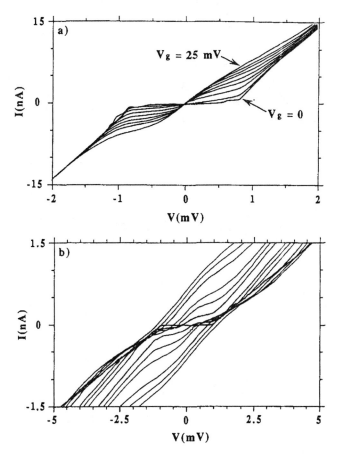

Figure 20. Current vs source-to-drain voltage for the resistively coupled SET-transistors where charge is injected via a gate resistor consisting of a 1D array of tunnel junctions. $T = 50\,\text{mK}$. (a) Sample A, the different curves are for the gate voltages $V_g = 0, 5, 7.5, 10, 12.5, 15, 17.5, 20$, and $25\,\text{mV}$. (b) Sample B, the different curves are for the gate voltages $V_g = 0, \pm 2.5, \pm 7.5, \pm 10, \pm 12.5, \pm 17.5, \pm 20$, and $\pm 22.5\,\text{mV}$. From Ref. [17].

Table I. Parameters of the three R–SET transisters

Sample	A	B	C
$2V_{\text{off}}(\text{mV})$	0.95	2.42	1.38
$R_D + R_S(\text{M}\Omega)$	0.11	1.8	0.33
$R_g(\text{M}\Omega)$	3.0	9.0	3.0
G_V	0.15	0.85	0.40
G_P	4.1	4.2	3.6
$N(\# \text{ gate jcns})$	13	11	13

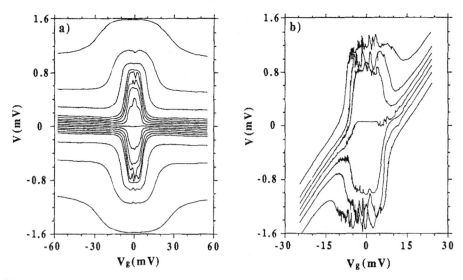

Figure 21. Source-to-drain voltage as a function of gate voltage for different bias currents. $T = 50\,\text{mK}$. (a) Sample A, curves are shown for $V_o = 0, \pm 0.5, \pm 1, \pm 1.5, \pm 2, \pm 2.5, \pm 5, \pm 10$, and $\pm 20\,\text{mV}$, $R_b = 1\,\text{M}\Omega$. A maximum voltage gain of 0.15 and power gain of 4.1 were estimated for this device. (b) Sample B, curves are shown for $V_o = 0, \pm 1, \pm 2, \pm 4\,\text{mV}$, $R_b = 10\,\text{M}\Omega$. A maximum voltage gain of 0.85 and power gain of 4.2 were estimated for this device. From Ref. [17].

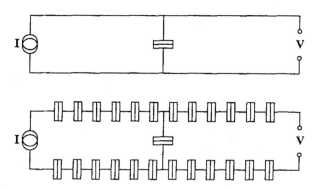

Figure 22. A schematic picture of the measurement circuits for the two junctions that were compared. The "solitary" junction was coupled directly to its measurement leads. The other junction was "protected" from the environment by arrays of similar tunnel junctions.

with the observed $I - V$ curves. However this can not explain the observations alone. If both superconductivity and SET effects are considered the simulations show good agreement with the observations (see Figs. 5 and 6 of Ref. [17]).

3.5. Decoupling of the electromagnetic environment

The environmental impedance is of utmost importance for the occurrence of SET effects, see Chap. 2. A tunnel junction inside an array may be effectively decoupled from the disastrous environment by neighboring junctions.

Here we will compare [18] a single unprotected junction (referred to as the solitary junction), coupled directly to the measurement leads with another single junction protected from the environment by arrays of similar junctions. The two different measurement circuits are shown in Fig. 22. Clear evidence for the decoupling of the environment was observed.

The "protected" junction was measured in a four point configuration with a six junction array close to the junction in each lead. The areas of all junctions were approximately $80 \times 80\,\text{nm}^2$. Both $I - V$ curves and derivative curves were taken at temperatures slightly above the superconducting transition temperature, $T_c \approx 1.2\,\text{K}$, of the aluminum electrodes.

Figure 23a shows the $I - V$ curve of the solitary junction where practically no offset voltage is observed. In the derivative plot there is a small dip in the conductance near the origin. In contrast to this, the protected junction shows a large Coulomb blockade as shown in Fig. 23b. The voltage offset is of the order of $400\,\mu\text{V}$, which corresponds to the estimated size of the junction. The derivative plot shows a clear reduction of the conductance at low voltages, an order of magnitude larger than for the solitary junction. Furthermore, the $I - V$ curve of the junction inside the array has the same form as the $I - V$ curve of the arrays.

This shows that the $I - V$ characteristic of the "protected" junction is much less affected by the environment than that of the solitary junction. The offset voltage is not suppressed in the "protected" junction in contrast to the situation in the solitary junction.

The comparison also made it possible to draw conclusions regarding the "tunneling time" for single electron tunneling events [18]. The problem of tunneling time is addressed in Chap. 1 of this volume.

4. Discussion

The results of the time correlation in arrays will be discussed from a fundamental point of view in subsection 4.1. They will be compared with other experiments. Two possible future experiments will be discussed briefly in subsections 4.2 and 4.3.

4.1. Self correlation

In the experiments described here the SET–oscillations have been observed in an indirect manner. The oscillations themselves have not been observed. Steps are observed in the $I - V$ characteristics at currents $I = ef$ by phase-locking the SET-oscillations to external microwaves. However, the inverse relation $f = I/e$, which would be the direct result of time correlation, has not been observed.

We may define the more strict concept of self correlation: Time correlation of tunnel events present in a system where no external hf-signal is applied. In the experiments with turnstiles [13] and pumps [14] (see also Chap. 3) the relation $I = ef$ has been observed with good accuracy, to about one part in 10^3. Time correlation exists in these experiments but it is induced by the external rf-signal which is applied to the gate. According to theory there is no self correlation of the tunnel events in such systems.

Theory does predict self correlation for the arrays. The experiments show that the tunneling is time correlated. However, one might argue that the observed steps in the $I - V$ characteristic of the arrays does not prove self correlation since external microwaves are applied. In other words the observed time correlation might be induced

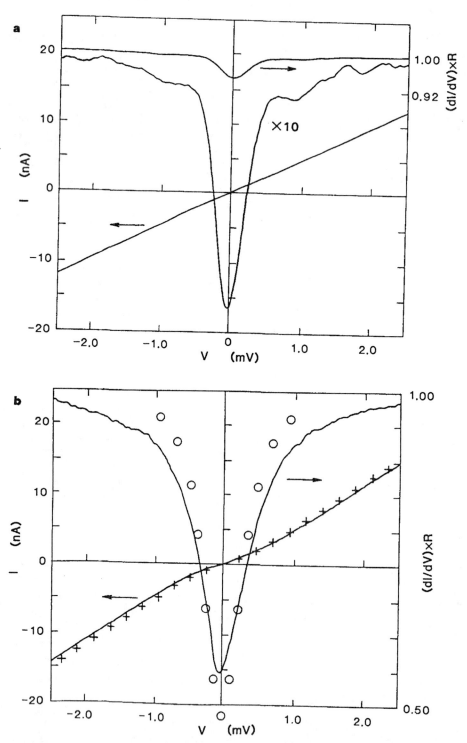

Figure 23. dc $I - V$ curves and differential conductances for two junctions embedded in different environments. (a) The solitary junction. A barely observable offset voltage is seen in the $I - V$ curve and a small ($\sim 5\%$) dip in the conductance is seen in the differential conductance. A blow up of the differential conductance is also shown. (b) The "protected" junction. A large offset voltage ($\sim 400\,\mu$V) is seen in the $I - V$ curve and a $\sim 50\%$ dip in the conductance is seen in the derivative plot. The rings and crosses are points from numerical simulations, for details see Ref. [18] .

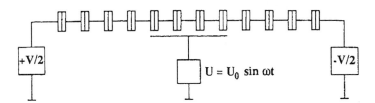

Figure 24. A schematic drawing of an array turnstile. The array is biased symmetrically at both ends and a gate electrode couples capacitively to one or several electrodes at the center.

by the external signal rather than being an inherent property of the array. An experiment where the relation $f = I/e$ is observed would be of great importance.

4.2. Array turnstiles

Comparing the array experiments with the experiments on pumps and turnstiles (Chap. 3 and Sec. 3.2 of this chapter) we note that, according to theory, the tunneling is self correlated in the arrays but not in the pumps and turnstiles. On the other hand the observed steps are much sharper in the pumps and the turnstiles. There are at least two obvious reasons for this. Firstly, the frequency (and therefore the current) was much higher in the array experiments. For the array experiments (Fig. 11) the ωRC product is about 0.03 whereas in Ref. [13] $\omega RC \approx 0.005$. A high ωRC product should smear the dc steps in the $I - V$ characteristic. Furthermore the common mode modulation or gate modulation seems to be superior to the differential mode modulation.

From these observations we can try to make better devices with sharper voltage steps. One possible suggestion would be an array which is modulated by a gate at the center (see Fig. 24). This can also be considered as a turnstile with a larger number of junctions. This device should have the advantage over the ordinary array that the common mode modulation could be fully used. The coupling between the gate and the different electrodes could be chosen individually for each electrode. Compared to the turnstile this device has the advantage of a larger number of junctions, which should decrease the amount of quantum leakage (see Chaps. 6, 9, and Refs. [7, 19]). The problem with background charge could possibly also be reduced. The background charge weakens the Coulomb blockades in the different junctions in a random way, and an electron (soliton) entering the array will see an average of the Coulomb blockades. This average will differ less from array to array if the number of junctions is large and therefore the device might be less sensitive to background charge. If this speculation is true, it will be easier to couple several of these devices in parallel, and thereby increasing the current.

4.3. Coupled arrays

Another experiment for the future is to investigate two parallel arrays, which are isolated from each other, but where each electrode of one array is capacitively coupled to an electrode in the other array. This has been treated theoretically in Ref. [20] for the case of three junctions in series and is discussed in Sec. 4 of Chap. 6.

If we look at longer arrays we can represent the arrays as shown in Fig. 25. As a soliton enters one array it will become very favorable for an antisoliton to enter in the

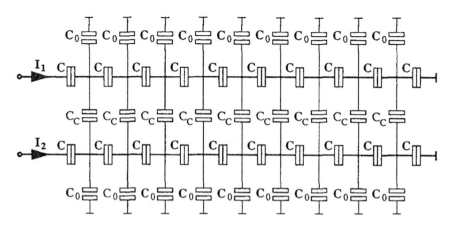

Figure 25. A schematic drawing of two arrays capacitively coupled to each other. C_C is the coupling capacitance between an electrode in one array and the corresponding electrode in the other array. If $C_C \approx C$ it should be possible to lock the currents in the two arrays so that $I_2 = -I_1$.

other array. This is true if the coupling capacitance between the arrays is fairly large, i.e., if $C_C \approx C$. As a current flows in one array the train of solitons in that array can phase-lock to a train of antisolitons in the other array giving rise to an equally large current of opposite sign, i.e., we have made a Quantum Current Mirror.

In a real experiment this could be proved by fixing a current I_1 in the first array and measuring the $I - V$ characteristic of the second array. A voltage step should occur at $I_2 = -I_1$. It might also be possible to lock a train of solitons in one array 180 degrees out of phase with a train of solitons in the second array. This means locking one soliton in array one between two solitons in array two. However, this kind of locking should be weaker than the one in phase between solitons and antisolitons. A variant of this coupled array experiment would be to allow tunneling between the arrays.

Using these soliton locking effects, it might be possible to series connect one side of a large number of these devices and to couple the opposite side of these devices in parallel. In this way we would get a Quantum Current Transformer, where a small, well defined current could be transformed to a larger (and more usable) current.

5. Conclusions

In this chapter we have discussed the properties of 1D arrays of ultrasmall tunnel junctions. We have reviewed experiments on 1D arrays of tunnel junctions, with high resistance and low capacitance, in a limit where the charging effects dominate over both the thermal energy and the Josephson coupling energy.

The offset voltage at high bias has been observed and is in good agreement with the estimated size in practically all samples. Another feature of the Coulomb blockade, the threshold voltage, has been observed at low temperature. The current below the threshold is zero within the accuracy of the measurement.

Time correlated single electron tunneling has been observed by phase-locking the SET-oscillations to external microwaves in the frequency range 0.3–5 GHz. First and second order steps were observed directly in the $I - V$ characteristics and up to fourth order peaks were observed in derivative measurements. The currents at which the steps occur agree well with the postulated $I = nef_{ext}$. The ωRC product and the way in which the microwaves are coupled to the array is found to be important. A lower ωRC product and combination of common mode modulation and differential mode modulation gives sharper steps than a pure differential mode modulation.

Space correlation of single electron tunneling in 1D arrays has been demonstrated in two different ways. A Coulomb staircase structure has been observed in an inhomogeneous 13 junction array. Transistor effects have been obtained in both homogeneous and inhomogeneous arrays. A voltage gain of 0.2 and a charge sensitivity of $2 \times 10^{-4}e/\sqrt{\mathrm{Hz}}$ have been achieved using capacitive coupling.

A number of R-SET transistors have been fabricated where a 1D array of junctions was used as the resistive element. It was shown that the 1D array can be used as a resistor for quasi-continous transport of charge. A voltage gain of 0.85 was obtained in one sample and real power gain of about 4 was shown in three different samples.

The effect of the high frequency electrodynamic environment on the charging effects (as measured by dc properties) of a single junction has been demonstrated. It has been shown that the low impedance environment of a single junction can be decoupled by arrays of similar junctions. All data presented in this chapter agree well with numerical simulation. A few smaller quantitative discrepancies can be explained by imperfections in the samples and the measurement system. However, it should be pointed out that only arrays of junctions where the resistance was high and the Josephson coupling low have been considered.

ACKNOWLEDGEMENTS. Discussions with T. Claeson, K. K. Likharev, D. Haviland, and L. S. Kuzmin as well as financial support from the Swedish Natural Science Research Council, the Swedish Research Council for Engeneering Sciencies and 'Carl Tryggers Stiftelse för Vetenskaplig forskning' is gratefully acknowledged.

References

[1]   A. Schmid, Phys. Rev. Lett. **51**, 1506 (1983).

[2]   K. K. Likharev, N. S. Bakhvalov, G. S. Kazacha, and S. I. Serdyukova, IEEE Trans. Magn. **25**, 1436 (1989).

[3]   N. S. Bakhvalov, G. S. Kazacha, K. K. Likharev and S. I. Serdyukova, Zh. Eksp. Teor. Fiz. **95**, 1010 (1989) [Sov. Phys. JETP, **68**, 581 (1989)].

[4]   M. Amman, E. Ben-Jacob, and K. Mullen, Phys. Lett. A, **142**, 431 (1989).

[5]   E. Ben-Jacob, K. Mullen, and, M. Amman, Phys. Lett. A, **135**, 390 (1989).

[6]   P. Delsing, Ph.D. thesis, Chalmers University of Technology (1990).

[7]   L .J. Geerligs, Ph.D. thesis, Delft University of Technology (1990).

[8]   P. Delsing, T. Claeson, K. K. Likharev and L. S. Kuzmin, Phys. Rev. B **42**, 7439 (1990).

[9]   M. Iansiti, M. Tinkham, A. T. Johnson, W. F. Smith and C. J. Lobb, Phys. Rev. B **39**, 6465 (1989).

[10] P. Delsing, K. K. Likharev, L. S. Kuzmin and T. Claeson, Phys. Rev. Lett. **63**, 1861 (1989).

[11] L. J. Geerligs, V. F. Anderegg, C. A. van der Jeugd, J. Romijn and J. E. Mooij, Europhys. Lett. **10**, 79 (1989).

[12] P. Delsing, K. K. Likharev, D. B. Haviland, A. N. Korotkov, and T. Claeson, to be published in the proceedings of SQUID '91.

[13] L. J. Geerligs, V. F. Anderegg, P. Holweg, J. E. Mooij, H. Pothier, D. Esteve, C. Urbina, and M. H. Devoret, Phys. Rev. Lett. **64**, 2691 (1990).

[14] H. Pothier, P. Lafarge, P. F. Orfila, C. Urbina, D. Esteve, and M. H. Devoret, Physica B **169**, 573 (1991).

[15] L. S. Kuzmin, P. Delsing, T. Claeson, and K. K. Likharev, Phys. Rev. Lett. **62**, 2539 (1989).

[16] K. K. Likharev, IEEE Trans. Magn. **23**, 1142 (1987).

[17] P. Delsing, T. Claeson, G. S. Kazacha, L. S. Kuzmin, and K. K. Likharev, IEEE Trans. Magn. **27**, 2581 (1991).

[18] P. Delsing, K. K. Likharev, L. S. Kuzmin and T. Claeson, Phys. Rev. Lett. **63**, 1180 (1989).

[19] D. V. Averin and A. A. Odintsov, Phys. Lett. A **140**, 251 (1989).

[20] D. V. Averin, A. N. Korotkov, and Yu. V. Nazarov, Phys. Rev. Lett. **66**, 2818 (1991).

Chapter 8

Single Charges in 2-Dimensional Junction Arrays

J. E. MOOIJ

Department of Applied Physics, Delft University of Technology
Lorentzweg 1, 2628 CJ Delft, The Netherlands

and

GERD SCHÖN

Institut für Theoretische Festkörperphysik, Universität Karlsruhe
Engesserstr. 7, D-7500 Karlsruhe 1, Germany

1. Introduction

Several circuits composed of small-capacitance tunnel junctions have been used to study single-electron and charging effects. Among them two-dimensional arrays take a special place due to the fact that for each junction the presence of all other junctions with high resistance provides an excellent decoupling from the environment. As a result many of the single-electron effects are very pronounced. In addition, properties special for the two-dimensional system are expected and observed [1]. Arrays built from normal junctions may show at low temperature insulating behavior, and only at higher temperature a Kosterlitz-Thouless-Berezinskii (KTB) [2, 3] phase transition to a conducting phase occurs [4]. In arrays built from Josephson junctions a phase transition separates an insulating from a superconducting state [5]. An interesting duality exists [6]–[10] between single charges, whose motion produces a current, and vortices, whose motion produces a voltage. The former dominate in arrays where the charging energy E_C exceeds the Josephson coupling E_J, the latter in arrays with E_J exceeding E_C. Moreover, in arrays with low dissipation the vortices can have a very special dynamics [11]–[16], associated with the electric field energy. They are found to show quantum as well as ballistic behavior.

Two-dimensional junction arrays are fabricated in the same way as single junctions, simple circuits and one-dimensional arrays. Very high quality can be achieved, even for

Single Charge Tunneling, Edited by H. Grabert and
M.H. Devoret, Plenum Press, New York, 1992

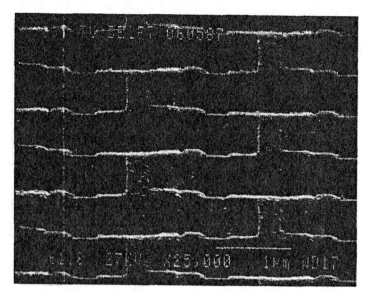

Figure 1. SEM picture of a fabricated array with junctions with $C \approx 10^{-15}F$.

large arrays of junctions smaller than 100 nm. Access is needed, however, to an electron beam writer with sufficient resolution and reproducibility over a large scan-field. With these techniques it is possible to fabricate arrays with junction capacitances in the range of 10^{-15} F or smaller, with a variation in the parameters such as capacitance, resistance, or the Josephson coupling in the 20% range or less. This means that in these arrays, in contrast to granular films, disorder plays a minor role.

In this chapter we first discuss the properties of the charge carriers, single electrons or Cooper pairs, in junction arrays (Sec. 2). We show that in suitable arrays the KTB transition of the 2D-Coulomb gas can occur. At the transition charge dipoles unbind, separating an insulating from a conducting phase. We then review in Sec. 3 the properties of vortices in classical Josephson junction arrays and their KTB transition where vortex-antivortex dipoles unbind, separating a superconducting from a resistive phase. The properties of both, charges and vortices, the competition between them and the general phase diagram of the junction array can be discussed in a coupled-Coulomb-gas model for the charges on the islands and the vorticity enclosed between islands [9] (Sec. 4). In Sec. 5 we study further properties of the vortices. We derive an effective action for vortices, obtain the vortex 'mass' and 'band energy' and discuss the Aharonov-Casher effect of vortices moving around a charge. Experimental results demonstrating the ballistic motion of vortices are presented. We also investigate the influence of dissipation on the vortex motion. Finally, we discuss further analogies between charges and vortices.

2. Charges in junction arrays

2.1. Normal state

The arrays considered consist of metallic regions, "islands", connected by tunnel junctions. A typical example is shown in Fig. 1. Most fabricated arrays have a regular

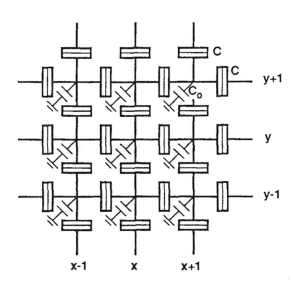

Figure 2. A schematic equivalent circuit for a two-dimensional junction array is shown.

square or triangular lattice, where identical junctions connect each island to their nearest neighbors. Each island has a capacitance to each of the other islands, to a far away ground, and to all other metallic regions in the neighborhood such as gates and leads. For overlapping tunnel junctions, the nearest neighbor capacitances are typically much larger than the others. The charging energy can be expressed in terms of a capacitance matrix C_{ij}

$$H_{\text{ch}} = \frac{1}{2} \sum_{i,j} Q_i C_{ij}^{-1} Q_j. \tag{1}$$

Here Q_i is the total charge on an island i. It can only be an integer multiple of the electronic charge e.

In order to discuss the physical effects we consider a specific example, a square array with junction capacitances C and island capacitances to ground C_0, but neglect all other capacitances. This network is indicated in Fig. 2. We further assume that C is much larger than C_0. We will return to a discussion of the more realistic case later. The charges $Q(x,y)$ and the potential $\Phi(x,y)$ on the islands $i = (x,y)$ satisfy a discrete Poisson equation

$$Q(x,y) = C_0 \Phi(x,y) + C[4\Phi(x,y) - \Phi(x-1,y) - \Phi(x+1,y)$$

$$- \Phi(x,y-1) - \Phi(x,y+1)].$$

Let us assume that all the islands have zero (total) charge except island $(0,0)$ which has a charge e. If the potential varies slowly on the scale of the lattice spacing, a quasi-continuous formulation can be used for distances (measured in units of the lattice spacing) far enough from the origin $r = (x^2 + y^2)^{1/2} \gg 1$

$$\nabla^2 \Phi(r) - \Lambda^{-2} \Phi(r) = 0 \quad \text{where } \Lambda = (C/C_0)^{1/2}. \tag{2}$$

The solution is $\Phi(r) = A K_0(r/\Lambda)$, where K_0 is the zero-order modified Bessel function and A is a constant. Application of Gauss's law around the charge yields $A = e/(2\pi C)$.

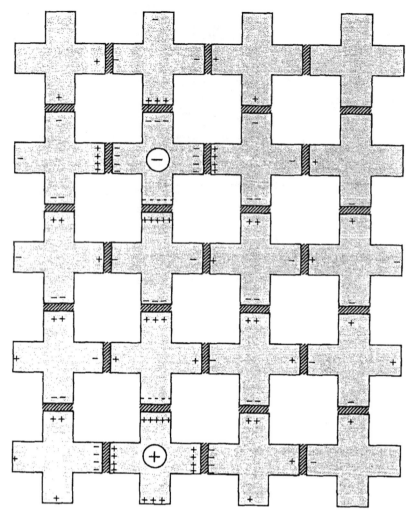

Figure 3. The polarization charges in a two-dimensional array of tunnel junctions induced by two opposite (integer) charges on two islands are shown for $C_0 \ll C$.

For $r \ll \Lambda$ the Bessel function is equal to $-\ln(r/\Lambda)$ and the potential around the charge e is

$$\Phi(r) = -\frac{e}{2\pi C}\ln\frac{r}{\Lambda} \quad \text{for } 1 \ll r \ll \Lambda. \tag{3}$$

In this region the field lines are drawn into the two-dimensional plane, which behaves as a 2D dielectric with effective dielectric constant C. The polarization charges on the islands accumulating at the junctions are shown in Fig. 3. A second charge in the array will be attracted or repelled by the first when the mutual distance is not much larger than Λ. Charges are repelled by an unconnected edge and attracted by a grounded edge. Compared with the situation in a one-dimensional chain of small junctions, the interaction is weaker. In 1D chains the potential does not fall off, whereas in 2D arrays the spreading of the polarization over the area leads to the logarithmic dependence on the scale of the screening length.

The charges in the array can move from island to island. According to the orthodox theory [17], the tunneling rate depends on the energy difference before and after the tunneling process. At low temperatures the rate is $\Gamma = |\Delta E|/(e^2 R_t)$ if the energy is lowered in the transition and zero otherwise. Here R_t is the high-temperature tunneling resistance of the junction. We will discuss the dynamics of the charges in the junction array and the resulting *I-V* characteristics further below.

2.2. Two-dimensional phase transition

At zero temperature no free charges will be present in an array, and all islands are neutral. As a result, the system is insulating for small voltages. At higher temperatures, one may expect free charges to be created by thermal activation, which leads to conducting behavior. The transition from one regime to the other is gradual for small Λ, but occurs as a phase transition of the special Kosterlitz-Thouless-Berezinskii type (KTB) [2, 3] when Λ is large. In this section we will assume that $C_0 = 0$, and the screening length Λ is infinite. A discussion of practical questions is postponed to Sec. 2.5. The interaction energy of a pair of charges e and $-e$ at mutual distance r, in the quasi-continuous approximation, is equal to $(E_C/\pi)\ln(r)$. Here and for later use we defined

$$E_C \equiv e^2/2C. \tag{4}$$

The interaction energy is exactly that of a pair of opposite charges in the "two-dimensional Coulomb gas", a model system which attracted much attention in statistical mechanics due to the peculiar nature of the two-dimensional topological phase transition. The junction arrays with dominant nearest-neighbor capacitance provide a physical realisation for this model system and show the mentioned phase transition.

Let us discuss the nature of this phase transition. In an overall neutral, infinitely large array the energy of a pair of charges e and $-e$ increases logarithmically with increasing separation, following the potential of Eq. (3). When the temperature is raised, more pairs are formed, at larger and larger separation. The binding force of a large pair is modified by the presence of smaller pairs, which act as a polarizable dielectric. The energy of the pairs is offset by the entropy term of the free energy. This is large because there are many ways of forming pairs. At the KTB transition, the first pair with infinite separation (free charges) is present in thermal equilibrium. The transition temperature is

$$T_{\text{cn}}^0 = E_C/(4\pi\epsilon_C). \tag{5}$$

In this expression ϵ_C is a non-universal constant, somewhat larger than 1. It is the effective dielectric constant experienced by the infinitely large pair at the transition temperature. Below T_{cn}^0, all charges are bound in pairs, and the conductance is zero. Above T_{cn}^0, the number of free charges increases. The density close to the transition $0 \leq (T - T_{\text{cn}}^0)/T_{\text{cn}}^0 \ll 1$ is given by

$$n_e(T) = K \exp\left[-\frac{b}{\sqrt{T/T_{\text{cn}}^0 - 1}}\right] \tag{6}$$

where K and b are constants of order 1. By fitting the linear conductance $G(T)$, which is proportional to the density of free charges $n_e(T)$, to the expression (6) the transition temperature can be determined.

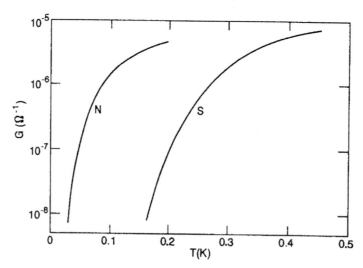

Figure 4. Conductance as a function of temperature for an array in normal (N) state and the same array in the superconducting (S) state (from [4]).

A clearer sign of the KTB transition can be obtained by studying the non-linear conductance. It is expected that below the transition the current I is proportional to a power of the voltage

$$I \propto V^{a(T)}. \tag{7}$$

The temperature dependent exponent is large at low temperatures $a(T < T_{cn}^0) \geq 3$. There a current can only flow at finite voltages, because bound pairs of charges have to be oriented and pulled apart by the voltage in order to produce free charges. Above T_{cn}^0, the current is proportional to V and the exponent $a(T > T_{cn}^0) = 1$. At the transition, in an ideal system, the exponent jumps from 3 to 1, which is a manifestation of what is known as the "universal jump of the superfluid density" [18]. The name is derived from the properties of a superfluid helium film, where the unbinding of vortex-antivortex pairs provides dissipation above the transition and the density of the superfluid is proportional to a similar exponent. Also in superconducting films and in Josephson junction arrays a vortex-antivortex unbinding transition [19, 20] occurs, as will be discussed in Sec. 3.

In a finite system, a pair with separation comparable to the size of the system already provides free charges at temperatures below T_{cn}^0. Thus the transition is rounded-off, showing thermal activation-like behavior at lower temperatures. A finite screening length has the same influence. It is important to note that in the theory of the phase transition, the logarithm of the separation of pairs $\ln(r)$ is the relevant quantity. The real phase transition is seen only when for the largest separation $\ln(r_{\max}) \gg 1$. This means that system size and screening length should be at least of the order of one hundred cells.

In Fig. 4 the linear conductance measured for an array of normal junctions is shown [4]. At low temperatures the array is insulating. Above a 'transition' temperature, which is of the order of the temperature T_{cn}^0 (5) where we expect a KTB transition, the conductance increases rapidly, roughly consistent with the predictions (6). However, the transition is strongly washed out. Indeed from the measured capacitances we can estimate the screening length to be only approximately 10 cells, which is too small to show a clear KTB transition. Similarly in the experiments of Ref. [21] the screening length is still too small.

2.3. Simulations of single electron tunneling in normal junction arrays

Several of the predictions of the response of the Coulomb gas are based on model assumptions about the dynamics of the system [22]. In the present case the dynamics of the single electron tunneling is known from the microscopic theory [17]. It has been studied extensively in the recent years, and many of the predictions derived from it, such as the Coulomb blockade, the Coulomb gap in the current-voltage characteristics, and the SET (single electron tunneling) oscillations of the voltage have been confirmed by experiments. The rate of tunneling of an electron (one out of many electrons on the mesoscopic island) from an island to a neighboring one, changing the charge configuration from $\{Q_i\}$ to $\{Q'_i\}$, depends on the energy difference ΔE and hence on the configuration of charges before and after the tunneling process

$$\Gamma(\{Q_i\} \to \{Q'_i\}) = \frac{\Delta E}{e^2 R_t}\left[\exp(\frac{\Delta E}{k_B T}) - 1\right]^{-1}. \tag{8}$$

The question whether the relevant energy change ΔE is that of the junction only where the tunneling occurs (local rule) or whether it refers to a larger system, possibly the whole array (global rule), had been discussed [23, 24, 25]. The answer depends on how much of the environment the tunneling electron sees, and hence depends on typical time scales in the problem. The work of Nazarov [24] demonstrated that the longest time, which in tunnel junction systems is the inverse of the energy difference $\Delta \tau = \hbar/\Delta E$, is the relevant time scale. The energy change is of order E_C or smaller. Hence typically the whole array is probed, and the change in the equilibrium energy $\Delta E = H_{ch}(\{Q_i\}) - H_{ch}(\{Q'_i\})$, given by Eq. (1), enters into the tunneling rate (8).

It is straightforward to simulate the stochastic charge dynamics of a junction array based on the rate (8) [23]. The dc I-V characteristic of an array (with N_s junctions in series) shows similar structure as single junctions with small capacitance. At low voltages and temperatures tunneling is prevented (Coulomb blockade) and no current is flowing. Beyond a certain threshold the current sets in, reaching a differential conductance which coincides with that of a classical array. However, the I-V characteristic is shifted relative to the classical linear I-V curve by what is called the 'Coulomb gap'. In the array this is of order $V_g \sim N_s e/C$. At finite temperature the Coulomb blockade is not perfect, a current flows already at small voltages, although for $k_B T \ll E_C$ it is still much smaller than the classical value. These features have been known for some time and have been confirmed in experiments. Extending the simulations to larger systems we found [26] that (i) the low voltage part of the I-V characteristic shows a nonlinear conductance as given by Eq. (7), (ii) the exponent $a(T)$ shows the temperature dependence predicted for a KTB transition (with finite size smearing), (iii) the transition temperature is given by Eq. (5), (iv) above T_{cn}^0 the conductance fits to the prediction (6), (v) at larger voltages the I-V characteristic shows the Coulomb gap. The results obtained for a 32×32 array are shown in Fig. 5. These results show that the single electron tunneling governed by the rate (8), which had been derived from the microscopic theory, yields both features, the KTB properties and the Coulomb gap. Both are fully consistent with one another.

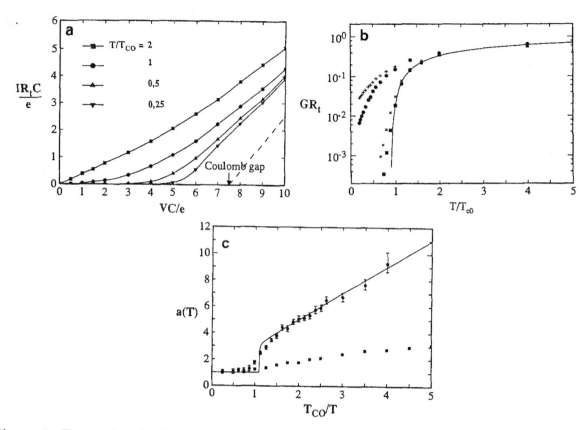

Figure 5. The results of a simulation of the single electron tunneling dynamics in a 32 × 32 array are shown (from Ref. [26]). (a) The *I-V* characteristic at different temperatures. (b) The linear conductance $G(T)$. The squares and symbols × show the result for two different voltages. The solid line is the theoretical prediction. The dots and crosses show the result in the presence of random offset charges. (c) The exponent of the nonlinear conductance $a(T)$ is shown by dots. The solid line is the theoretical prediction. The squares are the results obtained in the presence of random offset charges.

In Fig. 5 we show also results of a simulation where we allowed that on the island random quenched offset charges $Q_{x,i}$ were present at the beginning. These charges may be due to charged impurities which polarize the islands and bind a part of the charge. They need not to be integer and in the simulation where chosen to be equally distributed in the range $-e/2 \leq Q_{x,i} \leq e/2$. Also in the presence of offset charges the conductance still shows a strong temperature dependence, but the features characteristic for the KTB transition have completely disappeared.

2.4. Influence of quasiparticle tunneling on the charge KTB transition

The KTB phase transition described above follows from the electrostatic long range interaction of charges. We assumed that the tunneling of single electrons is weak. The tunneling is needed to establish an equilibrium distribution of the charges Q_i. Moreover, it determines the response of the system. On the other hand, if the tunneling is strong it may itself influence the KTB transition. In order to study this question we have to go beyond the simple stochastic tunneling picture based on the rate (8). Instead we start from a microscopic Hamiltonian which contains the charging energies but also accounts

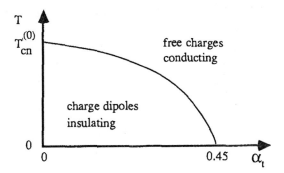

Figure 6. The phase diagram for the charge-unbinding transition of a normal junction array is plotted as a function of the tunneling conductance $\alpha_t = R_Q/R_t$.

for the tunneling of electrons across the tunnel barriers in a nonperturbative way. We find modifications of the KTB transition if the dimensionless tunneling conductance

$$\alpha_t \equiv \frac{R_Q}{R_t} \quad \text{where } R_Q = \frac{h}{4e^2} = 6.45 \, \text{k}\Omega. \tag{9}$$

is of order 1 or larger. The analysis of such a model is presented in Appendix A. There we show that the transition temperature of the KTB transition decreases with increasing α_t and vanishes for $\alpha_t \geq 0.45$ [4, 9]. The resulting phase diagram is shown in Fig. 6. In the region below the transition temperature the charges are bound in dipoles so that the array is insulating. Above, the free charges form a plasma phase and give a finite conductance.

2.5. Capacitances in real junction systems

In the previous discussion, we have employed the model circuit of Fig. 2. All elements of the capacitance matrix other than the nearest-neighbor and the direct island-to-ground terms were ignored. This is not a good representation for a typical, fabricated array, with metallic islands with dimensions of the order of 1 μm, on a substrate with a high dielectric constant. The calculation of the real field distribution in an actual array with specially shaped islands is beyond the scope of the present paper. However, when the potential varies slowly on the scale of the lattice spacing, it is a good approximation to view the array as a two-dimensional dielectric, and in such an approximation the potential around a single charge can be calculated. The analogous system is as follows: A thin dielectric layer, with thickness d and dielectric constant ϵ_r, is contained between media with dielectric constants ϵ_{r1} and ϵ_{r2} with $\epsilon_r \gg \epsilon_{r1}, \epsilon_{r2}$. At the origin of the cylindrical coordinate system, in the thin layer, a point charge e is present. The potential in the thin layer, for $\rho = (R/d)(\epsilon_{r1} + \epsilon_{r2})/\epsilon_r \gg 1$ is [27, 28]

$$V(\rho) = \frac{e}{4\epsilon_0 \epsilon_r d}[\text{H}_0(\rho) - \text{Y}_0(\rho)].$$

Here R is the radial coordinate, and H_0 is the zero-order Struve function and Y_0 the zero-order Bessel function of the second kind. To translate this to the junction array, we compare the polarization per unit length parallel to an electric field E in the dielectric layer, $d\epsilon_0(\epsilon_r - 1)E$, with the polarization charge in the array, CE. The latter quantity is found for a square array with lattice constant p by multiplying the number of junctions per unit length p^{-1} with the voltage drop per junction, pE, and the capacitance. For $\epsilon_r \gg 1$ we see that $d\epsilon_0\epsilon_r$ for the layer corresponds to C for the array. We also define a reference self-capacitance

$$C_{00} = 2p\epsilon_0(\epsilon_{r1} + \epsilon_{r2}) \tag{10}$$

which is the capacitance to a far away ground of a thin flat disc with radius $p/2$.

The potential in the array at distance $r = R/p$ is

$$V(r) = \frac{e}{4C}[\mathrm{H}_0(r/\lambda) - \mathrm{Y}_0(r/\lambda)] \quad \text{with } \lambda \equiv 2C/C_{00}.$$

In the limit $r \gg \lambda$ the potential reduces to that of a point charge e without an array. On the other hand, for $r \ll \lambda$ it is found that

$$V(r) \approx \frac{e}{2\pi C} \ln(r/\lambda)$$

which coincides with Eq. (3). However, notice that the screening length λ is not the square root but the first power of the ratio of the junction capacitance C and a self-capacitance C_{00}.

Arrays, fabricated for the study of charging effects, have typical junction capacitances in the range 10^{-16} to $10^{-15}\,\mathrm{F}$. Typical cell sizes are 0.5 to $2\,\mu\mathrm{m}$. The latter correspond, on Si substrates, to self-capacitances C_{00} of 1 to $5 \times 10^{-16}\,\mathrm{F}$. Clearly the effective screening length λ is still rather small. Typically the smallest junctions are combined with the smallest cells, so that the ratio C/C_{00} is not easily improved. It does not help to reduce the area of the metallic islands while leaving the cell size unchanged. In order to detect the charge KTB transition, one should fabricate arrays with the smallest possible cells and the largest acceptable junctions. The latter size is dictated by the transition temperature, which should not be below $50\,\mathrm{mK}$. This leaves, with practical present day technology, still a sufficiently wide temperature region for measuring below the transition. For a square array the dielectric constant in Eq. (5) is about 1.2, and the transition temperature is near $E_C/15$. This indicates that E_C should not be chosen below $1K$, or C not above $10^{-15}\,\mathrm{F}$, i.e. junctions should be about $100\,\mathrm{nm}$ square. With efforts a cell size of $300\,\mathrm{nm}$ should be obtainable, which yields $C_{00} \approx 7 \times 10^{-17}\,\mathrm{F}$. The screening length in such an array would be about 30 cells.

An important problem with experiments on single electronics is presented by offset, "frozen-in", charges on metallic islands. They are probably hard to avoid. They induce disorder, that will suppress the phase transition and other physical effects. In small circuits they can be tuned out by adjusting gate voltages, but this cannot be done for an array. It may be possible to reduce their density by careful preparation of surfaces and controlled cool-down.

It is interesting to consider the possibility of inducing a homogeneous "frustration"-charge to the charge-array, similar to frustration by a magnetic field in vortex-arrays. For this discussion we ignore random offset charges. By giving the whole array a certain potential with respect to a far-away ground, charges are induced on the metallic surfaces. For equal potential over the array, the islands near the edges will carry more charge than those in the middle. One can compare with the charge distribution on a flat circular metallic disc, which has about 3/4 of its total charge on the outer 1/4 of the radius. Maybe large metallic connecting banks can draw this edge charge, leaving a relatively flat distribution on the actual array. Calculations have to be made to see how well this might work. The other, more obvious method is to fabricate arrays with dominating capacitance to a ground plane, preferably between the array and the highly polarizable substrate. Again it is important to make the screening length as long as possible. When the capacitance to ground is large enough, the approximation of Sec. 2.1 is applicable and the screening length is $\Lambda = (C/C_0)^{1/2}$. Now it helps to reduce the size of the metallic region within the cell. For an island size of $0.2\,\mu m^2$, a distance between ground plane of $0.5\,\mu m$ with a dielectric with constant $\epsilon \approx 4$ gives $C_0 \approx 2 \times 10^{-17}\,F$. The screening length Λ could be about 7 cells. An array of about 10×10 cells, connected to metallic end electrodes at a common potential with respect to the ground plane would be rather homogeneously frustrated.

2.6. Charge transition in the superconducting state

In the superconducting state, at low temperatures and as long as typical voltages do not exceed the superconducting gap, Cooper pairs are the dominant charge carriers. As the Cooper pair charge is $2e$, all Coulomb energies are 4 times larger than for normal electrons. If the junction capacitances dominate, the Cooper pair charges interact logarithmically over sufficiently long distances. Hence one can expect again a 2D charge-unbinding transition separating an insulating from a conducting phase at a temperature [4]

$$T_{cs}^0 = E_C/(\pi\epsilon_C). \tag{11}$$

Below we will show that in a superconducting array in order for the Cooper pair charge-unbinding transition to occur we also need the requirement that the charging energy scale E_C is much larger than the Josephson coupling energy E_J. The transition temperature (11) of the superconducting array is larger by a factor 4 than the result for a normal array (5). This means, there exists a temperature regime where the superconducting array is insulating while a normal array already conducts!

A transition from an insulating to a conducting phase has indeed been observed in experiments on Josephson junction arrays. In Fig. 4 we show the linear conductance of a superconducting array, and compare it to the conductance of the same array in a magnetic field, which is strong enough to drive the array normal. It is obvious that the transition temperature of the superconducting array is indeed roughly 4 times larger than that of the normal array. Again the results are not yet a convincing proof of a KTB transition since the screening length of the array is only about 10 lattice spacings, too small to really observe a KTB transition.

Below we will demonstrate the duality between vortices in arrays where E_J is dominant and charges in arrays with large E_C. This analogy prompted Sugahara and Yoshikawa [6] to predict a charge-unbinding KTB transition in the superconducting state. Also Widom and Bajou [29] had indicated the phase transition and determined T_{cs}^0.

We have recently extended the simulations of the charge dynamics in junction arrays to include the tunneling of Cooper pairs (in the limit where it can be treated as incoherent) [30]. In contrast to the single electron tunneling rate (8) the tunneling rate of Cooper pairs shows a strong resonant structure when the energy is conserved. This leads to different overall *I-V* characteristics, in particular to a peak at voltages of the order of the Coulomb gap. But at low voltages the features typical for the KTB transition at the transition temperature (11) are recovered. The goal of these investigations is to include both Cooper pair and quasiparticle tunneling in the limit where quasiparticles are nearly frozen out and to study the influence of a low density of quasiparticles on the Cooper pair charge-unbinding transition.

3. Vortices in 2D junction arrays

In this short section we review some of the general properties of vortices in "classical" Josephson junction arrays. In later sections the emphasis will be placed on the properties of vortices in arrays of junctions with very small capacitance and very low subgap conductance. The aspects that are discussed here are shared by vortices in thin superconducting films, and by vortices in arrays of junctions with large ohmic shunt conductance, where charges on the islands are not well-defined.

In a classical 2D array of superconducting islands, coupled by "large" Josephson junctions, the electrostatic energy can be ignored, and the relevant energy is the sum of the Josephson coupling energies of the junctions

$$U = -\sum_{<i,j>} E_J \cos(\phi_i - \phi_j). \tag{12}$$

This means this system is a representative of the two-dimensional XY model of statistical mechanics, which is the standard example of a 2D system that shows the KTB transition where vortex-antivortex pairs unbind.

Actually the Josephson junction array can be more complicated than the model (12). A nonvanishing phase difference implies a supercurrent flow. This produces a magnetic field, which shifts in (12) the phase differences by a gauge term. Similarly, an applied magnetic field is shielded by supercurrents, the Meissner effect for the array. However, in the arrays of interest these currents are very small, and the magnetic screening length is very long. In fact at the temperature where the KTB transition occurs, the screening length is near 1 cm [19], much larger than practical samples. Another complication can arise if the currents through the junctions influence the magnitude of the order parameter in the islands. However, also this effect is usually small in the fabricated, mesoscopic junction arrays. Hence the phase of the order parameter on each island is the only relevant variable; and the description by the model (12) is sufficient, unless the charging energy needs to be taken into account.

Practical "vortex arrays" are strongly two-dimensional and hardly affected by the self-generated magnetic field in the third dimension. Hence it is the system size which limits the sharpness of the vortex-unbinding KTB transition. In contrast in "charge arrays" it is much more difficult to contain the electric fields generated by single charges in the two-dimensional system over reasonable lengths.

XY systems, and accordingly junction arrays, show two types of excitations. There are "spin waves", small amplitude wave-like variations of the phase over the array. For

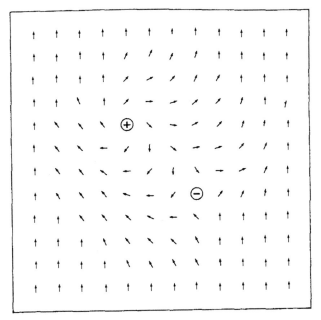

Figure 7. A vortex-antivortex pair in a junction array. The arrows indicate the value of the phase ϕ in the superconducting islands.

small wavevectors, their energy is small and they can easily be excited. At all non-zero temperatures, the spin waves destroy the correlation between phases at long distances. Spin waves in a superconducting array, as long as their amplitudes are small, do not lead to a resistance. Secondly there are the vortices. They can be detected in the model system with the following test: when all phase differences along a closed contour are added, the sum in units of 2π is equal to the net number of vortices within the contour. Even if spin waves destroy the phase coherence over long distances, they do not change this sum of phases. Vortices do, and are called topological excitations for this reason. There are vortices and also antivortices with opposite rotation. In Fig. 7 we show a vortex-antivortex pair in an array. The arrows indicate the values of the phase. A vortex and an antivortex attract each other. A pair of a vortex and an antivortex does not disturb the phases at long distances and therefore has a reduced energy. They interact logarithmically and hence are equivalent to the charges of a 2D Coulomb gas. Clearly there exists an analogy to the charges in ideal charge arrays.

From the analogy to the Coulomb gas it follows that the vortices undergo a KTB transition, where vortex dipoles unbind at a temperature

$$T_v^0 = \frac{\pi}{2\epsilon_v} E_J. \tag{13}$$

Here a dielectric constant ϵ_v appears. It differs in general from ϵ_C which enters the charge KTB transition temperature (5), but it also is of order 1. This transition separates a superconducting from a resistive phase.

The KTB transition of vortex arrays has been observed in experiments very clearly. It helps that the screening lengths of the vortex interaction are long, much longer than that of the charges. A discussion of physical properties can be found in Ref. [19]. A review of KTB vortex transitions in films and arrays is found in Ref. [20]. In Fig. 8 experimental data are shown for an array of shunted niobium tunnel junctions [31]. Figure 8a shows an

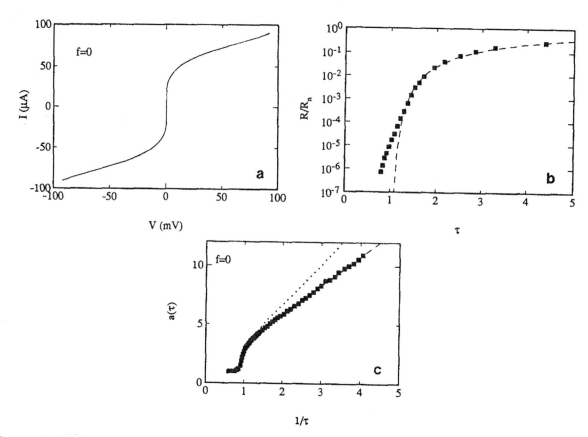

Figure 8. KTB transition of a vortex array [31]. (a) *I-V* characteristic near transition temperature. (b) Resistance as a function of temperature, normalized to the Josephson coupling energy. The dotted line is the theoretical prediction for the KTB transition. The deviation at low T is due to finite size effects. (c) The exponent of the nonlinear resistance ($V \propto I^{a(T)}$) is shown. KTB transition should show a jump from 3 to 1. The dotted line shows results from a Monte Carlo simulation.

I-V characteristic. Figure 8b shows the resistance as a function of temperature. It follows the predicted square-root-cusp dependence, similar to equation (6), over two orders of magnitude of the resistance. However, at the lowest temperatures, where the expression should best apply in an infinite system, the experimental data deviate. This is due to the finite size of the array and to the influence of a small residual external magnetic field. The transition is best seen in the non-linear resistance. For current-induced vortex pair unbinding, one expects V to be proportional to a power of the current, $V \propto I^{a}(T)$. The exponent $a(T)$ is 3 at the transition, jumping to 1 at temperatures above the transition temperature. As Fig. 8c shows, such a jump is well visible in an array with a width of 100 cells.

4. Charge-vortex duality

The charges and the phases of Josephson junctions are quantum mechanical conjugate variables. This has the consequence that charge order and phase order compete and exclude one another [17, 32]. In single junctions there exist a duality between the properties of the charges and phases. For instance the Bloch oscillations [33, 34] in small capacitance Josephson junctions (conveniently described in the charge picture) are in many respects analogue to the Josephson oscillations of classical junctions (described in the phase picture). In a Josephson junction array the charge order and the vortex order exclude one another. This leads to a superconductor-insulator transition at zero temperature [9]. In suitable systems there exists a nearly perfect duality between charges and vortices [9, 10] which implies a universal conductivity at the superconductor-insulator transition [7]. In order to study this we will first derive a coupled-Coulomb-gas description for the charges and the vortices. From it we will obtain the phase diagram of the junction array. In suitable systems we will find three phases: an insulating, a superconducting, and a conducting phase.

4.1. The coupled-Coulomb-gas description

In small capacitance Josephson junctions we have to include both the charging energy (1) and the Josephson coupling (12). The charges Q_i on the islands are quantum mechanical conjugates to the phases ϕ_i of the superconducting order parameter of the island. If we ignore dissipation, the Hamiltonian of this system is

$$H = \frac{1}{2}\sum_{i,j}(Q_i + Q_{x,i})C_{ij}^{-1}(Q_j + Q_{x,j}) - \sum_{<i,j>} E_J \cos[\phi_{ij} - A_{ij}(\tau)];$$

$$Q_i = \frac{\hbar}{i}\frac{d}{d(\hbar\phi_i/2e)} \cdot \qquad (14)$$

Here $\phi_{ij} = \phi_i - \phi_j$ refers to nearest-neighbors. Electromagnetic fields are accounted for by a vector potential $A_{ij} = 2\pi/\Phi_0 \int_i^j \vec{A} \cdot d\vec{l}$ where $\Phi_0 = h/2e$. We also allowed for "offset" or "external" charges $Q_{x,i}$ on the islands. They are fixed by external constraints and in general are noninteger. There origin and effect will be discussed further in Sec. 5.2. For the moment we included them to describe the most general model. In the following we will consider again the simple model of an array consisting of junction capacitances C and self capacitances C_0.

The partition function can be expressed as a path integral in imaginary times $0 \leq \tau \leq \beta = 1/k_B T$ (from now on we choose $\hbar = 1$)

$$Z = \sum_{\{n_i\}} \prod_i \int_{\phi_{i0}}^{\phi_{i0}+2\pi n_i} D\phi_i(\tau)\exp\{-S[\phi]\} \qquad (15)$$

which depends on the action

$$S[\phi] = \int_0^\beta d\tau \Big[\frac{C_0}{8e^2}\sum_i \dot\phi_i^2 + \frac{C}{8e^2}\sum_{<i,j>}\dot\phi_{ij}^2 - i\sum_i \frac{Q_{x,i}}{2e}\dot\phi_i - \sum_{<i,j>} E_J \cos[\phi_{ij}(\tau) - A_{ij}(\tau)]\Big].(16)$$

We consider junctions without Ohmic shunts and for the moment ignore single electron tunneling. Hence the charges on the islands can change only by Cooper pair tunneling. This means the Q_i are integer multiples of $2e$. This choice of charge states fixes the limits of the path integrals over the fields ϕ_i in Eq. (15) [32]. Values which differ by 2π are equivalent, and the integral in (15) includes a summation over winding numbers $\phi_i(\beta) = \phi_i(0) + 2\pi n_i$.

We (re-)introduce the charges on the islands in the path integral representation. In a mixed representation involving the phases $\phi_i(\tau)$ and charge trajectories $q_i(\tau) = Q_i(\tau)/2e = 0, \pm 1, \pm 2, \ldots$ the partition function is [32]

$$Z = \prod_j \int_{q_{j0}}^{q_{j0}} Dq_j(\tau) \sum_{\{n_i\}} \prod_i \int_{\phi_{i0}}^{\phi_{i0}+2\pi n_i} D\phi_i(\tau) \exp\{-S[q,\phi]\}. \tag{17}$$

which depends on the action

$$S[q,\phi] = \int_0^\beta d\tau \Big[2e^2 \sum_{i,j} [q_i(\tau) + q_{x,i}(\tau)] C_{ij}^{-1} [q_j(\tau) + q_{x,j}(\tau)]$$

$$+ i \sum_i q_i(\tau)\dot{\phi}_i(\tau) - \sum_{<i,j>} E_J \cos[\phi_{ij}(\tau) - A_{ij}(\tau)] \Big]. \tag{18}$$

Vortex degrees of freedom can be introduced by the Villain transformation [35], which can be generalized [9, 16] to the present problem with charges. It allows us to integrate out the phases at the expense of introducing at each (dual) space-time lattice point an integer-valued two-dimensional vector field $\vec{J}_i(\tau) = (J_i^x(\tau), J_i^y(\tau))$. For this purpose we introduce a lattice, with spacing $\Delta\tau$, also in the time direction. This spacing is of the order of the inverse Josephson plasma frequency, $(\Delta\tau)^{-1} \approx \omega_0 = \sqrt{8E_J E_C}$. Further details of the present derivation are given in Ref. [16]. The integration over the phase variables can now be performed, which yields

$$Z = \sum_{\{q_{i,\tau}\}} \sum_{\{\vec{J}_{i,\tau}\}} \exp\Big[-2e^2\Delta\tau \sum_{i,j,\tau} [q_i(\tau) + q_{x,i}(\tau)] C_{ij}^{-1} [q_j(\tau) + q_{x,j}(\tau)]$$

$$- \frac{1}{2\Delta\tau E_J} \sum_{i,\tau} |\vec{J}_{i,\tau}|^2 \Big]. \tag{19}$$

The summations are constrained by the 'continuity' equations at each lattice point

$$\partial_\tau q_{i,\tau} - \vec{\nabla} \cdot \vec{J}_{i,\tau} = 0. \tag{20}$$

Here, $\partial_\tau q_{i,\tau} = q_{i,\tau+\Delta\tau} - q_{i,\tau}$ denotes the time derivative on the lattice and is an integer. The constraint (20) can be solved by the parameterization

$$J_{i,\tau}^\mu = n^\mu (\vec{n} \cdot \vec{\nabla})^{-1} \partial_\tau q_{i,\tau} + \epsilon^{\mu\nu} \nabla_\nu M_{i,\tau}, \tag{21}$$

where $M_{i,\tau}$ is another integer field, $\epsilon^{\mu\nu}$ is the antisymmetric tensor and $\vec{n}$ a 2-D unit vector. The operator $\vec{n}(n \cdot \nabla)^{-1}$ represents a line integral. We now apply the Poisson resummation formula and introduce an integer field $v_{i,\tau}$, which is defined on the lattice dual to the original one. This field is the vorticity in the plaquette i. After performing the final Gaussian integral we obtain

$$Z = \sum_{\{q_i(\tau)\}} \sum_{\{v_i(\tau)\}} \exp\{-S_{\text{CCG}}[q,v]\}. \tag{22}$$

The action of the coupled-Coulomb-gas is (returning to a continuum notation for the times)

$$S_{\text{CCG}}[q,v] = \int_0^\beta d\tau \Big\{ \sum_{i,j} \Big[2e^2[q_i(\tau) + q_{x,i}(\tau)] C_{ij}^{-1}[q_j(\tau) + q_{x,j}(\tau)]$$

$$+ \frac{1}{4\pi E_J} \dot{q}_i(\tau) G_{ij} \dot{q}_j(\tau)$$

$$+ \pi E_J [v_i(\tau) + f_i(\tau)] G_{ij} [v_j(\tau) + f_j(\tau)]$$

$$+ i\dot{q}_i(\tau) \Theta_{ij}[v_j(\tau) + f_j(\tau)]\Big]$$

$$+ i \sum_i n_\mu (n \cdot \nabla)^{-1} \dot{q}_i(\tau) A_{i,i+\mu} \Big\}. \tag{23}$$

Here we introduced

$$f_i(\tau) = \frac{1}{2\pi} \epsilon_{\mu\nu} \nabla_\nu A_{i,i+\mu}(\tau) \tag{24}$$

which describes the magnetic flux through the plaquette i, measured in units of the flux quantum $\Phi_0 = h/2e$, but also an electric field. By $i + \mu$ we denote the nearest neighbor of i in μ-direction ($\mu = x, y$). Furthermore, we introduced the kernel $G_{ij} = G(\vec{r}_i - \vec{r}_j)$, which describes the interaction between vortices at site i and j

$$G(\vec{r}) = \frac{1}{2\pi} \int d^2q \frac{1}{q^2} [\exp(i\vec{q}\vec{r}) - 1] = -\ln\Big\{ \frac{1}{2}\big[1 + (1 + 2\pi r)^{1/2}\big] \Big\}. \tag{25}$$

The explicit result (25) is obtained for a convenient choice of the cutoff in q [14] and is defined also for small r. The kernel

$$\Theta_{ij} = \arctan\Big(\frac{y_i - y_j}{x_i - x_j} \Big) \tag{26}$$

describes the phase configuration at site i around a vortex at site j.

The first and third terms in (23) represent the classical action of the island charges and of the vortex gas. For $E_J = 0$ or $C^{-1} = 0$ the fields are constant in time τ, and the classical Coulomb gases of charges or vortices are recovered, respectively. In general the two 'charges' interact via the kernel Θ_{ij} as described by the fourth term. After a partial integration we recognize that this term is the interaction energy of a charge q_i with the voltage $\Theta_{ij}\dot{v}_j$ at site i due the the change of the vorticity at site j. The last term, also after a partial integration, describes the interaction of the charges with the line integral of the external electric field.

The action (23) shows a high degree of symmetry between the vortex and the charge degrees of freedom. If we consider the limit $C \gg C_0$ the inverse capacitance matrix becomes (for large distances)

$$e^2 C_{ij}^{-1} = \frac{E_C}{\pi} G_{ij} \qquad \text{for } C_0 = 0. \tag{27}$$

In this case charges and vortices are nearly dual. The duality is broken by the last term $\dot{q}_i G_{ij} \dot{q}_j$. This nonlocal kinetic contribution arises as the spin-wave contribution to the charge correlation function. The corresponding excitations in the charge gas are absent in the model defined by (14), so that an equivalent term $\dot{v}_i G_{ij} \dot{v}_j$ does not arise.

4.2. The phase transitions in the junction array

Here we discuss the phase transitions in the array without external fields $\vec{A} = 0$ and assuming that there exist no offset charges $Q_x = 0$. For $C^{-1} = 0$ or $E_J = 0$ the vortex and charge fields are constant in time, and (14) reduces to the Hamiltonian of classical Coulomb gases of vortices or charges, respectively. If the charging energy can be ignored, $E_C \approx 0$, the vortices undergo a KTB transition, where vortex dipoles unbind at a temperature T_v^0 given by (13). This transition separates a superconducting low temperature phase from a resistive high temperature phase. If the Josephson coupling is weak, $E_J \approx 0$, and the junction capacitance dominates, $C \gg C_0$, the charges interact logarithmically over sufficiently long distances. They also show a KTB transition where the dipoles, formed by a Cooper pair and a missing pair, unbind. The transition temperature T_{cs}^0 is given by (11). The transition separates a insulating from a conducting phase. At finite E_J and E_C charges and vortices need to be considered simultaneously. The charging energy provides a kinetic energy for the vortices, the Josephson coupling allows the tunneling of Cooper pairs and provides the dynamics for the charges.

If $E_C \ll E_J$, a perturbative approach can be used to show that the transition temperature of the vortex-unbinding KTB transition T_v is lowered below the classical value (13). Similarly, in the limit $E_J \ll E_C$, one can show [9], analogously to the treatment presented in Appendix A, that the charge unbinding transition temperature is reduced below the value (11)

$$T_{cs} = T_{cs}^0 \left(1 - 0.98 E_J/E_C\right). \tag{28}$$

The question remains what happens for $E_J \approx E_C$. If the duality between charges and vortices would be perfect, i.e. if the duality breaking last term in (23) would be absent, the transition temperatures would be symmetric around the self-dual point

$$(E_J/E_C)_{\text{self-dual}} = 2/\pi^2. \tag{29}$$

Assuming that at $T = 0$ there exists only one transition (below we will comment on this), we could immediately conclude [9, 10] that the critical value of E_J/E_C, separating the charge- from the vortex-ordered phase, at $T = 0$ is given by (29). But the duality breaking term, even if it becomes irrelevant at the fixed point, can lead to a shift of the critical value to

$$(E_J/E_C)_c = 2a/\pi^2, \tag{30}$$

which is larger than (29) by a numerical factor $a \geq 1$.

At $T = 0$ the system is effectively three-dimensional and the character of the phase transitions in general changes. In order to study the transitions in this limit we mapped the problem onto a different model which had been investigated by Korshunov [36] in

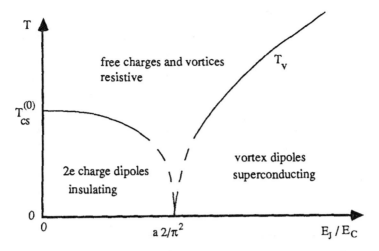

Figure 9. The phase diagram for the charge-unbinding transition and of the vortex-unbinding transition of a superconducting junction array is shown. Here quasiparticle tunneling is ignored. The two transitions meet at $T = 0$ at a critical value of $E_J/E_C = 2a/\pi^2$. The numerical coefficient a is larger but close to 1.

another context. He concluded that the system has only one transition. Combining this information with the perturbative results we arrive at the picture shown qualitatively in Fig. 9.

At the self-dual point $E_J/E_C \approx 2/\pi^2$, at $T = 0$ the phase transition separates a superconducting from an insulating phase. The charges are driven by an applied voltage, and their motion produces a current. On the other hand, the vortices are driven by an applied current, and their motion produces a voltage. From the duality between charges and vortices at the superconductor-insulator transition one can conclude [9] that the resistance of the array is given by the quantum resistance $R_Q = h/4e^2 = 6.45\,\text{k}\Omega$.

In Fig. 10 we also show the influence of quasiparticle tunneling. Its strength is characterized by the dimensionless conductance $\alpha_t = R_Q/R_t$ where R_t is the normal state resistance and $R_Q = h/4e^2 = 6.45\,\text{k}\Omega$ the quantum resistance. In the superconducting state at low temperature and as long as $E_C \ll \Delta$, where Δ is the superconducting gap, we have only virtual quasiparticle tunneling processes, which renormalize the junction capacitance by an amount proportional to the normal state conductance $1/R_t$ [37]

$$C \rightarrow C + 3\pi\hbar/(32\Delta R_t). \tag{31}$$

As a result the critical value for the insulator-superconductor transition at $T = 0$ is shifted

$$(E_J/E_C)_c = 2a/\pi^2 - 3\alpha_t^2/16.$$

On the other hand, in the limit $E_C \gg \Delta$, real single electron tunneling processes occur and, due to the lower activation energy, dominate over the Cooper pair tunneling. In this case the single-electron charges unbind at the lower KTB transition temperature (5). The combination of finite E_J and finite α_t enhances the dynamic screening effect discussed in Appendix A (see Fig. 6). As a result the critical value $\alpha_{t,c}$, beyond which even at $T = 0$ no charge ordered (insulating) phase exists, decreases.

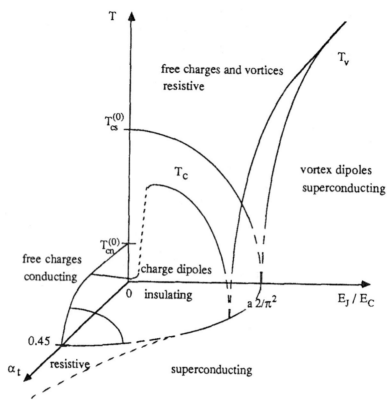

Figure 10. The phase diagram of a Josephson junction array as a function of E_J/E_C and the strength of the normal state tunneling $\alpha_t = R_Q/R_t$ is shown in the limit where the junction capacitances dominate $C \gg C_0$.

Charging effects in junction arrays and their influence on the phase transitions had actually been studied for some time. The research was stimulated by the experiments of the Minnesota group [38] on granular films, which showed that the low temperature properties of the films depend on the strength of the normal state dissipation. In the granular films the self-capacitance is probably dominant. The phase diagrams obtained for this limit [39] differ from the one in Fig. 10, most prominently by the absence of an insulating phase in the region of low E_J, α_t, and finite T. More precisely, at zero temperature, in ideal arrays also in the self-charging limit an insulating phase was predicted. The reason is that in the absence of dissipation any junction array, independent of the capacitances, can only be insulating or superconducting. However, at finite temperature and in the presence of dissipation it requires a mechanism like the long-range charge interaction to prevent the charges from moving. In other papers [40, 41, 42] the crucial dependence of the junction properties on the relative strength of the capacitances was recognized, but they missed the special properties of the charges and their KTB transition.

5. Quantum vortices

5.1. The vortex mass

The representation (23) displays the coupling between the charges and the vortices. In the extreme limits, $E_J = 0$ or $E_C = 0$, only one or the other needs to be considered.

In this subsection we focus on the limit $E_C \ll E_J$, where we obtain an effective action for the vortices, but the vortices are influenced by the charging effects. In the considered limit the charges are fluctuating strongly and can be treated as continuous variable. Hence they can be integrated out from the partition function. We first ignore external fields $f = 0$ and offset charges $Q_x = 0$. In the limit where the junction capacitance dominates $C \gg C_0$ the result is

$$S[v] = \int_0^\beta d\tau \sum_{i,j} \left[\frac{\pi}{8E_C} \dot{v}_i(\tau) G_{ij} \dot{v}_j(\tau) + \pi E_J v_i(\tau) G_{ij} v_j(\tau) \right]. \tag{32}$$

The vorticity at each space-time point can take the values 0 or 1. In general it can change discontinuously, subject only to the mentioned constraints. In particular vortices can be created and annihilated in pairs. Under certain conditions, however, we need to consider only vortices which move in a continuous fashion. If we label the vortices by their centre coordinate $\vec{r}_n(\tau)$ and the sign of the vorticity $v_n = \pm 1$ the vortex density at site $\vec{r}_i$ can be represented as

$$v_i(\tau) = \sum_n v_n \delta[\vec{r}_i - \vec{r}_n(\tau)]. \tag{33}$$

Substituting this expression into (32) we obtain the effective action

$$S[r] = \int_0^\beta d\tau \sum_{n,m} v_n v_m \left[\frac{\pi}{8E_C} \sum_{\mu,\nu=x,y} \dot{r}_n^\mu M_{nm}^{\mu\nu} \dot{r}_m^\nu + \pi E_J G\big(\vec{r}_n(\tau) - \vec{r}_m(\tau)\big) \right] \tag{34}$$

which contains the 'vortex mass' tensor obtained before by Simanek [11] and Eckern and Schmid [14]

$$M_{nm}^{\mu\nu} = \frac{\partial}{\partial r_n^\mu} \frac{\partial}{\partial r_m^\nu} G\big(\vec{r}_n - \vec{r}_m\big). \tag{35}$$

This expression converges rapidly for large $|\vec{r}_n - \vec{r}_m|$. Hence with sufficient accuracy we need to consider only $\vec{r}_n = \vec{r}_m$, and the action reduces to

$$S[r] = \int_0^\beta d\tau \left[\frac{1}{2} \sum_n \frac{\pi^2}{4E_C} \dot{r}_n^2 + \sum_{n,m} v_n v_m \pi E_J G\big(\vec{r}_n(\tau) - \vec{r}_m(\tau)\big) \right]. \tag{36}$$

This result demonstrates that in the considered limit the vortices can be viewed as particles with a mass

$$M_v^0 = \frac{\pi^2}{4E_C} \frac{\hbar^2}{p^2}. \tag{37}$$

In order to get a feeling for the magnitude we can compare the mass to that of an electron (and the lattice spacing p to the Bohr radius a_0). For $E_C \approx 0.1$ K and a cell size of $5\,\mu m^2$ the mass of the vortex is smaller than the electron mass by a factor 0.004. Hence we can expect strong quantum mechanical effects.

The idea of a 'vortex mass' and the expressions (35–37) have been developed before Ref. [11] and [14]. There is assumed that the phase configuration of a classical vortex is not affected by the potential barriers as the vortex moves through the lattice. This is the case only if the effective potential barriers are small, which according to Lobb et al. [19] have a height $0.2E_J$. Hence the bare vortex mass can be found only if the kinetic energy scale exceeds the barrier $E_C \geq 0.2E_J$. On the other hand, one can account at least partially for the complexity of the vortex motion, by reintroducing a potential

$$S_{\text{eff}}[r] = S[r] + \int_0^\beta d\tau \sum_n V(\vec{r}_n).$$ (38)

Here $S[r]$ is given by (36) and $V(\vec{r}_n)$ is the periodic potential with a modulation amplitude $0.1E_J$ as suggested in Ref. [19]. Using the simple picture, a vortex with bare band mass (37) moving in a periodic potential yields in the tight binding limit, $E_J \gg E_C$, the band mass

$$M_v \simeq \frac{1}{\sqrt{8E_J E_C}} \exp\left[\sqrt{0.41 E_J/E_C}\right].$$ (39)

Alternatively, in the limit $E_J \gg E_C$ we can use the action (32) without further approximations. The instanton action corresponding to a vortex moving in a time step $\Delta\tau$ from one site j to the neighboring one $j+1$ becomes [16, 43]

$$S_{\text{inst}} = -\frac{2\pi}{8E_C\Delta\tau}G(1) \simeq \sqrt{3.04 E_J/E_C}.$$ (40)

There remains some ambiguity (factors of order one) in the numerical coefficients in (40) due to uncertainties in the precise value of the time step $\Delta\tau$ and of $G(1)$. The explicit result given above is chosen to coincide with a result of a direct instanton calculation, based on the original action in the ϕ-representation [12, 43]. From the instanton action we obtain the 'band mass' of a vortex (again up to numerical coefficients of order one)

$$M_v \simeq \frac{1}{\sqrt{8E_J E_C}} \exp\left[\sqrt{3.04 E_J/E_C}\right].$$ (41)

The fact that this result differs from (39) — more than the ambiguity in the numerical coefficients — demonstrates that the vortex is not moving as a rigid object but adjusts during its motion its internal degrees of freedom.

At finite temperatures the vortices can move thermally activated from one site to another with a rate which depends on the barrier height. At low temperatures this process is still possible due to quantum mechanical tunneling. The tunneling rate is given by the instanton action (40)

$$\Gamma_q \propto \exp[-S_{\text{inst}}],$$ (42)

and provides a measure for the vortex mass. Furthermore, if the vortices have a mass one can expect them to move ballistically under suitable circumstances. This has recently been demonstrated in the experiments of van der Zant et al. [31, 44] and will be discussed in Sec. 5.5.

A non-vanishing self-capacitance $C_0 \neq 0$ provides a damping mechanism for the vortex dynamics [14]. (In this case the duality breaking term in (23) becomes important.) The difference between the effect of a self-capacitance and a junction capacitance can be understood by comparing the dispersion relations of the spin waves in the two limits.

If $C \gg C_0$ the spin waves have only an optical branch, whereas for $C_0 \neq 0$ they have an acoustic branch, which allows the generation of low energy spin waves, providing a mechanism for dissipation. The difference between these properties can be traced further back to the property of the system under Galilei transformations. It is invariant in the former limit, whereas the self-capacitance provides a frame of reference [14].

5.2. The Aharonov-Casher effect

The analogy between a vortex and a quantum mechanical particle goes even further. It has been suggested [45, 46] that charges act as a gauge field on vortices in the same way as a magnetic flux acts on a charged particle. This can be demonstrated from the coupled-Coulomb-gas description extended to include 'external charges'. An external charge $Q_{x,i}$ on the island i is fixed by external constraints and does not fluctuate. It can arise for example by coupling the island by means of the capacitance C_0 to an external voltage source. This binds a part of the charge Q_i on the island i at the capacitance C_0. If $C_0 \ll C$ the external charge remains approximately fixed even if the total charge on the island changes by tunneling. We can also imagine to couple some of the external islands of the array via a high-Ohmic resistor to the external circuit. In this case the external charge changes due to the externally imposed current. Finally we mention that charged impurities, e.g. in the substrate under the array, can create random external charges.

In the coupled-Coulomb-gas action (23) the external charges are included. In the limit $E_J \gg E_C$ we can integrate over the charges. This yields (for $f = 0$)

$$S[v; q_x] = S[v] + i \int_0^\beta d\tau q_{x,i}(\tau) \Theta_{ij} \dot{v}_j(\tau) \tag{43}$$

where the pure vortex action $S[v]$ is given by (32). Again we introduce continuous vortex trajecories $\vec{r}_n(\tau)$. Then the effective action becomes

$$S[r; q_x] = \int_0^\beta d\tau \left\{ \sum_n \left[\frac{M_v}{2} \dot{\vec{r}}_n^2 - i\vec{A}(\vec{r}_n) v_n \dot{\vec{r}}_n \right] + \sum_{n,m} v_n v_m \pi E_J G(\vec{r}_n - \vec{r}_m) \right\}. \tag{44}$$

Here $\vec{A}(\vec{r}_n) = \sum_i q_{x,i} \vec{a}(\vec{r}_n - \vec{r}_i)$ is a fictitious 'vector potential' created by the charges $q_{x,i}$ at the sites $\vec{r}_i$ and seen by the moving vortex at the site $\vec{r}_n$. The 'vector potential' of one unit charge is

$$\vec{a}(\vec{r}) = \vec{\nabla}\Theta(\vec{r}) = \hat{z} \times \vec{r}/r^2. \tag{45}$$

In the action (44) we see that the external charge create a gauge field (a 'fictitious' vector potential) which influences the vortex motion in the same way as a vector potential influences the motion of an ordinary charged particle. This implies that the quantum mechanical properties of a moving vortex depend on the enclosed charge. The influence of a charge on a magnetic particle was first studied by Aharonov and Casher [45]. Later van Wees [46] pointed out that the same effect applies for a vortex in a Josephson array, although in this case the magnetic flux of the vortex plays no role. The external charges can also create a 'fictitious' magnetic field '$\vec{B}$'= $\vec{\nabla} \times \vec{A}$, which creates a force, the Magnus force, on the classical level. It will be discussed further in the following section.

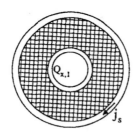

Figure 11. A ring shaped junction array with thick inner and outer electrodes. In this configuration the Aharonov-Casher effect can be observed. The quantum properties of a vortex enclosed between the thick electrodes depend on the charge $Q_{x,1}$ on the inner electrode.

We can study the consequences by considering a ring-shaped array as shown in Fig. 11. In this case we can fix the net number of vortices in the array by controlling the supercurrents in the inner and the outer ring-electrodes. They are assumed to be thick electrodes such that the fluctuations of the phase gradients are negligible. We, furthermore, can apply a magnetic field to put the minimum energy position of the vortex symmetrically between the two ring electrodes. Let us impose a phase gradient on the outer electrode; without loss of generality we can chose this phase as time independent. Then the value of the phase of the inner electrode ϕ_1, at least in the classical limit, determines the azimuthal position of the vortex core, and the band structure in the $q_{x,1}$ direction reflects the quantum motion around the ring. The band mass is given by the curvature of the ground state energy $[\partial^2 E_0/\partial q_{x,1}^2]^{-1}$ and (up to a factor $4e^2$) is equal to the effective capacitance. This in turn can be determined by measuring the voltage difference between the two electrodes, after a charge $q_{x,1}$ has been put onto one of it.

As shown by Geigenmüller [43, 16] the energy band structure differs noticably in the presence or absence of a vortex if $E_J > E_C$. He confirmed that the effective mass is equal to the bare vortex mass as long as $E_J \approx E_C$, but increases for larger E_J, consistent with (41). He also showed that the mass obtained from the instanton calculation is a better approximation than the simple expression (39).

5.3. Forces acting on vortices

An externally imposed current $\vec{I}_i(\tau)$ can be accounted for by adding to the action $S[\phi]$, given in (16), a term

$$S_I[\phi] = \int_0^\beta d\tau \frac{1}{2e} \sum_i \vec{I}_i(\tau) \cdot \vec{\nabla}\phi_i(\tau). \tag{46}$$

In Appendix B we analyze this extra term in the framework of our present discussion (following Ref. [16]). We find that the vortex action $S[v]$ acquires a new contribution

$$S_I[v] = -\int_0^\beta d\tau \frac{1}{2e} \sum_{i,j} (\vec{\nabla} \cdot \vec{I}_i(\tau))\Theta_{ij}v_j(\tau) = \int_0^\beta d\tau \frac{1}{2e} \sum_{i,j} \vec{I}_i(\tau) \cdot \frac{\hat{z} \times \vec{r}_{ij}}{r_{ij}^2} v_j(\tau),$$

coupling the current at site i to the vorticity at site j. Let us now assume that the current is uniform, then $S_I[v]$ reduces to

$$S_I[v] = -2\pi \int_0^\beta d\tau \frac{1}{2e} \sum_j \vec{I}(\tau) \cdot (\hat{z} \times \vec{r}_j) v_j(\tau). \tag{47}$$

This means that a vortex with vorticity v experiences a force

$$\vec{F}_I = -\vec{\nabla}\left[-2\pi \frac{\vec{I}}{2e} \cdot (\hat{z} \times \vec{r})\right] v = \frac{2\pi}{2ep}(I_\tau^y \hat{x} - I_\tau^x \hat{y}) v \tag{48}$$

which is perpendicular to the applied current.

The question whether vortices experience a Magnus force perpendicular to their velocity $\dot{\vec{r}}$ has been a controversial issue, and the answer depends on details of the model considered [47, 48, 49]. We have seen in the previous section that exteral charges create a 'fictitious' vector potential $\vec{A}(\vec{r})$ for the vortex motion. From this we obtain a fictitious 'magnetic' field '$\vec{B}u$'$= \vec{\nabla} \times \vec{A}$, which in turn gives rise to a ficticious 'Lorentz' force

$$\vec{F}_M(\vec{r}) = v_n q_x(\vec{r}) \hat{z} \times \dot{\vec{r}} \tag{49}$$

(For a discussion of some subtleties in the derivation of (49) see Ref. [50].) The force (49) can be called Magnus force. As a result of the Magnus force the vortex velocity gets a component parallel to the direction of the current. This in turn implies a nonvanishing Hall voltage.

Notice that the sign of Magnus force and Hall voltage depend on the charge profile. The result (49) coincides with that obtained by Fisher [49] for continuous films. In the junction array we find an additional periodicity. The properties of the system are invariant if the charge on an island changes by multiples of the Cooper pair charge. Hence also the Magnus force depends $2e$-periodically on the local charges, which implies that it depends only on the offset charges q_x and not on the charges created by the tunneling of Cooper pairs [10].

5.4. Dissipation by quasiparticle tunneling

An imposed current through the array creates a force on the vortices. As a result the vortices will be accelerated until dissipation limits their motion. It is, therefore, essential to include the dissipation in our description. In an ideal array the most important source of dissipation is the tunneling of quasiparticles. This can be described in a compact form by an effective action [37]

$$S_{\text{eff}}[\phi] = S[\phi] + \int_0^\beta d\tau \int_0^\beta d\tau' \sum_{<i,j>} \alpha(\tau - \tau')\{1 - \cos[\phi_{ij}(\tau) - \phi_{ij}(\tau')]\}. \tag{50}$$

The first part $S[\phi]$, describing the charging and the Josephson coupling, is given by (16), the second part, describing the tunneling of quasiparticles, involves a kernel which in the normal state is given by

$$\alpha(\tau) = \alpha_t \frac{(1/\beta)^2}{\sin^2(\pi\tau/\beta)}, \quad \alpha_t = \frac{R_Q}{R_t} \tag{51}$$

where R_t is the normal state tunneling resistance. In the superconducting state $\alpha(\tau)$ depends on the superconducting gap. At $T = 0$ in ideal junctions it is short-ranged, and for small frequencies the second term in (50) reduces to a renormalization of the nearest neighbor capacitance (31). A finite subgap conductance $1/R_{\rm qp}$ can be accounted for by a kernel of the form (51), but α_t is replaced by $\alpha_{\rm qp} = R_Q/R_{\rm qp}$.

Also for the more general model (50) we can proceed along the lines described above. We introduce the charges on the islands and express the phase configuration in terms of spin waves and vortices. The details of the derivation will be presented elsewhere [51, 52]. We find again a coupled-Coulomb-gas description for the charges and the vortices; however, we have to sum over all the events where a single electron tunnels. In the limit $E_J \gg E_C$ we can eliminate again the charges and find an effective action for the vortices only. In leading order in the frequencies it is [51]

$$S_{\rm eff}[v] = S[v] - \int_0^\beta d\tau \int_0^\beta d\tau' \sum_i \sum_{\mu=x,y} \alpha(\tau-\tau') \cos\left\{\frac{1}{2}\sum_j \nabla_\mu \Theta_{ij}[v_j(\tau) - v_j(\tau')]\right\} \tag{52}$$

where $S[v]$ is given by (32). If we label again the vortices by their vorticity v_n and their trajectory $\vec{r}_n(\tau)$, as given in Eq. (33), we find

$$S_{\rm eff}[r] = S[r] + \int_0^\beta d\tau \sum_n V[\vec{r}_n(\tau)] - \int_0^\beta d\tau \int_0^\beta d\tau' \int d^2r \sum_\mu \alpha(\tau-\tau')$$

$$\times \cos\left\{\frac{1}{2}\sum_n v_n \frac{\partial}{\partial r_\mu}[\Theta(\vec{r}-\vec{r}_n(\tau)) - \Theta(\vec{r}-\vec{r}_n(\tau'))]\right\}. \tag{53}$$

The first term, which accounts for the kinetic energy and interaction of vortices, is given by Eq. (36). We added the periodic potential (as given e.g. by Lobb et al. [19]) in order to account for the fact that the vortex motion is not necessarily a smooth one, rather in many situations a hopping motion [see the discussion of Eq. (38)]. The dissipative term involves the phase configuration $\Theta(\vec{r}-\vec{r}_n(\tau))$ at position $\vec{r}$ due to a vortex at site $\vec{r}_n(\tau)$. Assuming that the vortices move slowly we can expand this nonlinear term and obtain

$$S_{\rm eff}[r] = S[r] + \int_0^\beta d\tau \sum_n V[\vec{r}_n(\tau)]$$

$$-\frac{\pi}{2}\int_0^\beta d\tau \int_0^\beta d\tau' \sum_{n,m} v_n v_m \alpha(\tau-\tau') G[\vec{r}_n(\tau) - \vec{r}_m(\tau')]. \tag{54}$$

This expression shows that each vortex, apart from the logarithmic interaction with the other vortices, has properties which coincide with that of a Fröhlich polaron in two dimensions [53, 14, 15].

The last term in (54) describes the dissipation of the vortex motion. It is of a similar structure as the last term in Eq. (50) which describes the dissipation of the phase dynamics of a Josephson junction. In order to estimate the strength of the vortex damping we expand the function $G(r)$ in (54). Then the action of one vortex at position $\vec{r}(\tau)$ becomes

$$S_{\text{eff}}[\vec{r}(\tau)] = \int_0^\beta d\tau \left[\frac{M_v}{2}\dot{r}^2 + V(\vec{r})\right] + \frac{\pi^2}{4}\int_0^\beta d\tau \int_0^\beta d\tau' \alpha(\tau - \tau')[\vec{r}(\tau) - \vec{r}(\tau')]^2. \qquad (55)$$

If we compare (55) with the expansion of (50), we realize that the vortex coordinate is influenced by dissipation in the same way as the phase of a single junction. However, the dissipation of the vortex motion (the effective α) is weaker by a factor 2 than that of the phase dynamics of a single junction. This result also been found by Orlando et al. [54] on a more phenomenological level. In this argumentation we have to keep in mind that the effective resistance of good quality Josephson junctions — and hence also of arrays of such junctions — is strongly voltage dependent. If the conductance of the junctions is finite (say $1/R_{\text{qp}}$) we can parameterize the vortex dissipation by the parameter $\alpha_{\text{eff}} = R_Q/2R_{\text{qp}}$. The resulting equation of motion for a vortex with vorticity v and coordinate $r(\tau)$ (in units of the lattice spacing) then can be written as

$$M_v\ddot{\vec{r}} + \alpha_{\text{eff}}\dot{\vec{r}} = -(\pi/e)v\hat{z} \times \vec{I} + \vec{F}_M, \qquad (56)$$

where the forces on the r.h.s. had been discussed in the preceeding section. In general α_{eff} depends strongly on the velocity of the vortices [14].

One consequence of the dissipation is the possibility of a phase transition of individual vortices, similar to the transition discussed by Schmid [55] for the phase of a single Josephson junction. This transition may occur in arrays at low temperatures where a magnetic field produces excess vortices with a low density, such that their interaction can be ignored. The phase transition separates a localized and a delocalized phase. It occurs at a critical strength of the dissipation $\alpha_{\text{eff,c}} = 1$. For strong dissipation $R_t < R_Q/2$ the individual vortices are localized in the minima of the potential and the system is superconducting, whereas for $R_t > R_Q/2$ the vortices are free and the system is resistive.

5.5. Experimental observation of ballistic vortices

An experiment has been performed recently by van der Zant et al. to demonstrate the ballistic nature of vortices in arrays of strongly underdamped tunnel junctions [31, 44]. A special aluminium sample was fabricated for this purpose, as shown in Fig. 12. It contains three separate regions: a 2D array on the left where vortices, generated by a small magnetic field, are accelerated by a current, a narrow channel through which vortices can only pass in a straight line through the middle, and a detector array on the right where no currents flow. The idea is to detect where the vortices that pass through the channel leave the array. If the motion is diffusive, the vortices from the channel push each other towards the edge, and few vortices reach the edge opposite the channel. This is in fact seen at high temperatures. At low temperatures, however, it is found that in a certain regime of field values and driving currents in the left array, virtually all vortices

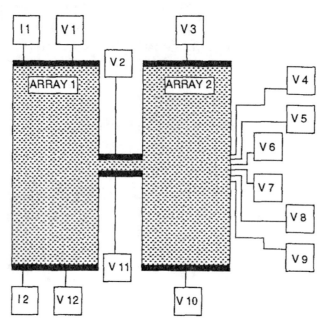

Figure 12. Layout of sample to detect ballistic vortices. A current is imposed in the left array, accelerating the vortices. A narrow channel connects to the right, where voltage probes are placed.

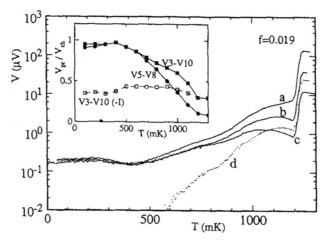

Figure 13. Measurements on the sample of Fig. 12. Plotted are the voltages across different sets of probes as a function of temperature: (a) the voltage across the channel, (b) the voltage across the right array $V3 - V10$, (c) the voltage drop between $V5 - V8$, which counts the vortices which have propagated ballistically, d) the voltage drop between $V5 - V3$, which is a measure of the vortices which have deviated from a ballistic trajectory. The inset shows the ratio of voltages to the channel voltage. The open squares give the results after the current in the left array has been reversed.

from the channel reach the opposite edge in a narrow beam. Fig. 13 shows the voltages over different probes as a function of temperature, for the same driving current in the left array. In the inset the ratio of various probe voltages to the channel voltage is given. The inertial mass of vortices is clearly demonstrated. In so far as the velocity of the vortices in the channel can only be estimated, no good quantitative conclusions can be drawn from the data. It was possible to slow down the vortices in the right hand array by applying a decelerating current there.

6. Charge dynamics

In junction arrays where the charging energy dominates the most important degrees of freedom are the charges on the islands. In the limit $E_C \gg E_J$ we can disregard the discrete nature of the vortices and integrate them out, in analogy to what we did with the charges in Sec. 4. Here we also include the effect of electromagnetic fields. The result is an effective action for the charges

$$S[q] = \int_0^\beta d\tau \sum_{i,j} \Big[\frac{1}{2\pi E_J} \dot{q}_i(\tau) G_{ij} \dot{q}_j(\tau) + \frac{2E_C}{\pi} q_i(\tau) G_{ij} q_j(\tau) + i f_i(\tau) \Theta_{ij} \dot{q}_j(\tau) \Big]. \tag{57}$$

This result shows that the 'mass' of the charge is $M_q = 1/E_J$. [Technically the kinetic energy term arises as a sum of two terms, one from the integration, the other from the duality breaking term explicit in (23).] Actually these results could be obtained much more directly [10]. The Josephson coupling describes the hopping of a Cooper pair from one island to the next with a hopping matrix element of strength E_J. In a periodic array this implies the formation of a band structure and the concept of a band mass.

The third term in (57) describes the Aharonov-Bohm effect. Notice again the similarity to the last term in (43), demonstrating the duality between magnetic frustration and the 'charge frustration' due to exteral charges. If the magnetic field, which is responsible for the magnetic frustration is homogeneous, it produces a force on the charges, give by

$$\vec{F} = q\vec{E} + q\dot{\vec{r}}_q \times \vec{B}. \tag{58}$$

Here $\dot{\vec{r}}_q$ is the velocity of a charge and $\vec{B}$ is the static field related to f. The electric field describes a force on charges due to an applied voltage along the array, dual to the force (48) on vortices exerted by an imposed current. The magnetic field produces a Lorentz force on the Cooper pair charges moving in a Josephson junction array in the same way as for ordinary free charges. Hence we expect a Hall effect for the Cooper pairs in these arrays.

Quasiparticle tunneling again provides a source of dissipation. In view of the analogy to the dynamics of a single junction [32] we expect that the strength of the dissipation for the charge dynamics is governed by a parameter $\bar{\alpha} \approx R_t/R_Q$.

The analogy of the action for the charges (57) and that for the vortices (32) suggests further physical effects. At this stage we can speculate for instance about 'ballistic charge motion' in arrays. It will be interesting to see whether the charge-vortex duality stimulates further investigations in this direction.

ACKNOWLEDGEMENTS. Much of what is covered in the present article is the result of a stimulating collaboration with R. Fazio and H. S. J. van der Zant. We, furthermore, want to acknowledge many fruitful discussions about the issues of this article with U. Eckern, L. J. Geerligs, U. Geigenmüller, T. P. Orlando, A. van Otterlo, A. Schmid, and B. J. van Wees.

A. Arrays of normal junctions with arbitrary strength of the tunneling

In order to study arrays with general strength of the tunneling we start from a microscopic Hamiltonian which contains the charging energies but also accounts for the tunneling of electrons across the tunnel barriers. The partition function of an array of normal tunnel junctions can be expressed as [37, 32] a path integral over a macroscopic field ϕ_i, defined as the integral of the voltage

$$\phi_i(\tau) = \int^\tau d\tau' \, 2eV_i(\tau') \tag{59}$$

on the island i. (The factor 2 is introduced for a unified description of the normal and superconducting case). The partition function reads (for $\hbar = 1$)

$$Z = \prod_i \int D\phi_i \exp\{-S_n[\phi]\}. \tag{60}$$

It involves the action $S_n[\phi] = S_{\rm ch}[\phi] + S_t[\phi]$

$$S_{\rm ch}[\phi] = \int_0^\beta d\tau \Big[\frac{1}{16E_0} \sum_i \dot\phi_i^2 + \frac{1}{16E_C} \sum_{<i,j>} \dot\phi_{ij}^2 \Big], \tag{61}$$

$$S_t[\phi] = \sum_{<i,j>} \int_0^\beta d\tau \int_0^\beta d\tau' \alpha(\tau - \tau') \Big\{ 1 - \cos\Big[\frac{\phi_{ij}(\tau) - \phi_{ij}(\tau')}{2} \Big] \Big\}. \tag{62}$$

The two terms in $S_{\rm ch}$ represent the charging energy due to the self-capacitance $E_0 = e^2/2C_0$ and the nearest-neighbor capacitance $E_C = e^2/2C$, respectively. The action S_t describes the dissipation due to the single electron tunneling. The subscripts i label the islands on the lattice, and $\phi_{ij} = \phi_i - \phi_j$ refers to nearest neighbors. The dissipative kernel for normal junctions is

$$\alpha(\tau) = \alpha_t \frac{(1/\beta)^2}{\sin^2(\pi\tau/\beta)} \tag{63}$$

where the dimensionless tunneling conductance is $\alpha_t = R_Q/R_t \approx 6.45\,{\rm k}\Omega/R_t$.

We have not yet specified the limits of the path integrals over the fields ϕ_i in (60). They depend on the allowed charge states of the system [32]. We consider junction arrays without Ohmic shunts. This allows us to assume that the total charges q_i on the islands are integer multiples of e. The discreteness of the charge implies that the conjugate phases ϕ_i have to be defined on a unit circle. Values which differ by 4π are equivalent. (The 4π, rather than 2π, is again a consequence of the factor 2 in the definition of ϕ_i.) As a consequence the integral in Eq. (60) includes a summation over winding numbers $\phi_i(\beta) = \phi_i(0) + 4\pi n_i$.

For the present problem it is convenient to formulate the problem in terms of the winding numbers. For this purpose we decompose the phases as

$$\phi_i(\tau) = \phi_i(0) + 4\pi n_i \frac{\tau}{\beta} + \vartheta_i(\tau) \qquad \text{where } \vartheta_i(0) = \vartheta_i(\beta) = 0. \tag{64}$$

Accordingly, the charging part of the action is decomposed into two terms, one $S_{ch}[\vartheta]$ depending on $\vartheta_i(\tau)$, the other $S_{ch}[n]$ depending on the winding numbers

$$S_{ch}[n] = \frac{\pi^2}{\beta E_0} \sum_i n_i^2 + \frac{\pi^2}{\beta E_C} \sum_{<i,j>} (n_i - n_j)^2. \tag{65}$$

This action (65) coincides with the 'Discrete Gaussian Model' (DGM), which has been studied in connection with the roughening transition of solid-solid interfaces [56]. For $1/E_0 = 0$ it is known to be analogue to the Coulomb gas problem and to have a KTB transition at a critical temperature, which, of course, coincides with the result (5).

Next we study the effect of the dissipative tunneling. As long as it is weak, i.e. α_t is small, we can treat it perturbatively. To first order in α_t the system is described by

$$Z \approx Z_{ch,\vartheta} \sum_{\{n\}} \exp\{-S_1[n]\}.$$

The winding number part of the effective action is

$$S_1[n] = S_{ch}[n] - \sum_{<i,j>} \int_0^\beta d\tau \int_0^\beta d\tau' \alpha(\tau - \tau') g_{ij}(\tau - \tau') \cos[2\pi n_{ij}(\tau - \tau')/\beta] \tag{66}$$

with $n_{ij} = n_i - n_j$, and the correlation function is

$$g_{ij}(\tau) = \exp\{-\frac{1}{8}\langle[\vartheta_{ij}(\tau) - \vartheta_{ij}(0)]^2\rangle_{ch,\vartheta}\} = \exp\{-2E_C|\tau|(1 - |\tau|/\beta)\} \tag{67}$$

The average in $\langle\ldots\rangle_{ch,\vartheta}$ and the partition function $Z_{ch,\vartheta}$ are taken with the charging action $S_{ch}[\vartheta]$.

The correlation function $g_{ij}(\tau)$ decays exponentially in time, which allows us to expand the second term of Eq. (66) in n_{ij}. As a result we recover the model (65) but with a renormalized nearest-neighbor capacitance C. This implies that in the limit of vanishing C_0 the transition is again of the KTB type but with a renormalized transition temperature [4, 9]

$$T_{cn}/T_{cn}^{(0)} = 1 - 0.11\alpha_t \qquad \text{for } \alpha_t \ll 1. \tag{68}$$

For weak tunneling the picture of charges on the island is still meaningful, but the tunneling leads to a 'dynamic screening'. This screening effectively decreases the strength of the Coulomb interaction and lowers the KTB transition temperature.

For stronger dissipation the fluctuations of ϑ_{ij} are suppressed, and we can expand in these fields. In this limit our parameterization (64) is an expansion in terms of saddle point solutions plus fluctuations. It is related to the description presented by Korshunov [57] and by Zaikin [58], simplified for the normal junction (where the instantons reduce to straight lines) and generalized to an array. (It should be noted that the 'instantons' of Zaikin are not saddlepoint solutions.) In second order the dissipative term of the action reduces to

$$S_t[\vartheta, n] \approx 2\alpha_t \sum_{<i,j>} |n_{ij}|$$

$$+\frac{1}{8} \sum_{<i,j>} \int_0^\beta d\tau \int_0^\beta d\tau' \alpha(\tau - \tau')[\vartheta_{ij}(\tau) - \vartheta_{ij}(\tau')]^2 \cos[2\pi n_{ij}(\tau - \tau')/\beta]. \quad (69)$$

The combination $\alpha_t|n_{ij}|$ arises from Fourier-transforming $\alpha(\tau)$. For very strong dissipation it is even sufficient to consider only the first term in (69), and we arrive at

$$S[n] = \frac{\pi^2}{\beta E_0} \sum_i n_i^2 + \frac{\pi^2}{\beta E_C} \sum_{<i,j>} (n_i - n_j)^2 + 2\alpha_t \sum_{<i,j>} |n_i - n_j|. \quad (70)$$

The first two parts have been already recognized as the DG model, the third term is known as the 'absolute solid on solid' (ASOS) model, which has also been studied in connection with the roughening transition of interfaces. As the temperature approaches zero only this third term survives. Monte Carlo simulations [59] show that the ASOS model also undergoes a transition of the KTB type at the critical value of coupling constant [4, 9]

$$\alpha_{t,c} \approx 0.45 \qquad \text{at } T = 0. \quad (71)$$

A comment is appropriate concerning the validity of the result (71). The effective action (70) was obtained under the assumption of strong dissipation. This implies, strictly speaking, that we can use it only to discuss the plasma phase of the system. However, all the considered step models belong to the same universality class. Hence the transition, if it exists, is of the KTB type. From the fact that at $T = 0$ in the two limits, weak or strong dissipation, we are in the ordered or disordered phase we can conclude that a KTB transition exists at a critical value of α_t. We, therefore, expect that the expansion (70), even when applied to small values of α_t, only leads to quantitative errors, e.g. the numerical value of $\alpha_{t,c}$ may change, but the existence of the phase transition and its properties are not affected.

The effect of the temperature is to decrease the critical value of dissipation. We can evaluate the phase boundary by mapping (69) onto a generalized XY model. The details of this calculation are presented in Ref. [9]. The models are dual in the sense that the weak coupling region is mapped into the strong coupling region of the XY action. We learn from it that at low temperatures the shift of the critical value of dissipation is [4, 9]

$$\alpha_{t,c} \approx 0.45(1 - \frac{\pi}{4} \frac{T}{T_{cn}^{(0)}}) \quad (72)$$

The phase diagram for an array of normal junctions is shown in Fig. 6.

B. Effect of an imposed current

An externally imposed current $\vec{I}_i(\tau)$ can be accounted for by adding to the action S, given in (16), the term S_I given in (46). We can proceed as in Sec. 4.1 and eliminate the phases in favor of the vortices. Let us temporarily replace $\vec{I}_{i,\tau}$ by $i(2e/\Delta\tau)\vec{I}_{i,\tau}$. Then the extra term in S implies that the continuity equation (20) is replaced by

$$\partial_\tau q_{i,\tau} + \vec{\nabla} \cdot \vec{I}_{i,\tau} - \vec{\nabla} \cdot \vec{J}_{i,\tau} = 0 \tag{73}$$

which can be solved by (21), however, we have to add to the right hand side the term $\tilde{I}^\mu_{i,\tau}$. Moreover, we find the constraint $\partial_\tau q_{i,\tau} + \vec{\nabla} \cdot \vec{I}_{i,\tau} = $ integer, which can be enforced by introducing the integer field $\lambda_{i,\tau}$

$$\sum_{\{\lambda\}} \exp\left[2\pi i \sum_{i,\tau} \lambda_{i,\tau}(\partial_\tau q_{i,\tau} + \vec{\nabla} \cdot \vec{I}_{i,\tau})\right].$$

With these changes we find the following modified coupled-Coulomb-gas action

$$S_{\text{CCG}}[q; v] = S_0[q] + S_0[v] + 2\pi i \sum_{i,\tau} \lambda_{i,\tau}(\partial_\tau q_{i,\tau} + \vec{\nabla} \cdot \vec{I}_{i,\tau})$$

$$+ \sum_{i,j,\tau}\Big\{ -i(\partial_\tau q_{i,\tau} + \vec{\nabla} \cdot \vec{I}_{i,\tau})\Theta_{ij}v_j(\tau)$$

$$+ \frac{1}{4\pi\Delta\tau E_J}\big[\partial_\tau q_{i,\tau} + \vec{\nabla} \cdot \vec{I}_{i,\tau}\big]G_{ij}\big[\partial_\tau q_j(\tau) + \vec{\nabla} \cdot \vec{I}_j(\tau)\big]\Big\} \tag{74}$$

where the first two terms are the same as those in (23). The charge-vortex interaction term has acquired a new contribution, which can be written in terms of $\vec{I}_i(\tau)$ as

$$S_I[v] = \int_0^\beta d\tau \frac{1}{2e} \sum_{i,j}[\vec{\nabla} \cdot \vec{I}_i(\tau)]\Theta_{ij}v_j(\tau) = -\int_0^\beta d\tau \frac{1}{2e} \sum_{i,j} \vec{I}_i(\tau) \cdot \frac{\hat{z} \times \vec{r}_{ij}}{r_{ij}^2}v_j(\tau) \tag{75}$$

where $\vec{r}_{ij} = \vec{r}_i - \vec{r}_j$.

References

[1] see for instance: *Coherence in Superconducting Networks*, Proceedings of the NATO Advanced Study Institute, ed. by J. E. Mooij and G. Schön, Physica B **152**, 1–302 (1988).

[2] J. M. Kosterlitz and D. J. Thouless, J. Phys. C **6**, 1181 (1973).

[3] V. L. Berezinskii, Zh. Eksp. Teor. Fiz. **59**, 907 (1970) [Sov. Phys. JETP **32**, 493 (1971)].

[4] J. E. Mooij, B. J. van Wees, L. J. Geerligs, M. Peters, R. Fazio, and G. Schön, Phys. Rev. Lett. **65**, 645 (1990).

[5] L. J. Geerligs, M. Peters, L. E. M. de Groot, A. Verbruggen, and J. E. Mooij, Phys. Rev. Lett. **63**, 326 (1989).

[6] M. Sugahara and N. Yoshikawa, in: *Extended Abstracts of 1987 International Superconductivity Electronics Conference (ISEC'87)*, p. 341; N. Yoshikawa, T. Akeyoshi, M. Kojima, and M. Sugahara, Jpn. J. Appl. Phys. **26**, 949 (1987).

[7] M. P. A. Fisher, G. Grinstein, and S. M. Girvin, Phys. Rev. Lett. **64**, 587 (1990); M. P. A. Fisher, B. P. Weichman, G. Grinstein, and D. S. Fisher, Phys. Rev. B **40**, 546 (1989); M.-C. Cha, M. P. A. Fisher, S. M. Girvin, M. Wallin, and A. P. Young, preprint.

[8] X. G. Wen and A. Zee, Int. J. Mod. Phys. B **4**, 437 (1990).

[9] R. Fazio and G. Schön, Phys. Rev. B **43**, 5307 (1991); R. Fazio and G. Schön, in: *Transport Properties of Superconductors — Progress in High T_C*, ed. by R. Nikolsky, Vol. 25, p. 298 (World Scientific, 1990).

[10] B. J. van Wees, Phys. Rev. B **44**, 2264 (1991).

[11] E. Simanek, Solid State Comm. **48**, 1023 (1983).

[12] S. E. Korshunov, Physica B **152**, 261 (1988).

[13] A. I. Larkin, Yu. N. Ovchinnikov, and A. Schmid, Physica B **152**, 266 (1988).

[14] U. Eckern and A. Schmid, Phys. Rev. B **39**, 6441 (1989).

[15] U. Eckern, in: *Applications of Statistical and Field Theory Methods to Condensed Matter*, ed. by D. Baeriswyl, A. R. Bishop, and J. Carmelo, NATO ASI series B Vol. 218, p. 311 (Plenum, 1990).

[16] R. Fazio, U. Geigenmüller, and G. Schön, in: *Quantum Fluctuations in Mesoscopic and Macroscopic Systems*, ed. by H. A. Cerdeira, F. Guinea Lopez, and U. Weiss, p. 214, (World Scientific, 1991).

[17] see for instance D. V. Averin and K. K. Likharev in: *Mesoscopic Phenomena in Solids*, ed. by B. L. Altshuler, P. A. Lee, and R. A. Webb, Ch. 6 (North-Holland, 1991).

[18] B. I. Halperin and D. Nelson, J. Low Temp. Phys. **36**, 599 (1979).

[19] C. J. Lobb, D. W. Abraham, and M. Tinkham, Phys. Rev. B **27**, 150 (1983).

[20] J. E. Mooij, in: *Percolation, Localization and Superconductivity*, ed. by A. M. Goldman and S.A. Wolf, (Plenum, 1984).

[21] P. Delsing et al., in: *SQUID'91*, Springer Proceedings in Physics, ed. by H. Koch, to be published.

[22] V. Ambegaokar, B. I. Halperin, D. R. Nelson, and E. D. Siggia, Phys. Rev. B **21**, 1806 (1980); K. K. Mon and S. Teitel, Phys. Rev. Lett. **62**, 673 (1989).

[23] U. Geigenmüller and G. Schön, Europhys. Lett. **10**, 765 (1989).

[24] Yu. V. Nazarov, Zh. Eksp. Teor. Fiz. **96**, 240 (1989).

[25] L. S. Kuzmin, P. Delsing, T. Claeson, and K. K. Likharev, Phys. Rev. Lett. **62**, 2539 (1989).

[26] P. Bobbert, R. Fazio, U. Geigenmüller, and G. Schön, in: *Macroscopic Quantum Phenomena*, ed. by T. D. Clark et al., p. 119, (World Scientific, 1990).

[27] L. V. Keldysh, JETP Letters **29**, 658 (1979).

[28] D. Chklovskii, unpublished.

[29] A. Widom and S. Badjou, Phys. Rev. B **37**, 7915 (1988).

[30] A. van Otterlo, P. Bobbert, and G. Schön, in: *Proceedings of the Conference 'Physics in Two Dimensions'*, Neuchâtel (1991), to be published in Helv. Phys. Acta.

[31] H. S. J. van der Zant, H. A. Rijken, and J. E. Mooij, J. Low Temp. Phys. **79**, 289 (1990).

[32] G. Schön and A. D. Zaikin, Phys. Rep. **198**, 237 (1990).

[33] A. Widom, G. Megaloudis, T. D. Clark, H. Prance, and R. J. Prance, J. Phys. A **15**, 3877 (1982); K. K. Likharev and A. B. Zorin, J. Low Temp. Phys. **59**, 347 (1985).

[34] L. S. Kuzmin and D. Haviland, Phys. Rev. Lett. **67**, 2890 (1991).

[35] See for example J. V. Jose et al., Phys. Rev. B **16**, 1217 (1977).

[36] S. E. Korshunov, Europhys. Lett. **11**, 757 (1990).

[37] U. Eckern, G. Schön, and V. Ambegaokar, Phys. Rev. B **30**, 6419 (1984).

[38] B. G. Orr, H. M. Jaeger, A. M. Goldman, and C. G. Kuper, Phys. Rev. Lett. **56**, 378 (1986); H. M. Jaeger, D. B. Haviland, B. G. Orr, and A. M. Goldman, Phys. Rev. B **40**, 182 (1989); see also [1].

[39] S. Chakravarty, G.-L. Ingold, S. Kivelson, and A. Luther, Phys. Rev. Lett. **56**, 2303 (1986); A. Kampf and G. Schön, Phys. Rev. B **36**, 3651 (1987); M. P. A. Fisher, Phys. Rev. B **36**, 1917 (1987); W. Zwerger, J. Low Temp. Phys. **72**, 291 (1988); R. Fazio, G. Falci, and G. Giaquinta, Sol. St. Comm. **71**, 275 (1989); for further references see also [1].

[40] K. B. Efetov, Sov. Phys. JETP **51**, 1015 (1980).

[41] R. M. Bradley and S. Doniach, Phys. Rev. B **30**, 1138 (1984).

[42] B. Mirhashem and R. A. Ferrell, Phys. Rev. B **37**, 649 (1988); U. Eckern and G. Schön, in: *Advances in Solid State Physics*, ed. by U. Rössler, Vol. 29, p. 1, (Vieweg, 1989).

[43] U. Geigenmüller, in: *Macroscopic Quantum Phenomena*, ed. by T. D. Clark et al., p. 131 (World Scientific, 1990).

[44] H. S. J. van der Zant, F. C. Fritschy, T. P. Orlando, and J. E. Mooij, Phys. Rev. Lett. **66**, 2531 (1991); H. S. J. van der Zant, F. C. Fritschy, W. E. Elion, L. J. Geerligs, and J. E. Mooij, submitted to Phys. Rev. Lett.

[45] Y. Aharonov and A. Casher, Phys. Rev. Lett. **53**, 319 (1984); B. Reznik and Y. Aharonov, Phys. Rev. D **40**, 4178 (1989).

[46] B. J. van Wees, Phys. Rev. Lett. **65**, 255 (1990).

[47] P. Nozieres and W. F. Vinen, Philos. Mag. **14**, 667 (1966).

[48] J. Bardeen and M. J. Stephen, Phys. Rev. **140**, 1197 (1965).

[49] M. P. A. Fisher, preprint.

[50] R. Fazio, A. van Otterlo, G. Schön, H. van der Zant, J. E. Mooij, in: *Proceedings of the Conference 'Physics in Two Dimensions'*, Neuchâtel 1991, to be published in Helv. Phys. Acta.

[51] R. Fazio and G. Schön, in: *SQUID'91*, Springer Series in Electronics and Photonics, ed. by H. Koch and H. Lübbig, to be published.

[52] R. Fazio and G. Schön, unpublished.

[53] see R. P. Feynman, *Statistical Mechanics*, Chap. 8, (Benjamin/Cummings, 1972).

[54] T. P. Orlando, J. E. Mooij, and H. S. J. van der Zant, Phys. Rev. B **43**, 10218 (1991).

[55] A. Schmid, Phys. Rev. Lett. **51**, 1506 (1983).

[56] S. T. Chui and J. D. Weeks, Phys. Rev. B **14**, 4978 (1976).

[57] S. E. Korshunov, Pis'ma Zh. Eksp. Teor. Fiz. **45**, 342 (1987) [JETP Lett. **45**, 434 (1987)].

[58] A. D. Zaikin, in: *Quantum Fluctuations in Mesoscopic and Macroscopic Systems*, ed. by H. A. Cerdeira, F. Guinea Lopez, and U. Weiss, p. 255 (World Scientific, 1991); and this volume.

[59] Y. Saito and H. Müller-Krumbhaar, in: *Applications of the Monte Carlo Method*, ed. by K. Binder, (Springer, 1984).

Chapter 9

Possible Applications of the
Single Charge Tunneling

D. V. AVERIN and K. K. LIKHAREV

Department of Physics, State University of New York, Stony Brook
NY 11794, USA,
and Department of Physics, Moscow State University
Moscow 119899 GSP, USSR

1. Introduction

We have had a temptation to call this chapter traditionally, something like "Single-Electronic Devices", but eventually decided to change the title. The reason is that the present-day applied physics literature is full of descriptions of various electronic "devices" which have no real chance to be used in practice, because their possible performance is inferior to that of commercially available devices. (Most amusing is that in most cases this comparison is never being made, with silent mutual understanding of the grim facts by the writing, reading, and possibly even financing parties). In some situations, this sort of research work is justified (provided that its authors do not depart too far from what Rolf Landauer calls the "truth in advertising" [1]). In fact, operation of many new "devices" involves very complex physics which sometimes can be studied in this way alone.

In our present field (which we personally prefer to call Single-Electronics [2]), the situation is significantly different. Since its formulation in the mid-80's, the so-called "orthodox" theory of the correlated single charge tunneling (for its comprehensive review, see Ref. 3) was shown to be extremely successful in description of most experimental results (see Ref. 4 as well as preceding chapters of the present volume). More exactly, recent experiments with systems of few tunnel junctions (see, e.g., Refs. 5–9), like uniform 1D arrays [6], "turnstiles" [7], "pumps" [8], and "dots" [9], demonstrated that the orthodox theory gives a *quantitative* description of the correlated tunneling in all known tunnel junction systems with metallic electrodes of linear dimensions a from some 5 nm [5] to a few hundreds nm (the latter boundary is apparently determined just by the lowest temperatures available, with the scaling $a_{max} \propto T^{-1/2}$). This range comfortably covers that (crudely, 10 to 100 nm) of the junctions which are both interesting for practice and can be fabricated reproducibly using existing nanofabrication techniques.[1] As a result, one can rely on the orthodox theory in invention, simulation, and performance analysis of possible applications of the correlated tunneling.

We believe that this is the time to stop playing with toy-like few-junction "devices" [6–9] (which were very important at the initial stage of development of the field) and to start the real applied work by development of more complex, typically very non-uniform, multi-junction devices which really could have a considerable practical potential. This chapter is devoted to the basic principles of such a development. We will restrict ourselves to devices using correlated tunneling of single electrons, because those using the tunneling of single Cooper pairs (in ultrasmall Josephson junctions) are more complex and their practical potential is still far from being clear.

We will start with a brief description (in Sec. 2) of the "rules of the game" as prescribed by the orthodox theory, and a demonstration of their application to a particular system (Sec. 3). In Sec. 4–6 we will apply these rules to description of relatively simple analog devices. In Sec. 7 we discuss the possible digital integrated circuits which are the most ambitious goal of the single electronics. The concluding Sec. 8 is devoted to a discussion of the most urgent problems of the applied single electronics.

2. Single electronics: The rules of the game

General operating principle of single-electronic devices is to control the tunneling of single electrons in systems of small tunnel junction. The orthodox theory prescribes the rules of this single electron tunneling. Its predictions are especially simple in the quasiclassical limit when tunnel conductances G_i of all the junctions of the system under consideration are small enough:[2]

$$G_i \ll R_Q^{-1}, \qquad R_Q \equiv \pi\hbar/2e^2 \simeq 6.5\,k\Omega, \tag{1}$$

so that quantum fluctuations of the electric charge on the electrodes of the junctions are negligible. In this limit, the theory says that tunneling of an electron through the junction has the probability rate

$$\Gamma_i = \frac{1}{e}I_i(V_i)\left[1 - \exp(-\frac{eV_i}{k_BT})\right]^{-1}, \qquad V_i \equiv \Delta W_i/e. \tag{2}$$

Here $I_i(V)$ is the dc $I-V$ curve of the i-th junction (when biased by a fixed dc voltage V, so that the single electron charging effects vanish), while ΔW_i is the decrease of the potential (electrostatic) energy of the system, resulting from this particular tunneling event.

In the low temperature limit

$$k_BT \ll \Delta W_i, \tag{3}$$

---

[1] Semiconductor structures with 2D electron gas can exhibit the correlated single charge tunneling (see Refs. 10-12, and Chap. 5), but are still too irreproducible to be considered seriously for practical applications.

[2] To relate formulas given in this chapter to results of earlier chapters use $R_Q = R_K/4$.

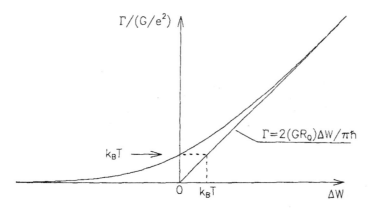

Figure 1. Rate Γ of the single electron tunneling as a function of the electrostatic energy reduction ΔW, for a normal metal tunnel junction with the linear dc $I - V$ curve ($I(V) = GV$).

the function $\Gamma(\Delta W)$ virtually vanishes at $\Delta W < 0$ (Fig. 1) so that Eq. (2) is reduced to a very simple

Crude Rule: A particular single electron tunneling event takes place as soon as it decreases the electrostatic energy of the system (including the external energy sources).

Nevertheless, for many practical devices the exponentially small tail of the function $\Gamma(\Delta W)$ at $\Delta W < 0$ caused by the thermal fluctuations, should be taken into account even in this limit, because it can yield a small but still unacceptable rate of undesirable tunneling events (for example, errors in single electron logic circuits, see Sec. 7).

The second important source of undesirable events is the macroscopic quantum tunneling of charge (q-mqt, see Ref. 13 and Chap. 6 which can provide a nonvanishing rate Γ even for $\Delta W < 0$ and $T \to 0$. In essence, this effect is nothing more than the single electron tunneling through $n > 1$ junctions at a time, provided that the process as a whole leads to $\Delta W_\Sigma > 0$, even if its partial steps give $\Delta W_i < 0$. The rate Γ_Q of q-mqt is described by relatively complex formulas of Sec. 3.2 of Chap. 6 but crudely scales as

$$\Gamma_Q \propto \frac{1}{\hbar}(GR_Q)^n \frac{\Delta W_\Sigma^{2n-1}}{|\Delta W_i|^{2n-2}}. \tag{4}$$

Thus, design of each single electronic device can be separated in two stages: an approximate classical analysis using the Crude Rule, and a more quantitative analysis of undesirable effects due to the thermal activation and macroscopic quantum tunneling.

3. Rules of the game: An illustration

As the simplest example, consider a possible structure of the single-bit memory cell. The starting idea would be, of course, to store the information as an extra single electron charge of an (almost) insulated metallic electrode (Fig. 2a). In order to insert and extract the charge (i.e., perform writing in and reading out the information), one would need to connect this electrode to external circuits by at least one tunnel junction.

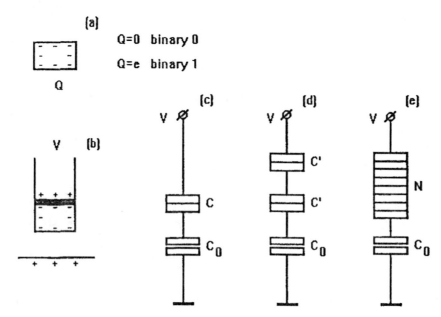

Figure 2. Development of the single electron memory cell: (a) the basic idea of the charge storage; (b) and (c) single-junction system (no internal memory); (d) double-junction system (internal memory vulnerable to q-mqt); (e) more stable multi-junction system.

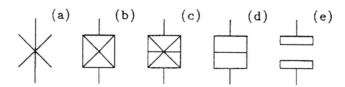

Figure 3. "Saclay Notation" for basic elements of the single electronic equivalent circuits: (a) large ("classical") Josephson junction, (b) ultrasmall Josephson junction with Cooper-pair tunneling ("Bloch junction"), (c) similar junction with additional tunneling of single electrons, (d) ultrasmall tunnel junction with single electron tunneling ("SET junction"), and (e) capacitor.

Thus one arrives at the system shown in Fig. 2b. Using the notation accepted at the Saclay meeting in April 1990 and shown in Fig. 3 (this notation will be used through all this chapter), we obtain an equivalent circuit of Fig. 2c, where C_0 is own capacitance of our metallic electrode (capacitance to the ground), while C is that of the tunnel junction.

If external voltage V is fixed (independent of the cell contents), it is straightforward to write down the electrostatic energy of the system:

$$W(n) = \frac{(ne + Q_0)^2}{2C_\Sigma} + neV\frac{C_0}{C_\Sigma}, \qquad C_\Sigma \equiv C + C_0, \tag{5}$$

where Q_0 is the (possible) background charge of the electrode (for the sake of simplicity we will accept $Q_0 = 0$ until discussion in Sec. 7) and n is an integer number of extra electrons on the electrode. Applying the Crude Rule, one immediately obtains that the system would be in its binary-zero state at

$$-\frac{e}{2C_0} < V < +\frac{e}{2C_0} \qquad (n = 0) \tag{6}$$

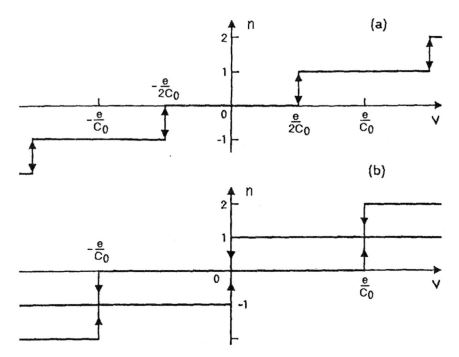

Figure 4. Static numbers of electrons in (a) single-junction cell (Fig. 2c) and (b) multi-junction cells (Fig. 2d, e) as a function of the applied voltage V.

and in its binary-unity state at

$$+\frac{e}{2C_0} < V < +\frac{3e}{2C_0} \qquad (n = 1). \tag{7}$$

Our goal seems to be achieved, but it is not: the regions (6) and (7) *do not overlap* (see Fig. 4a). Hence, contents of the cell is completely determined by the external voltage; in other words, it lacks *internal* memory (operation of such a system without memory was experimentally demonstrated recently, see Chap. 3 and Ref. 9).

In order to get a real memory cell, parameter regions corresponding to $n = 0$ and $n = 1$ should be partly overlapped. The simplest way to realize it is to use two tunnel junctions connected in series instead of the single one, see Fig. 2d. In fact, this circuit can be characterized by two integer numbers, n and n', of electrons passed through each of the junctions, and its electrostatic energy is

$$W(n, n') = \frac{e^2 n^2}{2C_\Sigma} + \frac{e^2 (n')^2}{2C'_\Sigma} - \frac{e^2 nn'}{2\overline{C}} + (\alpha n + \beta n')eV, \tag{8}$$

where

$$C_\Sigma = D/(C + C'), \quad C'_\Sigma = D/(C + C_0), \quad \overline{C} = D/2C_0, \quad D \equiv CC' + CC_0 + C'C_0$$

and

$$\alpha = C'C_0/D, \quad \beta = CC_0/D.$$

Figure 4b shows n as a function of V for a simple case when $C = C' = C_0$ (and as a consequence $C_\Sigma = C'_\Sigma = \overline{C} = (3/2)C_0$, $\alpha = \beta = 1/3$), as calculated using the Crude

Rule. One can see that this system does exhibit bistability (i.e., internal memory) within the range $0 < V < e/C_0$. Moreover, thresholds for undesirable events ($2e/C_0$ for $n \to 2$, and $-e/C_0$ for $n \to -1$) are far enough from the basic operation range, so that the memory cell has quite big parameter margins.

This completes the approximate analysis of classical dynamics of electron tunneling in the memory cell circuit. Now we can pass to the second stage of the analysis, and consider the effects of thermal and quantum fluctuations. Let the cell be in the binary 0 state ($n = n' = 0$) and biased in the middle of the bistability region ($V = e/2C_0$); then Eq. (2) yields the following estimate for the probability rate of the undesirable events $n' \to 1$:

$$\Gamma_T \simeq \frac{G}{C_0} \exp\left[-\frac{e^2}{6C_0 k_B T}\right], \qquad \text{for} \quad k_B T \ll e^2/C_0. \tag{9}$$

After this thermally activated passage through the first junction, the electron has an equal probability to tunnel further ($n \to 1$) or come back ($n' \to 0$), so that the rate of the former event (giving parasitic switching of the memory cell to the binary 1 state) is equal to half of that indicated by Eq. (9). It is straightforward to check that opposite switching (binary 1 $\to$ binary 0) has the similar probability in this bias point.

In order to evaluate the error probability, let us assume that we are using typical state-of-the-art $Al/Al_2O_3/Al$ junctions of area $S = 60 \times 60$ nm^2, which yields capacitance C about 2×10^{-16} F. In order to keep the characteristic time scale $\tau = C/G$ of the cell not very long (say, below 100 ps), the junction conductance G should be not less than $(500 \text{ k}\Omega)^{-1}$. Using Eq. (9), one obtains that in order to keep the probability of the thermally induced errors small (say, below 10^{-30} during one clock cycle period $\tau \sim$ 0.6 ns), one should make the operation temperature quite low, below 0.02 K. (The only realistic way to higher temperatures is scaling the junctions down: $T \propto S^{-1}$).

Coming to the second source of the digital errors, the macroscopic quantum tunneling, one can use general formulas of Chap. 6 (Sec. 3.2) to get the following result for characteristic value of the tunneling rate:[3]

$$\Gamma_Q \simeq \frac{G R_Q}{6\pi^2} \frac{G}{C}. \tag{10}$$

Substituting the same parameters as above, one gets a very high error rate $\Gamma_Q \simeq 2 \times 10^{+6}$ s^{-1}. Any serious attempt to reduce Γ_Q by a drastic reduction of G would result in unacceptably large values of the time constant $\tau = C/G$. One can see that our simplistic approach has failed: the macroscopic quantum tunneling of charge makes the memory cell (shown in Fig. 2d) virtually unstable. It is easy to comprehend that this drawback cannot be cured by a slight modification of the circuit: all few-junction single-electronic systems have relatively high q-mqt rates, so that their possible practical applications are quite restricted.

---

[3] In the middle of the bistability region, $V = e/2C_0$, electron transfer through both junctions does not change electrostatic energy of the circuit, so that the q-mqt rate in this bias point is proportional to temperature T, and vanishes at $T \to 0$ (see Chap. 6). Hence, this point is not representative for calculation of the q-mqt rate, and we evaluate it at $V = 3e/4C_0$, i.e., in the middle of the interval between $V = e/2C_0$ and the boundary of the bistability region.

Nevertheless, the q-mqt can be made exponentially small by using a considerable number of tunnel junctions in one device. Figure 2e shows an evident generalization of our previous memory cell. Analysis of this circuit (similar to that carried out above) shows that the bistability region is limited by the following margins

$$\frac{e}{2C}[c - N + 1] \le V \le \frac{e}{2C}[c + N - 1], \qquad c \equiv \frac{C}{C_0} \tag{11}$$

so that at $C = C_0(N - 1)$ the state diagram of the system is similar to that shown in Fig. 4b. In the middle of the range (11) ($V = e/2C_0$) the rate of the thermal events can be evaluated as

$$\Gamma_T \sim \frac{G}{C_0} \exp\left[-\frac{e^2}{16C_0 k_B T}\right], \qquad \text{for } N \gg 1, \ k_B T \ll e^2/C_0, \tag{12}$$

so that for fixed C the probability of thermally induced errors can be much less than for the two-junction cell. For example, at $N = 8$ and the same junction parameters as above, we get $C_0 = 3 \times 10^{-17}$ F, and operation temperature can be raised to ~ 0.1 K.

The characteristic rate of macroscopic quantum tunneling for $N \gg 1$ can be estimated as follows (the bias point for this estimate is chosen similarly to that for Eq. (10), $V = e/2C_0 + e/4C$):

$$\Gamma_Q \sim \frac{G}{2C} \left(\frac{GR_Q}{\pi^2}\right)^{N-1} \left(\frac{N}{2}\right)^{2N} \frac{1}{(2N - 1)! \, [(N - 1)!]^2}. \tag{13}$$

For $N = 8$ and the above parameters this expression yields a reasonably low value $\Gamma_Q \sim 10^{-20}\,\text{s}^{-1}$.

This example shows that the one-dimensional (1D) arrays of several junctions allow one to increase stability of the single electron devices with respect to both thermal and (especially) quantum fluctuations. Another possible way is to use high-ohmic metallic resistors ($R \sim 10^6$–$10^7\,\Omega$) with a small stray capacitance $C_s \le C_0$ (for a junction-resistor system, the q-mqt rate scales as $\exp(-R/R_Q)$ [13], i.e., can be exponentially low). The requirement of small stray capacitance, however, limits the useful resistor length to few tens μm, so that its sheet resistance should be as high as $\sim 10^6\,\Omega/\square$. Present-day technologies (see, e.g., Ref. 14) can provide values not more than $\sim 10^3$–$10^4\,\Omega/\square$. This is why we will not discuss devices using the resistors in this chapter (although such a use would extend considerably the variety and flexibility of the single electron devices).

4. DC current standards

The most evident practical application of the correlated tunneling are the dc current standards (first suggested [15] for the Cooper pairs and then [16] for the single electrons.) The basic idea is that the Coulomb repulsion of the single charges (e or $2e$) allows one to organize a transfer of exactly one (or a fixed integer number n) of the charges during one cycle of an external rf drive. In this case the dc current carried by the charges is

$$I = nqf, \qquad q = \begin{cases} e, & \text{for electrons} \\ 2e, & \text{for Cooper pairs} \end{cases} \tag{14}$$

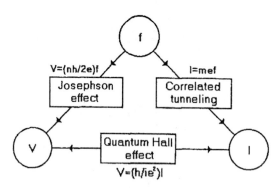

Figure 5. "Quantum Metrology Triangle" formed by three major variables related through three fundamental physical effects (i, m, n, are integers).

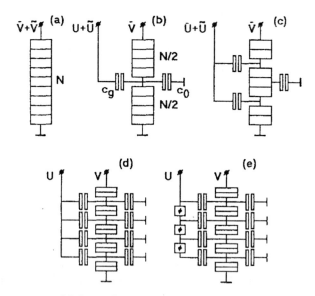

Figure 6. Multi-junction arrays which can be used for dc current standards (for a discussion, see text).

where f is the rf drive frequency. If this frequency is well characterized (present-day metrology allows such a characterization accurate up to $\sim 10^{-14}$), one gets a very accurate source of the dc current, provided that relation (14) is precise. Such a high-precision current standard would close the "quantum metrology triangle" (Fig. 5) [17].

The most apparent way to the dc current standards is the phase locking of the SET or Bloch oscillations (see Chap. 7) by n-th harmonic of the external rf drive. The first candidate for such a system is merely a 1D array of tunnel junctions (Fig. 6a) in which the SET oscillations were observed for the first time [18]. Numerical simulations using the quasiclassical approximation (2) of the orthodox theory show [19] that a nearly complete phase locking can really be implemented in the array, provided that stray capacitance C_0 of its electrodes is small ($C_0 < 0.01C$), the number N of junctions is large ($N > 3(C/C_0)^{1/2} > 30$), and temperature is low ($k_B T < 0.01e^2/C$).

This simplest system, however, suffers from a major drawback: a small phase locking range. In fact, in order to achieve a precise phase locking of such a intrinsically random process as the SET oscillations (see Sec. 2 above), the rf drive period $1/f$ should be much larger than the characteristic time constant $\tau = RC$ of each junction; i. e.

$$fRC \ll 1. \tag{15}$$

However, for the relaxation-type systems (like overdamped Josephson junctions, SET oscillators, etc.), the phase locking parameter range always scales as f at $fRC \to 0$, and in the limit (15) is rather small.

Another system seems therefore more practical (Fig. 6b). Here the rf drive is applied to the middle point of the 1D array with a relatively small number of junctions: $N < (C/C_0)^{1/2}$. Operation of this system is close to that of the "single electron turnstile" [7] (see also Chap. 3): during the negative half-cycle, the rf drive causes injection of one electron into the middle point of the structure through one of its arms, while the positive half-cycle pushes the electron out through the second arm. This system (at a proper choice of its parameters) does not exhibit the autonomous SET oscillations, and the periodic transfer of the single electrons is induced (rather than phase locked) by the rf drive. This is why the parameter range where Eq. (14) holds, is constant rather than vanishing at $fRC \to 0$.

Note that such a turnstile-type device can be presented as a memory cell with two (rather than one) ports, and its dynamics is very similar to that described above (Sec. 3). A similar analysis of thermally induced and quantum tunneling "errors" (transfer of an extra electron or hole per cycle) shows that for the parameters assumed in Sec. 3, $N/2 = 8$ junctions in each arm of the device is quite sufficient for approaching accuracy $\sim 10^{-10}$ of the dc currents below $\sim 10^{-13}$ A ($f < 50$ MHz).

The same estimate remains valid for the device shown in Fig. 6c, where the rf drive is used to modulate energy barriers between the memory capacitor and the array edges. A more exact comparison of these two devices is still to be carried out, but presumably the really optimum system is their synthesis, shown in Fig. 6d. Here the rf drive changes potential of each of the array electrodes with a weight determined by individual gate capacitance $(C_g)_i$. Generally speaking, distributions of stray capacitances $(C_0)_i$ and even the junction capacitances C_i should be also optimized, although in all cases the net electrode capacitance $(C_\Sigma)_i = (C_g)_i + (C_0)_i$ should be much less than the junction capacitance

$$(C_\Sigma)_i/C_i < N^{-2} \ll 1 \tag{16}$$

in order not to screen the electron-electron interaction [19].

Even better parameters can be presumably obtained if one allows certain phase shifts φ_i between the rf drives of the electrodes (Fig. 6e). The simplest possibility here is organization of the travelling wave drive: $\varphi_i = \text{const} \simeq 2\pi/N$. (A three-junction prototype of such a "single electron pump" has already been tested [8], see Chap. 3). This device, although more complex for implementation, can be apparently somewhat more stable than the turnstile-type devices (with the same total number N of junctions), especially if special (pulse-like) non-sinusoidal drive waveforms are used.

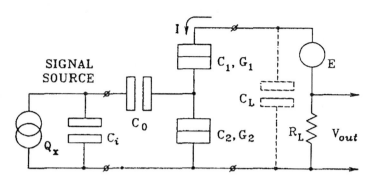

Figure 7. The single electron transistor, capacitively coupled to the signal source, as an electrometer. Output capacitance $C_L \gg C_0$, C_1, C_2 and resistance $R_L \sim R_1$, R_2 do not affect the system sensitivity.

A comprehensive analysis of the devices described above is still to be carried out, but their general features are already clear:

- very high potential accuracy (expectedly, better than 10^{-10}),
- relatively large number of junctions ($N > 10$),
- low current values (below $\sim 1\,\mathrm{pA}$).

The last drawback can be presumably corrected by either using the single-Cooper-pair tunneling (see Sec. 6), or parallel connection of a large number of 1D-array devices, or both. Note that the parallel connection of the devices should not create any problem with their dynamics, so that the possible integration scale (and hence the current multiplication factor), is only limited, first, by fabrication technology (which should provide random parameter scattering within the proper operation margins, typically $\sim \pm 10\%$), and, second, by the residual background charge (see Sec. 7).

If these problems are solved successfully, one can expect that well characterized dc currents up to $\sim 0.1\,\mu\mathrm{A}$ can be obtained using single-chip single-electronic devices. A further increase of the current (by at least two orders of magnitude) can be achieved using the superconducting dc current transformers [20, 21].

5. Supersensitive electrometry

The second straightforward application of the correlated single electron tunneling is the supersensitive electrometry. Really, already the first experiments [22, 23] with the simple double-junction system (the "Single Electron Transistor", see Chap. 2) have confirmed the theoretical prediction [24, 25] that the dc current I through such a system should be sensitive to sub-single-electron changes of the charge Q injected to its middle electrode. In the further experiments [26, 27], charge resolution of the order of $10^{-4}e/\sqrt{\mathrm{Hz}}$ has been demonstrated.

Recently, Korotkov et al. [28] have completed a theoretical analysis of the single electron transistor, capacitively coupled to the signal source (Fig. 7). In order to be realistic, this analysis takes into account the non-vanishing intrinsic capacitance C_i of the source, which can be comparable with or larger than the junction capacitances C_1 and C_2. In the analysis, C_i should be considered as a fixed quantity, while the coupling capacitance C_0 may be considered as an adjustable parameter. (Too small C_0 makes

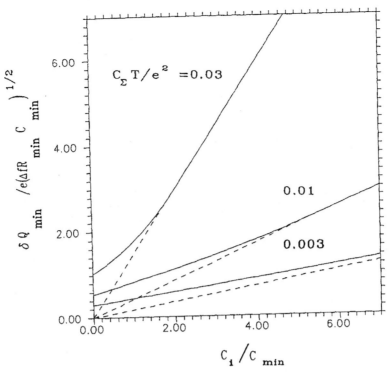

Figure 8. Charge sensitivity of the capacitively coupled single electron transistor as a function of the signal source capacitance C_i for several levels of the reduced temperature.

the signal source coupling weak, while too large C_0 can make the effective capacitance $C_\Sigma = C_1 + C_2 + C_0'$, where $C_0' = (C_i^{-1} + C_0^{-1})^{-1}$, too large in comparison with $e^2/k_B T$ and thus allow the thermal fluctuations to destroy the effect of correlated tunneling.) Tunnel junction capacitances $C_{1,2}$ can also be considered as adjustable parameters, but nevertheless limited from below by some fixed value $C_{\min}$ determined by the available technology.

Figure 8 shows the main result of the quasiclassical analysis, the signal charge resolution $\delta Q_{\min}$ of the electrometer[4] as a function of the signal source capacitance C_i. One can see that the resolution becomes worse with an increase of C_i:

$$\delta Q_{\min} \simeq \begin{cases} 5.4 C_{\min}(k_B T R \Delta f)^{1/2} & \text{for} \quad C_i \ll C_{\min}, \\ A C_i (k_B T R \Delta f)^{1/2} & \text{for} \quad C_i \gg C_{\min}, \end{cases} \tag{17}$$

where $A \to 5.4$ for $k_B T/(e^2/C_{\min}) \to 0$.

According to these semi-classical formulas, the charge resolution can be arbitrary high at $R \to 0$ and/or $T \to 0$. This divergency, however, is removed by an account of the quantum fluctuations inducing the q-mqt events. A quantitative theory of this limitation is still to be completed; a semi-quantitative evaluation [28] yields

$$\delta Q_{\min} = B(\hbar C_{\min} \Delta f)^{1/2}, \quad B \simeq (R_Q/R)^{1/2}, \quad \text{for} \quad T \to 0, \quad C_i \ll C_{\min}. \tag{18}$$

---

[4] The charge resolution is defined in a standard way, $\delta Q_{\min} \equiv (S_I(0) \Delta f)^{1/2}/|dI/dQ|$, where $S_I(0)$ is the low-frequency spectral density of the current fluctuations at the electrometer output.

Equations (17), (18) imply that at finite temperature T there is an optimal junction resistance $R_{min} \simeq (\hbar R_Q / C_{min} k_B T)^{1/2}$ for which δQ_{min} is minimum. At this resistance and $C_i \ll C_{min}$

$$\delta Q_{min} \simeq B'(\hbar C_{min} \Delta f)^{1/2}, \quad B' \simeq (C_{min} k_B T / e^2)^{1/4}. \tag{19}$$

For typical present-day parameters ($S_{min} \simeq 60 \times 60 \text{ nm}^2$, $C_{min} \simeq 3 \times 10^{-16}$ F, $T \simeq 50$ mK) Eq. (19) yields $\delta Q_{min} \sim 10^{-6} e / \sqrt{\text{Hz}}$. This figure is a factor of ~ 100 smaller than that demonstrated experimentally [25, 26]; the reason for this difference is not clear yet. Note, however, that even the "poor" experimental resolution is some six orders of magnitude better than sensitivity of the best commercially available semiconductor-transistor electrometers ($\delta Q_{min} \sim 10^{+2} e / \sqrt{\text{Hz}}$).

As the intrinsic capacitance of the signal source becomes larger, this performance gap shrinks. If for example, $C_i = 1$ pF, the charge resolution according to Eq. (17) should be close to $10^{-2} e / \sqrt{\text{Hz}}$ (for the parameter set specified above). The electrometer working in this mode ($C_i \gg C_{0,1,2}$) is more adequately considered as a voltmeter with the voltage resolution

$$\delta V_{min} = \delta Q_{min} / C_i \tag{20}$$

independent of the source capacitance. Equation (17) shows that the voltage resolution of the single charge electrometer is not particularly brilliant ($\sim 10^{-9} V / \sqrt{\text{Hz}}$ for our above example). It means that the real strength of the single charge electrometry is in measurement of charges arriving from low-capacitive (high-impedance) sources.

This statement is even more valid for the resistively-coupled single electron transistor [24, 28]. This device can be only valuable for measurement of charge/current signals from very high resistive sources ($R_i \gg R_0 \gg R$) and in quasiclassical regime has the current sensitivity

$$\delta I_{min} \simeq (8 k_B T R_{min} \Delta f)^{1/2}. \tag{21}$$

It is very tempting to compare the present-day situation in low-noise electrometry with one which appeared in the low-noise magnetometry with the advent of the SQUIDs in the late 1960s. A SQUID itself is capable of high resolution measurements of signals from very low impedance sources (say, inductances L_i below $\sim 10^{-9}$ H). However, a use of superconducting dc transformers allows one to match the SQUIDs with signal sources of a relatively high impedance ($L_i \sim 10^{-7}$ H). The electrometry, unfortunately, lacks this remarkable opportunity because the electrostatics is described by one-component electric potential φ rather than three-component vector potential $\vec{A}$ as the magnetostatics. (The superconducting dc transformers are essentially based on the possibility to twist the magnetic vector potential $\vec{A}$ in space.) Nevertheless, this conclusion is strictly valid only for very low frequencies. At nonvanishing frequencies, several opportunities for charge multiplication in capacitive systems do exist, though their applications to the single charge tunneling devices still is to be explored.

Another important issue for the new electrometry is reduction of the tunnel junction area to at least $\sim 20 \times 20 \text{ nm}^2$ ($e^2 / 2 C_\Sigma \sim 30$ K) which would allow the single electron transistor to operate reliably at usual helium temperatures, convenient for many applications. Unless it happens, new electrometry is limited to unique physical experiments (see, e.g., Ref. 9) inside dilution refrigerators.

6. Infrared radiation receivers

One more possible application of the single electron tunneling is a sensitive detection of high-frequency electromagnetic radiation [29]. Such detectors can be presumably most competitive in the far-infrared (IR) frequency range, where there is a considerable lack of sensitive receivers of other kinds (see, e.g., Ref. 30).

Principle of operation of such a detector can be based on the high nonlinearity of the dc $I - V$ curve of any system of ultrasmall tunnel junctions which exhibits the Coulomb blockade: the current is fairly small for dc voltages V below Coulomb blockade threshold V_t, and arises sharply beyond the threshold. This is exactly the vicinity of V_t where the junction is most sensitive to weak electromagnetic radiation: absorption of the radiation quanta creates a non-vanishing current I ("photon-assisted tunneling") within the interval

$$V_t - \frac{\hbar\omega}{e} < V < V_t. \tag{22}$$

In a single junction, this response can be described by the usual formula [31] characteristic for all devices based on the single electron tunneling

$$I(V, A) = \sum_n J_n^2(\frac{eA}{\hbar\omega})I(V + n\hbar\omega/e), \tag{23}$$

where A is proportional to the radiation field amplitude, J_n are Bessel functions of the first kind, and $I(V)$ is the dc $I - V$ curve of the junction in the absence of radiation. Noise properties of the single-junction receivers are also similar to those of other devices involving single electron tunneling, in particular, the celebrated S-I-S detectors and mixers (see, e.g., Ref. 32).

The Coulomb blockade in a single junction is, however, suppressed, so that the dc I-V curve $I(V)$ does not exhibit considerable nonlinearity, until the junction is isolated from the electrodynamic environment by high-value resistors (see Chap. 2). These resistors would absorb most of the radiation, and the sensitivity of the resulting receiver should be relatively low. As a result, one comes again to multi-junction arrays. Unfortunately, for the multi-junction systems Eq. (23) is *not* generally valid, and the analysis of their response to electromagnetic radiation is more complicated. Nevertheless, for some important particular cases it can be carried out [29]. The results indicate that the sensitivity of a one-dimensional array as a quadratic videodetector can be limited by the background radiation alone, provided that temperature is low enough ($k_B T \ll e^2/C, \hbar\omega$) to suppress the thermal activation of tunneling, number of junctions is large enough ($N \geq \ln(R/R_Q) \gg 1$) to suppress the macroscopic quantum tunneling of charge, and random spread of the background charge Q_0 is low enough ($\Delta Q_0/e \leq \hbar\omega/(e^2/C)$) to provide a sharp edge of the Coulomb blockade on the voltage scale of the photon-assisted-tunneling region.

Coming to more practical limitations, the arrays should be impedance matched with the radiation source (reduced to the array terminals). Because of a high impedance of the array ($NR \gg \rho_0$, where $\rho_0 = 120\pi \ \Omega$ is the free-space wave impedance) the matching can be presumably achieved only in a small frequency band ($\Delta\omega/\omega \sim \rho_0/NR$). Another problem is that for most present-day receiver applications, millikelvin temperatures are mostly impractical, and one should reach higher operation temperatures (say, 1.5 K) by using smaller junctions (areas like 20×20 nm^2 should be quite sufficient). As a bottom line, the IR receivers based on arrays of ultrasmall tunnel junctions do certainly look as an interesting possibility, but several fabrication and design problems should be solved to transform this possibility to reality.

7. Digital circuits

The most remarkable opportunities for the single electronics, however, hopefully reside in the field of digital circuits. There can be two major ways for implementation of these hopes. The first possible way is to mimic the semiconductor integrated circuits, using the single electron transistors as analogs of their semiconductor counterparts. An analysis [25] has shown that despite some drawbacks (like relatively low voltage gain) this way is feasible. However, another way seems much more attractive. Here the single digital bits are coded by single electrons (rather than by dc voltage/current levels as in the transistor circuits). Such an approach, using the natural quantization of physical variables, has proved to be extremely useful [33] in superconductor digital circuits, where the digital bits can be presented by single quanta of magnetic flux. Because of a deep (though incomplete) duality between the Josephson junction devices and single charge tunneling devices (see Chap. 2) one can make an attempt of a literal translation of the single flux quantum circuits to the single electronic language.

The first attempt of this kind was made as early as in 1987 [34], but was only partly successful. The main drawback of this first version of the "Single Electron Logic" (SEL) was that the duality mentioned above dictates changing all series connections of the circuit elements to parallel ones. It means in particular that all elementary circuits ("gates") which were very comfortably connected in parallel for the dc supply voltage in Josephson junction circuits [33] turn out to be connected in series in the single electronic circuits. This fact creates numerous technical problems, and makes the whole logic family hardly feasible.

After several unsuccessful attempts, a new version of the SEL was suggested very recently [35]. In this version, the direct duality was abandoned for the sake of performance. Figure 9 shows the simplest gate of this SEL family, the clocked buffer/inverter. One can see that this (internally bistable) circuit is basically similar to that shown in Fig. 6c, except for the fact that the upper and lower arrays are controlled by single electron outputs of the preceding gates rather than by an external voltage.

The clock cycle of the SEL circuit starts with arrival of a single hole to the clock input $T^{(+)}$. This signal is passed through the coupling capacitance as a change of the electric potential which opens the lower array for the single electron tunneling and restores zero extra charge (a higher initial potential) of the middle point of the gate, or does nothing if there was no charge during the preceding clock cycle.

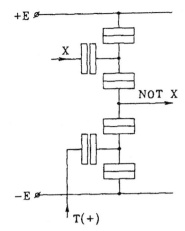

Figure 9. Simplified version of the simplest stage (buffer/invertor) of the Single Electron Logic family. In real circuits, each junction should be replaced by a one-dimensional array of several (typically, 4 to 7) junctions.

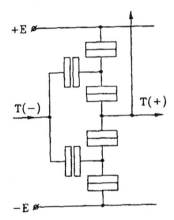

Figure 10. A possible modification of the circuit shown in Fig. 9, suitable for inversion/reproduction of the clock pulses.

During the new clock cycle, a single hole may appear or not appear at the input terminal X. In the first case (which would mean the binary unity) the upper array opens and passes a single electron to the middle point, thus switching the device to its alternative state with the negative extra charge (a lower electric potential) of the middle electrode. This potential can now be used to control similar gates (again via coupling capacitances). The clock period is terminated by arrival of the new clock pulse $T^{(+)}$ which restores the gate to its initial state with the zero charge (higher potential) of the middle electrode.

Figure 10 shows how simply the clock pulses can be generated. The SEL circuits generally need a local self-timing which means that the initial clock pulse ($T^{(-)}$ in Fig. 10) as a rule arrives from the neighboring clock circuit. Of course, this cell can be controlled by external voltage as well (thus performing the conversion of information from dc-voltage to single-electron form). It is important that this clock gate can control more than one gate, for example a similar clock gate and a signal gate (like that shown in Fig. 9).

Figure 11 shows that the similar principle can be used to design more complex

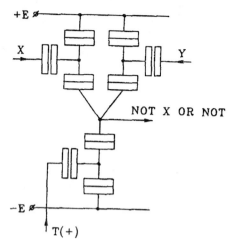

Figure 11. A possible structure of the $\bar{X} \wedge \bar{Y}$ gate of the SEL family.

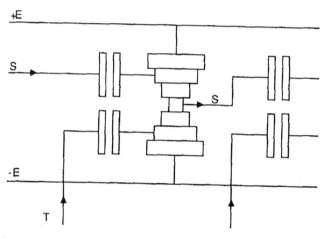

Figure 12. A version of the buffer/invertor gate with larger parameter margins (cf. Fig. 9).

gates as well. In all these circuits, a single digital bit is coded by passing of a single electron/hole between poles of the dc supply source. Preliminary estimates show that the basic gates of the Single Electron Logic family can have reasonable parameter margins (of the order of ± 10 %) provided that their parameters are chosen in an optimum way. It implies, as a rule, a considerable decrease of the area S (and hence of the capacitance C) of the junctions near the middle point (Fig. 12). This measure increases energy (and electric potential) of the electron/hole carrying the signal and allows it to control "weaker" (larger) junctions of the next gates located closer to the dc supply lines.

Table I should explain why the SEL circuits are considered as an important possible line of development of the digital electronics. It gives crude estimates of the main parameters of these circuits for various levels of the fabrication technology.

Starting from the smallest present-day junctions ($\sim 30 \times 30$ nm^2, see Ref. 36) with capacitance $C \simeq 5 \times 10^{-17}$ F, we note that their energy scale (defined as $e^2/2C$) corresponds to ~ 20 K. However, the real upper temperature limit is much smaller than this figure, because the probability Γ_T of the thermally activated tunneling events (leading to digital errors) must be extremely small: $\Gamma_T < 10^{-50}$ sec^{-1} is a reasonable requirement if we really want to run a VLSI circuit for a long time. Using Eq. (12) one arrives at the

Table I. Single Electron Logic Family: Scales of basic parameters (adopted from [3, 25]). The single junction resistance was assumed to be 600 kΩ ($\simeq 10^2 R_Q$), which yields a sufficiently low rate of the macroscopic quantum tunneling of charge, see Eq. (13).

Fabrication Technology Level	Junction Area (nm^2)	Energy Scale (J/bit)	Temperature Limit (K)	Time Scale (ps)	Integration Scale (gates/cm^2)
State-of-the-art	30×30	$3 \cdot 10^{-22}$	0.3	30	$3 \cdot 10^8$
Nanolithography Limit	10×10	$3 \cdot 10^{-21}$	3	3	$3 \cdot 10^9$
Macromolecular Devices (?)	3×3	$3 \cdot 10^{-20}$	30	0.3	$\simeq 10^{10}$

estimate $T_{max} \simeq 0.3$ K given in the Table I. Thus present-day circuits of this type are confined to very low temperatures (practically, to helium dilution refrigerators), which is of course not encouraging for most prospective users.

Time scale (defined as C/G) is also not very fascinating, because the modern electronics offers several room-temperature devices with few-ten-picosecond switching times. Even worse, one should remember that due to intrinsically random character of the single electron tunneling, the real minimum clock period should take many relaxation times. A good estimate of the additional factor is again $\ln(10^{50}) \simeq 10^2$, so that one arrives at minimum clock period of the order of three nanoseconds.

Much more encouraging is the last column of the Table I which yields a conservative estimate of the number of SEL gates (not junctions!) which can be placed on a 1 cm^2 chip. One can see that if the present-day technology is extended to VLSI (which is of course *not* a simple task) we would have the most dense electronic circuits ever available with 3×10^8 gates/cm^2, while the most optimistic projections for the semiconductor-transistor logic circuits yield the maximum density around 3×10^7 gates/cm^2. The former figure is quite compatible with the energy consumption of the circuits: for the first line of the table, switching of even all 300 million gates on the chip each 3-ns period would dissipate just some 0.1 mW heat which can be readily removed even at sub-1-K temperatures.

A further improvement of the fabrication technology to, say 10×10 nm^2 junction area would bring us to fabulous integration scales of the order of 3 billion gates per chip. Simultaneously, the time scale and hence the maximum clock frequency would become much better, and a simple He^4 cooling (quite acceptable for some prospective users) becomes possible. Despite a 100-fold increase of the total dissipation power (to about 10 mW) it is still not critical, because of a better heat removal capability of He^4. We believe these figures do really indicate that the circuits of the Single Electron Logic family (and/or other probable single-electronic circuits of the same kind) are the only reasonable prospect for future digital nanoelectronics.

It would be certainly very hard to scale the devices down below the level indicated in the second line of Table I. If we speculate for a minute that it is really possible (see the third line) we arrive at figures like ~ 10 billions gates per chip. Probably, such an advance is only possible if one uses some self-assembling macromolecular structures as the building blocks of the SEL circuits. Despite the speculative character of this proposal, we think that there is nothing unbelievable in it: there is a very broad class of macromolecules (like proteins) which can be self-assembled in a predetermined order. What about the single electron transfer, the requirements are quite limited: it presumably suffice that the molecules consist of a conducting core and an insulating shell (thin enough to enable the electron tunneling). This simplicity gives us a hope that the single electronics is the first available realistic physical basis for molecular electronic devices performing the digital information processing on nanometer space scale.

Moreover, if the self-assembly of such devices is really implemented, there is no reason to stay in two dimensions, and one can think about the three-dimensional integration. According to Table I, the space density of these devices should be as high as $\sim 10^{15}$ gates/cm^3 (the figure to be compared with some $\sim 10^7$ gates/cm^3 density of neurons in the human brain), operation speed would become quite high, and operation temperature could be in the region attainable with simple close-cycle refrigeration. The power dissipation at this stage would, however, present quite a problem, so that only a fraction of gates could be used during a given clock cycle (which is typical for very complex systems).

Another major disadvantage of the SEL circuits is a big impedance mismatch between the gates ($R \sim 10^6 \Omega$) and typical transmission lines ($\rho \sim 10^2 \Omega$). It means that ballistic propagation of the signals (so typical for the Josephson junction prototypes [33] of the SEL circuits) is hardly feasible here, and that long interconnections will produce an additional increase of the logic delays. In order to constrain this drawback, VLSI SEL circuits will require special system architectures which would minimize the long-distance communications. A large class of algorithms presumably allows such a conveyor-type ("parallel-pipeline") implementation.

8. Discussion

We believe that the list of possible applications of the single charge tunneling is by no means exclusive, and that new applications will be found as soon as the field of single electronics matures. Nevertheless, even the examples presented in this paper may justify the following strong claim: *the single charge tunneling is the only evident basis for future nanoelectronics* (if the last term means devices with linear sizes below ~ 30 nm). It does not mean, however, that the applied single-electronics has no problems. Let us list the most important of them.

8.1. Background charge relaxation

In Sec. 3 we have already met the background charge Q_0, and for the sake of simplicity have assumed $Q_0 = 0$. The reality is more complex; in several experiments with single electronic "toys" a considerable background charge was registered, and sometimes it should be compensated from outside in order to ensure a proper work of the device. For large-scale integrated circuits with their millions of junctions, such an individual trimming of Q_0 is of course impractical. Fortunately, there is an implicit but considerable experimental evidence [22, 26, 37, 38] that in at least some junctions the background charge exhibits a natural relaxation to a value $|Q_0| \ll e$. Let us argue why this evidence is not quite surprising.

The simplest model of a tunnel barrier as a pure insulating layer yields [25, 39] some fixed value Q_{00} of the background charge. However, a larger contribution to Q_0 can come from charged impurities embedded into the barrier. With the simple assumption of a constant dielectric susceptibility of the matrix material, one can write

$$\frac{Q_0}{e} = \mathrm{frac}[\frac{Q_{00}}{e} + \sum_i \frac{x_i}{d}], \tag{24}$$

where x_i is the distance of the i-th impurity from surface of one of the electrodes, and d is the thickness of the tunnel barrier. Let the junction be cooled down starting from high temperatures where the impurities can diffuse easily, in the absence of external electric fields. The electrochemical energy of the junction is then $Q_0^2/2C$, so that $Q_0 = 0$ corresponds to its minimum that is lower than the maximum (at $Q_0 = \pm e/2$) by $\Delta W = e^2/8C$. Let the temperature $T_0 = \Delta W/k_B$ be higher than the temperature T_D below which the impurity diffusion is effectively suppressed. (The time scale of impurity diffusion below T_D is larger than duration of the experiment.) Then in the range $T_D < T < T_0$ the impurities should reach an equilibrium distribution with $Q_0 \simeq 0$ in order to minimize the energy, and the further reduction of temperature (below T_D) will just freeze this distribution. (On the contrary, if $T_D > T_0$ one can expect a nearly uniform probability distribution of Q_0 within the interval $[-e/2, +e/2]$.)

One can see that with a proper choice of material (with smaller T_D) and with smaller junctions (smaller C and hence larger T_0) one can hope to provide a virtually perfect relaxation of Q_0 in all junctions of a single electronic circuit. There is an evidence that for such a "classical" barrier material of single electronics as Al_2O_3 the relaxation is quite effective even for junctions as large as $0.1 \times 0.1\mu m^2$, and that the non-vanishing Q_0 observed in several experiments was due to the charge impurities trapped in the silicon oxide layer covering the substrate. This gives a hope for a simple solution of the background charge problem, since removal of the silicon oxide layer does not do any harm (moreover, this operation would allow a natural shunting of the junctions by the silicon substrate during their fabrication and handling at elevated temperatures, helping to avoid their blow up by occasional electric discharges). Nevertheless, a need of more detailed and explicit measurements of the Q_0 relaxation is evident.

8.2. Fabrication technology

The second important issue of the applied single electronics is the progress of the fabrication technology, which should be pursued along at least three complementary directions:

  – extension of the present-day shadow-mask e-beam technology to smaller junction
    sizes (down to at least 20 nm level, see Table I);
  – its extension to VLSI circuits (up to $\sim 10^9$ junctions per $1\,cm^2$ chip);
  – search for alternative technologies which would allow a further progress in both
    directions.

For the first direction, the most promising approach is seemingly a use of non-organic
resist materials like AlF_3 for the shadow masks. In the second direction, the largest
problem may be a big time of the direct e-beam writing, and a transfer to X-ray lithography may become necessary. Finally, for more complex circuits, several (more than
two) metallic layers may become necessary, and one should replace the shadow mask
technique with a more comprehensive technology, possibly a direct patterning of each
layer.

8.3. Design automation

One more problem is automation of numerical simulation and design of the single
electronic devices. Such a simulation presents a problem even for simple few-junction
devices within the simple quasiclassical approach described by Eq. (2). Quantitative calculations of rate of the macroscopic quantum tunneling of charge still present a problem
even for relatively simple systems. There is no hope to create complex single electronic
devices until advanced software packages (which would allow an automatic calculation
of a device performance, error rate, and parameter margins) are developed.

Besides modeling on the equivalent-circuit level, at least two other type of software support are necessary. Programs of the first type should allow calculations of
intercapacitance matrices for a given circuit layout. For the single electronic digital circuits, upper-level programs are also necessary, which would use simplified (but adequate)
models of the SEL gates and thus would provide faster simulation of VLSI circuits and
systems. Strange as it could seem just a couple years ago, the computer-aided analysis
and design may become the largest problem of applied single-electronics.

We believe that the problems listed above (and those which would inevitably arise
in future) will be overcome, and at the next conference on the single charge tunneling
we will be able to speak not about "possible", but about real important applications of
these wonderful effects.

ACKNOWLEDGEMENTS. Many useful discussions with our colleagues from the Soviet
Union, Europe, United States, and Japan are gratefully appreciated. The authors of
Refs. 8, 9, 11, 12, 14, and 21 were very kind to allow us to use their results prior to
publication.

References

[1] R. Landauer, Physica A **168**, 75 (1990).

[2] K. K. Likharev, *Dynamics of Josephson Junctions and Circuits* (Gordon and Breach, New York, 1986), Chap. 16.

[3] D. V. Averin and K. K. Likharev, in: *Mesoscopic Phenomena in Solids*, ed. by B. L. Altshuler, P. A. Lee, and R. A. Webb (Elsevier, Amsterdam, 1991), Chap. 6.

[4] K. K. Likharev, in: *Physics of Granular Nanoelectronics* (Plenum, New York, to be published).

[5] R. E. Wilkins, E. Ben-Jacob, and R. C. Jaklevic, Phys. Rev. Lett. **63**, 801 (1989).

[6] P. Delsing, K. K. Likharev, L. S. Kuzmin, and T. Claeson, Phys. Rev. Lett. **63**, 1180 (1989); P. Delsing, K. K. Likharev, L. S. Kuzmin, and T. Claeson, Phys. Rev. Lett. **63**, 1861 (1989).

[7] L. J. Geerligs, V. F. Anderegg, P. A. M. Holweg, J. E. Mooij, H. Pothier, D. Esteve, C. Urbina, and M. H. Devoret, Phys. Rev. Lett. **64**, 2691 (1990).

[8] H. Pothier, P. Lafarge, C. Urbina, D. Esteve, and M. H. Devoret, Europhys. Lett. **17**, 249 (1992).

[9] P. Lafarge, H. Pothier, E. R. Williams, D. Esteve, C. Urbina, and M. H. Devoret, Z. Phys. B (1991).

[10] U. Meirav, M. A. Kastner, and S. J. Wind, Phys. Rev. Lett. **65**, 771 (1990).

[11] P. L. McEuen, E. B. Foxman, U. Meirav, M. A. Kastner, Y. Meir, N. S. Wingreen, and S. J. Wind, Phys. Rev. Lett. **66**, 1926 (1991).

[12] B. Su, V. Goldman, and J. E. Cunningham, Bull. Am. Phys. Soc. **36**, 400 (1991) and to be published.

[13] D. V. Averin and A. A. Odintsov, Phys. Lett. A **140**, 251 (1989).

[14] L. S. Kuzmin and D. Haviland, Phys. Rev. Lett. **67**, 2090 (1991).

[15] K. K. Likharev and A. B. Zorin, in: *Proceedings of 17th Int. Conf. on Low Temp. Phys. (Contributed Papers)*, ed. by U. Eckern et al. (Elsevier, Amsterdam, 1984), p. 1153.

[16] D. V. Averin and K. K. Likharev, in: *SQUID'85*, ed. by H.-D. Hahlbohm and H. Lübbig (Walter de Gruyter, Berlin, 1985), p. 197.

[17] K. K. Likharev and A. B. Zorin, J. Low Temp. Phys. **59**, 347 (1985).

[18] P. Delsing, K. K. Likharev, L. S. Kuzmin, and T. Claeson, Phys. Rev. Lett. **63**, 1861 (1989).

[19] K. K. Likharev, N. S. Bakhvalov, G. S. Kazacha, and S. I. Serdyukova, IEEE Trans. Magn. **25**, 1436 (1989).

[20] R. F. Dzuiba and D. B. Sullivan, IEEE Trans. Magn. **11**, 716 (1975).

[21] E. R. Williams, J. Martinis, M. H. Devoret, D. Esteve, C. Urbina, H. Pothier, P. Lafarge, and P. Orfila, to be published.

[22] L. S. Kuzmin and K. K. Likharev, Pis'ma Zh. Exp. Teor. Fiz. **45**, 289 (1987) [JETP Lett. **45**, 496 (1987)].

[23] T. A. Fulton and G. J. Dolan, Phys. Rev. Lett. **59**, 109 (1987).

[24] D. V. Averin and K. K. Likharev, J. Low Temp. Phys. **62**, 345 (1986).

[25] K. K. Likharev, IEEE Trans. Magn. **23**, 1142 (1987).

[26] L. S. Kuzmin, P. Delsing, T. Claeson, and K. K. Likharev, Phys. Rev. Lett. **62**, 2539 (1989).

[27] L. J. Geerligs, V. F. Anderegg, and J. E. Mooij, Physica B **165/166**, 973 (1990).

[28] A. N. Korotkov, D. V. Averin, K. K. Likharev, and S. A. Vasenko, report presented at SQUID'91 (Berlin, June 1991), and to be published.

[29] I. A. Devyatov and K. K. Likharev, to be published.

[30] T. G. Blaney, in: *Infrared and Millimeter Waves*, ed. by K. J. Button, vol. 3 (Acad. Press, New York, 1980), p. 2.

[31] P. K. Tien and G. P. Gordon, Phys. Rev. **129**, 647 (1963).

[32] I. A. Devyatov, L. S. Kuzmin, K. K. Likharev, V. V. Migulin, and A. B. Zorin, J. Appl. Phys. **60**, 1808 (1986).

[33] K. K. Likharev and V. K. Semenov, IEEE Trans. Appl. Supercond. **1**, 3 (1991).

[34] K. K. Likharev and V. K. Semenov, in: *Ext. Abstr. of Int. Supercond. Electron. Conf. (ISEC'89)*, (Tokyo, 1989), p. 182.

[35] K. K. Likharev, S. F. Polonsky, and S. V. Vyshenskii, to be published.

[36] G. J. Dolan and J. H. Dunsmuir, Physica B **152**, 7 (1988).

[37] J. Lambe and R. C. Jaklevic, Phys. Rev. Lett. **22**, 1371 (1969).

[38] J. B. Barner and S. T. Ruggiero, Phys. Rev. Lett. **59**, 807 (1987).

[39] I. O. Kulik and R. I. Shekhter, Zh. Exp. Teor. Fiz. **68**, 623 (1975) [Sov. Phys - JETP **41**, 308 (1975)].

Index